C0-BEF-279

SETON HALL UNIVERSITY
3 3073 10189171 9

STEM CELLS AND TISSUE HOMEOSTASIS

THE SECOND SYMPOSIUM OF
THE BRITISH SOCIETY FOR CELL BIOLOGY

STEM CELLS AND TISSUE HOMEOSTASIS

EDITED BY

B. I. LORD
Senior Research Officer, Paterson Laboratories
Christie Hospital & Holt Radium Institute, Manchester

C. S. POTTEN
Senior Research Officer, Paterson Laboratories
Christie Hospital & Holt Radium Institute, Manchester

R. J. COLE
Professor of Developmental Genetics
University of Sussex

CAMBRIDGE UNIVERSITY PRESS
CAMBRIDGE
LONDON · NEW YORK · MELBOURNE

Published by the Syndics of the Cambridge University Press
The Pitt Building, Trumpington Street, Cambridge CB2 1RP
Bentley House, 200 Euston Road, London NW1 2DB
32 East 57th Street, New York, NY 10022, USA
296 Beaconsfield Parade, Middle Park, Melbourne 3206, Australia

First published 1978

Printed in Great Britain at the
University Press, Cambridge

Library of Congress Cataloguing in Publication Data

Main entry under title:

Stem cells and tissue homeostasis

(The Symposium of the British Society for Cell Biology; 2)
'Symposium organised by the British Society for Cell Biology'
Includes bibliographies and index
1. Stem cells – Congresses. 2. Homeostasis – Congresses. 3. Tissues – Congresses. 4. Cell proliferation – Congresses. 5. Cell differentiation – Congresses. I. Lord, Brian Iles, 1936– II. Potten, C. S., 1940– III. Cole, R. J. IV. British Society for Cell Biology. V. Series: British Society for Cell Biology. Symposium – British Society for Cell Biology; 2.
QH573.S73   574.8'761   77-80844
ISBN 0 521 21799 7

CONTENTS

PREFACE

Under the general title of Stem Cells and Tissue Homeostasis, specialist speakers were invited to review the current understanding of the control of cell proliferation and differentiation in a wide variety of tissues and animal species. The result was this collection of papers which were presented at the second Symposium organised by the British Society for Cell Biology and held at the University of Manchester in April 1977.

The selection of a wide range of disciplines with a common theme reflected the equally wide-ranging interests of the Society's members and it was felt that this broad overview, brought together in a single volume, would be of interest to research workers and students of cell biology throughout the world. The kinetic parameters of many cell systems have been studied extensively over the past twenty years and the pattern of development of the tissues is in many cases quite well understood. The identity and location of the stem cells, however, have frequently remained elusive and it is only recently that it has become possible to fit them into the picture of the tissues' organisation and function. The time was appropriate, therefore, that research workers interested in stem cells should be brought up to date. It was noticeable that the chairmen of the various sessions were never pressed to stimulate discussion and this should be attributed as much to the standards of presentation and interest generated by the speakers as to the still controversial nature of some of the reports.

About one hundred participants at the Symposium were indebted to members of the Paterson Laboratories at the Christie Hospital, in particular Dr Raymond Schofield, for the organisation and smooth running of the meeting. The presence of Dr E. A. McCulloch of the Ontario Cancer Institute, Toronto, as principal guest and Flow lecturer was made possible by the generosity of the Flow Laboratories.

July 1977 B.I.L.

Cell lineages, stem cells and the 'quantal' cell cycle concept

H. HOLTZER

Department of Anatomy, School of Medicine, University of Pennsylvania, Philadelphia, Pennsylvania

A question that can be asked of all emerging systems of some degree of complexity is: what are the roles of extrinsic, as opposed to intrinsic, factors? This question is as relevant to the generation of diversity during evolution as it is to the generation of diversity in (1) the embryo, and (2) the mature animal.

With regard to the emergence of different species, the answer has been unequivocal. Lamarckian versions of the causative role of the environment in directing and determining diversification have been rejected. Evolutionists begin with the assumption that the environmental input is limited to a permissive one; organisms are allowed or not allowed to express options generated by genetic mechanisms. Equally important from the point of view of mechanism is that the options open to any one organism for evolving are strictly limited in number. Selective pressures can not, in one step, induce a flatworm to evolve into a crab, or into a starfish, or into a frog. In an evolutionary lineage the most proximal precursor is unique, stable and programmed; many such species link primitive to advanced organisms. The emergence of terminal forms without lineages is inconceivable. To follow the obligatory sequence of stepwise changes which occur during evolution requires an understanding of those endogenous mechanisms which generate and transmit novel metabolic options in an autonomous, unique organism.

In contrast to the thinking in the field of evolution, concepts and experimentation in cell differentiation are strongly coloured by neo-Lamarckian notions of the didactic role of exogenous molecules in the generation of cell diversity (Holtzer, 1963, 1968; Holtzer *et al.*, 1975*b*; Holtzer & Holtzer, 1976). Most biochemically oriented investigators describe how exogenous proteins, nucleic acids, glycosaminoglycans, cyclic nucleotides, etc., instruct 'undifferentiated' cells to transform into 'differentiated' cells. The purpose of this review is to present evidence that although no cell exists independently of its microenvironment, the role of exogenous molecules in the *specification* of phenotypic diversification is minimal. Exogenous influences such as hormones, growth factors, light, or cell–cell interactions merely permit the responding cells to select one of their alternative phenotypic choices. What cannot be overemphasised is that (1) all cells, embryonic as well as mature, are equally restricted with respect to the number of phenotypic programmes they can activate, and (2) this limited repertoire is predetermined by the phenotypic programme of the mother cell. The central problem in cell diversification is but a variation of the central problem in evolution. It is to define the endogenous mechanisms that insure that a 'mother' cell, having *inherited* one set of phenotypic options from its 'mother', will transmit to its daughters a different, but equally predetermined and restricted set of phenotypic options.

The solution to many of these issues lies in understanding the pivotal role of stem cells and their symmetrical and asymmetrical cell divisions in the generation of cell lineages. The problems of the stem cell are a paradigm for the basic problems of cell diversification. For example, are there 'uncommitted' stem cells? Are there classes of stem cells that have larger phenotypic repertoires, greater degrees of phenotypic 'freedom', than other classes? If there are, then instructive exogenous molecules must play decisive roles in cell diversification. If there are no genetically uncommitted stem cells, and if stem cells are maximally *bipotent*, then the distinction between stem cells and many replicating precursor cells will be hard to define. If most stem cells generate a predetermined, binary-set of phenotypic options, then the influence of exogenous molecules is limited to determining the number and/or survival of the alternative phenotypes emerging from a given lineage.

INDUCTION, UNDIFFERENTIATED CELLS AND MULTIPOTENT CELLS

The dramatic grafting experiments of Spemann and his school (1938) have had a profound impact on concepts of cell diversification. That work led to the notion that biochemically undifferentiated cells, or cells without circumscribed phenotypic programmes, became determined or differentiated as a result of inductive interactions (Saxén & Toivonen, 1962). Whether the undifferentiated cells differentiated into nerve, gut, muscle or cartilage was postulated to be a function of specific, exogenous inducing molecules. Subdivision into different kinds of nerve, digestive tract, muscle and cartilage cells was likewise effected by unique, exogenous molecules released by secondary and tertiary inducers. This view has led many to look for molecules that 'cause' differentiation. This concept of instructive, cell–cell interactions conforms to the conventional view that embryonic cells are more versatile metabolically than their descendants, can be more readily switched into new phenotypic programmes by variations in their microenvironment. Most schemes for cell diversification assume that embryonic cells, as a class, are subject to regulatory controls or possess properties no longer functional in mature cells.

Fig. 1 presents an extreme form of this view in the context of the generation of five distinct phenotypes. An aliquot of the putative undifferentiated cells exposed to inducing molecule A should, without cell division, transform directly into phenotype A. Alternatively, aliquots of equivalent cells exposed to inducing molecules B, C or D should, without cell division, transform directly into phenotypes B, C and D. This scheme, it is worth stressing, is the only operational definition which justifies the concept of an 'undifferentiated' cell. To my knowledge there are no experiments which remotely support this view. Neither the zygote or blastula cells on the one hand, nor the most anaplastic tumour cells on the other, respond in this fashion.

Similarly, there are no experiments which demonstrate that 'embryonic' cells are more readily reprogrammed, or that they have, as cells, a greater number of phenotypic choices than 'mature' cells. No data support the contention that it is easier to transform a given blastula cell directly into a myoblast or chondroblast than to so transform a given haematocytoblast or to so transform a given stratum germinativum cell. Pending such experiments, there is no reason to assume that 'embryonic' cells are more readily reprogrammed by exogenous molecules than 'mature' replicating cells.

For these and other reasons (Holtzer *et al.*, 1975*b*) the concept of the undifferentiated cell must be rejected. Without undifferentiated cells to instruct, the concept of inducing molecules must be reconsidered. At most, inducing molecules act as hormones – they

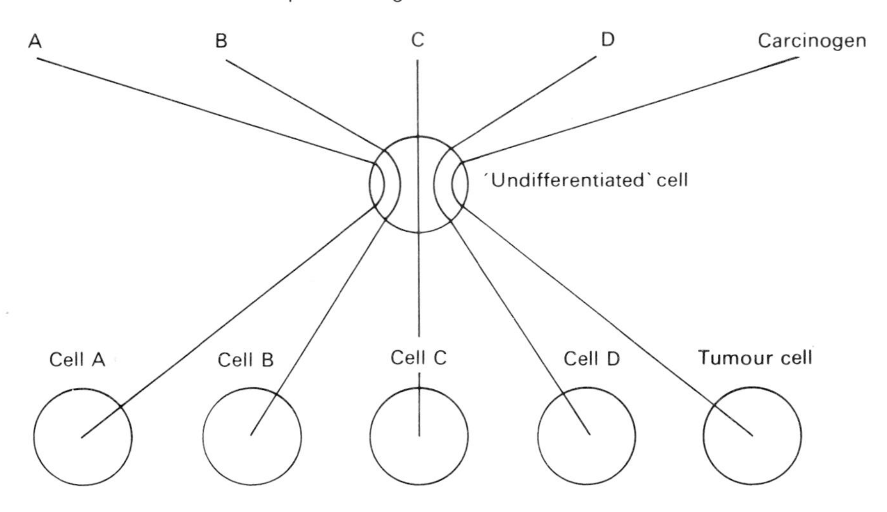

Fig. 1. A scheme accounting for the generation of five terminal phenotypes that stresses the concept of embryonic induction by specific exogenous molecules. It is based on two assumptions: (1) some cells are biochemically 'undifferentiated' and genetically 'totipotent', or at least, 'pluripotent'; and (2) there are large numbers of specific, inducing molecules that 'cause' undifferentiated cells to differentiate. If molecule A impinges on the totipotent or the pluripotent cell, the cell itself directly transforms into cell A; if molecule B impinges on the cell, the cell itself directly transforms into cell B, etc. Note that (1) there is no requirement for passage through a cell cycle in this transformation from undifferentiated to differentiated state, and (2) it is the information supplied by the exogenous molecule that determines the phenotypic programme of the responding cell, not the phenotypic information stored within the cell. According to this scheme, any cell can be induced to transform directly into a neoplastic cell and anaplastic cells derived from different normal cells converge phenotypically (see text for further details).

can act only on already-differentiated target cells, i.e. cells already programmed to respond.

The experimental design in Fig. 1 is equally applicable to ascertaining the biological reality of multipotent cells. The issue is not whether a specific precursor, after two or more generations, yields progeny with three or more phenotypic programmes. The issue is whether, in the absence of replication, a population of equivalent cells can be induced to transform into three or more phenotypes. The issue of the reality of multipotent cells is enlarged on in the next section of this article and by others in this volume. Here it suffices to say that a distinction must be made between a multipotent *system* and a multipotent *cell*. Before attempting to make a molecular model of a multipotent cell, it would be wise first to establish its existence. The phenotypic versatility imputed to the multipotent cell is similar to that attributed to the undifferentiated cell, and with as little experimental justification. In brief, though a single cell obviously can generate many phenotypes, little has been done to determine the role of the successive generations in the emergence of those phenotypes. Such data as there are lead to the conclusion that intervening generations are extremely important and that in fact there is no compelling evidence demonstrating the existence of multipotent cells as defined in Fig. 1.

THE GENERATION OF CELL LINEAGES BY BIPOTENT CELLS

A very different scheme which also accounts for the emergence of different phenotypes, but which assigns a permissive rather than instructive role to exogenous molecules, is based on the old concept of cell lineages and is illustrated in Fig. 2. In this scheme all cells, including zygote, blastula, embryonal carcinoma cells and anaplastic tumour cells are fully differentiated, for each occupies a particular compartment in a particular lineage and, as such, is endowed with a set of circumscribed phenotypic options. The central feature of Fig. 2 is the postulate that each cell within any given compartment is equivalent, and that as such cells pass through S and G_2 they may generate, *at the most*, the next scheduled binary-set of phenotypic options. Each daughter receives one member of this newly generated binary-set. Which set of phenotypic options a cell receives from its mother, which *new binary-set* it is scheduled to generate and which it will transmit singly to its daughters, is a function of which compartment of the lineage the precursor occupies. A specific phenotypic reprogramming occurs only when a particular precursor replicates. On a molecular level the most provocative implications of this scheme are: (1) exogenous molecules have very little influence in the generation of the set of phenotypic options of a specific cell; (2) sets of phenotypic options appear and are lost in successive generations; and (3) with respect to the number of sets of phenotypic options that can be generated, no precursor is ever more than *bipotent*. For example, precursor cell IIA in Fig. 2 cannot, itself, be induced by exogenous molecules to transform directly into cell

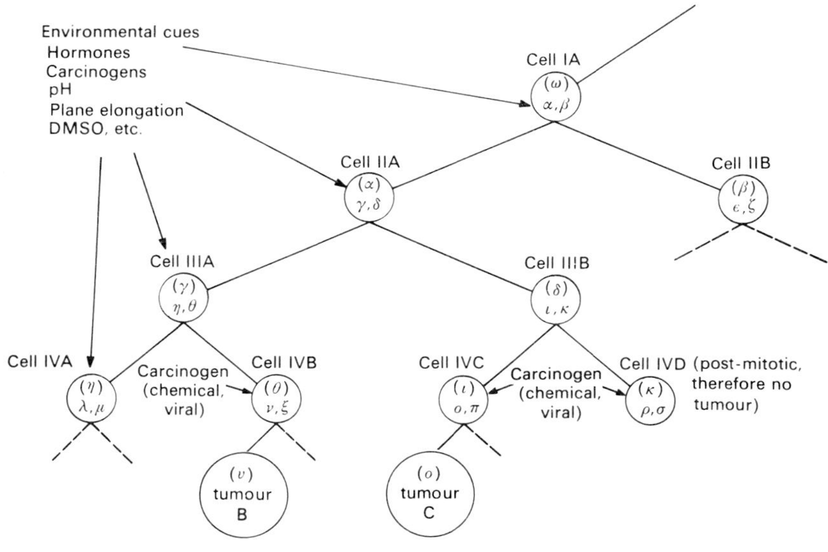

Fig. 2. A lineage to generate terminal phenotypes based on a sequence of cells each capable of binary decisions only (Abbott, Schiltz & Holtzer, 1974; Dienstman, Biehl, Holtzer & Holtzer, 1974). Cells in any compartment of any lineage are maximally bipotent with respect to phenotypic diversification. Cells IVA, IVB, IVC and IVD constitute terminal compartments; cells IIIA and IIIB penultimate compartments; cells IIA and IIB antepenultimate compartments, etc. Transition from compartment to compartment requires passage through a *quantal cell cycle* and cells in successive compartments generate a different phenotypic programme from that of their mother cell. Each cell, from oocyte to anaplastic tumour, inherits a circumscribed set of cell-unique options (symbolised by a Greek letter within parentheses) from its mother cell. This maternal inheritance determines the new phenotypic options of its daughters, which in turn synthesise their cell-unique molecules (symbolised by two Greek letters at bottom of each cell). There are no 'undetermined', no 'undifferentiated', no 'pluripotent' cells. The maternal contribution of cell IA to its daughters IIA and IIB is different from the maternal contribution it received. It is this maternal contribution that determines the daughters' new phenotypic options. Environmental influences may alter rate of transit through the lineage and even select against the survival of some phenotypes; but exogenous molecules do not impose, or define, a phenotypic programme. All type IV cells are 'terminal', but only IVD is post-mitotic and so can not be transformed into a neoplastic cell. In this scheme even anaplastic tumour cells are members of some lineage. Proliferative cell cycles which only increase the numbers of cells within a compartment are not shown (see text for additional details).

IIIA or cell IIIB. The option to generate cell IVA or cell IVD is only available to precursors IIIA and IIIB, respectively. In this scheme each cell is a member of a particular compartment, in a particular lineage, and is endowed with one set of phenotypic options. Cells proximal in a lineage do not have more degrees of 'freedom' than their descendants. Cells 'moving' through a lineage acquire as many phenotypic options as they lose. No embryonic cell is more amenable to environmentally induced reprogramming than are the cells in the haemopoietic systems, stratum geminativum, or gut in a 99-year-old man. This concept of lineages of bipotent cells predicts that the differences that do obtain between 'embryonic' and 'mature' cells will not be referable to the former having more of its single-copy genes available for transcription than the latter.

BIPOTENT STEM CELLS IN THE MYOGENIC AND CHONDROGENIC LINEAGES

Many have claimed that the early limb bud or somite consists of equivalent, multipotent mesenchyme (Ms) cells and that specific inducing molecules can transform a given Ms cell directly into a definitive myoblast, or chondroblast, or fibroblast (Lash, 1968; Zwilling, 1968; Searls & Janners, 1969; Caplan & Koutroupas, 1973; Flickinger, 1974; Caplan & Rosenberg, 1975). It has even been claimed that Ms cells actually synthesise the cell-unique myosins and sulphated proteoglycans which characterise the definitive myoblasts and definitive chondroblasts respectively (Medoff & Zwilling, 1972; Orkin, Pollard & Hay, 1973). This widely accepted view of the Ms cell is schematised in Fig. 3(*a*).

To reassess these claims, we cloned single cells from 8-day chick leg 'muscles' and from 3-day chick limb buds. If a single cell from 8-day leg 'muscle' formed a clone of 50 cells or more, invariably that clone contained both definitive myoblasts and definitive fibroblasts (Abbott, Schiltz & Holtzer, 1974). When such single cells were reared under conditions known to permit presumptive chondroblasts to differentiate terminally, neither they, nor their progeny, formed chondroblasts. Clearly, in 8-day leg 'muscles' the cloneable cell that is capable of bifurcating into myogenic and fibrogenic cells is only bipotent; that cell does not have the option to found a chondrogenic lineage. Accordingly these cells, designated presumptive myoblast–fibroblast cells (PMbFb in Fig. 3*b*), must constitute a compartment at least one generation distal to the mesenchyme (Ms) compartment (Dienstman & Holtzer, 1975).

Table 1 and Fig. 3(*b*), summarising comparable data from early 3-day limb buds, also indicate the relationship between mesenchyme cells and their distal progeny of myoblasts, fibroblasts and chon-

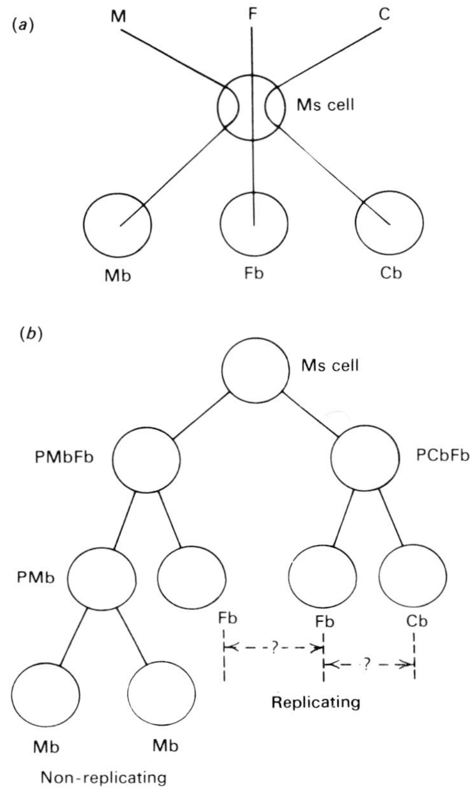

Fig. 3. The mesenchyme cell (Ms) by definition is the common progenitor of myoblasts, fibroblasts and chondroblasts. In scheme (*a*) the Ms cell is viewed as being pluripotent. In one microenvironment it can be induced to transform directly into a myoblast (Mb), whereas in other microenvironments it can be induced to transform into a fibroblast (Fb) or chondroblast (Cb). This model, favoured by embryologists, stresses the contribution of inducing molecules and is consistent with the notion that differences among mesodermal cells are quantitative rather than qualitative.

Scheme (*b*) depicts the Ms cell as bipotent. A minimum of two quantal cell cycles is required for a given Ms cell to yield a chondroblast, a minimum of three to yield a myoblast. Proliferative cell cycles are not represented. There is at least one compartment, the PMbFb compartment, between the Ms cell and the presumptive myoblast (PMb). This antepenultimate compartment in the myogenic lineage also channels cells into the fibrogenic lineage. Similarly, the PCbFb compartment consists of cells ancestral to the chondrogenic and fibrogenic lineages. This scheme predicts that fibroblasts have more than one origin, and that their relationship to myoblasts is different from their relationship to chondroblasts (Chacko, Abbott, Holtzer & Holtzer, 1969; Holtzer & Mayne, 1973; Mayne, Schiltz & Holtzer, 1973). Adding molecules such as cAMP, glycosaminoglycans, nicatinomide, acetylpyridine, collagen, mesodermal growth factors, etc., to cultures of these cells could: (1) prevent terminal differentiation of Mb, Fb or Cb cells; (2) induce PMbFb cells

Table 1. *Occurrence of terminal cell types in limb mesoderm cultures*

	Explants		Monolayers	
	8:1:1	F−10	8:1:1	F−10
Stage 17–18				
Muscle	+	±	−	−
Cartilage	+	+	−	−
Stage 21–22				
Muscle	++	+	+	−
Cartilage	++	++	+	+
Stage 24–25				
Muscle	+++	++	++	−
Cartilage	+++	+++	++	++

Key: −, absence of tissue; ±, exceedingly sparse tissue; +, tissue present; ++, moderate amount of tissue; +++, abundant tissue.

8:1:1 and F−10 refer to the medium in which these chick limb buds were grown. Note that the precursor myogenic cells known to be present in stage 17–18 limb buds simply do not survive or terminally differentiate in F−10. The addition of complex molecules to F−10 that would permit these myogenic cells to survive and terminally differentiate does not prove that such molecules establish the myogenic nature of such cells.

droblasts. Clones were not readily established from limb buds younger than stage 18 (Dienstman, Biehl, Holtzer & Holtzer, 1974). Clones from older limb buds consisted *either* of definitive myoblasts and fibroblasts, *or* of definitive chondroblasts and fibroblasts. Of the hundreds scored, a mixed clone of both myoblasts and chondroblasts was never observed. From these and related observations it was concluded that while the penultimate compartment in the myogenic lineage might consist of myogenic cells only – presumptive myoblasts or PMb in Fig. 3(*b*) – the antepenultimate compartment consists of PMbFb cells. PMbFb cells from 3-day limb buds, like those from 8-day leg 'muscles' are maximally bipotent: they do not yield progeny for the chondrogenic lineage. Conversely, there are distal

to divide symmetrically yielding two Fb cells rather than one PMb and one Fb cell; (3) induce PCbFb cells to divide, yielding two Fb cells rather than one Cb and one Fb cell; (4) induce PMbFb or PCbFb to undergo proliferative rather than quantal cell cycles. According to this scheme, exogenous molecules do not induce multipotent mesenchyme cells to become myoblasts, or fibroblasts, or chondroblasts.

compartments in the chondrogenic lineage that consist of bipotent cells with the option of funnelling only into the chondrogenic and fibrogenic lineages: they do not have the option of yielding myogenic cells. By any definition, the PMb cell is the final stem cell in the myogenic lineage. Newly 'born' myoblasts in a 48-hour chick embryo and the newly 'born' myoblasts in the regenerating muscles of a 99-year-old man are the daughters of PMb cells and are subject to the same intracellular controls. The PMb cells in mature muscle have been termed 'satellite' cells (Mauro, 1961; Mauro, Shafiq & Milhorat, 1970).

There is evidence that some structural genes which are actively transcribed in Mb cells are not transcribed in PMb cells. Antibody to mature skeletal myosin will precipitate myosin heavy chains extracted from Mb cells, but not the myosin heavy chains extracted from PMb cells (Holtzer, Marshall & Finck, 1957; Holtzer, 1961; Okazaki & Holtzer, 1966; Chi, Fellini & Holtzer, 1975; Fellini & Holtzer, 1976). The myosin light chains in Mb cells have molecular weights of 25 000 and 18 000, whereas those of light chains of PMb cells are 20 000 and 16 000. Equally striking qualitative differences have been reported between the sulphated proteoglycans synthesised by Cb cells and those synthesised by PCb cells (Okayama, Pacifici & Holtzer, 1976).

These results are incompatible with claims that (1) there are cells in limb buds or somites which are multipotent, or (2) the precursors to the definitive myoblasts or chondroblasts synthesise the same myosins or sulphated proteoglycans respectively as those synthesised by their daughters, the definitive myoblasts or chondroblasts.

Much in Fig. 3(*b*) remains obscure. It is unclear what exogenous factor(s) triggers the decision of a given PMb to enter a quantal rather than a proliferative cell cycle. Nor is it known whether all PMb cells are equally facultative with respect to the symmetry of their quantal cell cycles, i.e. whether they always produce one myoblast and one presumptive myoblast or may, depending upon microenvironmental conditions, produce two myoblasts. However, simply by influencing such decisions, all of the dramatic effects on cultured myogenic cells which have been attributed to specific molecules in 'conditioned medium' (Konigsberg, 1971; White, Bonner, Nelson & Hauschka, 1974), such as insulin (de la Haba, Cooper & Etling, 1966), cAMP (Zalin & Montague, 1975), mesodermal growth factor (Gospodarowicz, Wiseman, Moran & Lindstrom, 1976), BudR (Stockdale, Okazaki, Nameroff & Holtzer, 1964; Okazaki & Holtzer, 1966; Bischoff & Holtzer, 1970), or the phorbol ester, PMA (Cohen *et al.*, 1977) can be accounted for (see Yeoh & Holtzer, 1977). In none of these instances is there evidence that the exogenous molecules acted on those sites of the genome which commit these cells to the myogenic lineage. They simply permitted, or did not permit, the cell to *express* its particular myogenic options; they did not *determine* those myogenic options. With the possible exception of the BudR experiments (see below), it is not likely that this type of molecular approach to cell differentiation will reveal how the PMb cell, the mother cell, initiates those processes which result in the functioning of a new set of phenotypic options in its daughter, the Mb cell.

Fig. 3(*b*) provides a simple explanation for the many claims that myogenic cells in limb buds or somites can be shunted into the chondrogenic lineage (or vice versa) by alteration of cell density or oxygen tension, or the addition of glycosaminoglycans, cAMP or 3-acetylpyridine to the culture medium (Caplan & Rosenberg, 1975; Solursh & Reiter, 1975; Slavkin & Greulich, 1975). In a population of PCbFb and PMbFb cells, alterations in culture conditions could augment the frequency of cell divisions of the former, thus accounting for the eventual dominance of chondroblasts. It is more likely, however, that the altered culture conditions simply favoured the survival and terminal differentiation of Cb cells, whereas the Mb cells that emerged in equal numbers did not differentiate terminally or even survive (see Table 1). These considerations of survival and/or terminal differentiation also can account for the effect of collagen on myogenesis. The observation that collagen 'stimulates' myogenesis (Hauschka & Konigsberg, 1966) is not to be attributed to a unique and specific interaction of this molecule with myogenic cells. The stimulation observed is simply due to the fact that presumptive myoblasts replicate on, and the myoblasts elongate and migrate on, the collagen substrate. In the absence of collagen these cells do not adhere to the substrate, do not replicate, do not migrate or elongate, and shortly die (Yeoh & Holtzer, 1977). Collagen and glycosaminoglycans are as essential to myogenesis as are Ca^{2+} ions, proper pH, etc. However, such experiments provide no information about why only some descendants of Ms cells enter the myogenic lineage.

Table 2 summarises the known phenotypic programmes of some

Table 2. *Phenotypic programmes of cells in different compartments of the myogenic, chondrogenic and fibrogenic lineages*

	Mesenchyme cell; presumptive myoblast; presumptive chondroblast; presumptive fibroblast	
	+DNA synthesis −Fusion −Ach R −M CPK +'c' myosin heavy chains +'c' light-meromyosin +'c' myosin light chains −Type IV sulphated proteoglycan ? Collagen chains	
Definitive myoblast	*Definitive chondroblast*	*Definitive fibroblast*
−DNA synthesis[a] +Fusion[b] +Ach R[c] +M CPK[d] +Fibrillar myosin heavy chains[e] +Fibrillar light-meromyosin[f] +Fibrillar myosin light chains[g] 'c' myosin heavy chains ? 'c' light-meromyosin ? 'c' myosin light chains −Type IV sulphated proteoglycan ? Collagen chains	+DNA synthesis[h] +Type IV sulphated proteoglycan[j] +Type II collagen chains[k] +'c' myosin light chains +'c' light-meromyosin +'c' myosin light chains	+DNA synthesis −Type IV sulphated proteoglycan +Type I collagen chains +'c' myosin heavy chains +'c' light-meromyosin +'c' myosin light chains

[a] Stockdale & Holtzer (1961); [b] Holtzer, Abbott & Lash (1957); [c] Fambrough & Rash (1972); [d] Turner *et al.* (1976); [e] Holtzer *et al.* (1973, 1975*b*); [f] Fellini & Holtzer (1976); [g] Chi, Fellini & Holtzer (1975); [h] Holtzer *et al.* (1961); [j] Okayama, Pacifici & Holtzer (1976); [k] Miller & Matukas (1974).

This table contrasts the phenotypic programmes of cells in the antepenultimate, penultimate and terminal compartments of the myogenic, chondrogenic and fibrogenic lineages. Although the cells in the terminal compartments of these lineages have strikingly different phenotypic programmes, with current biochemical techniques it is impossible to identify, or to distinguish between, (1) their 'mother' cells in the penultimate compartments, or (2) their 'mother' and 'grandmother' cells in the antepenultimate compartment. Note also that the phenotypic options cells in the myogenic, chondrogenic and fibrogenic lineages. It illustrates, in terms of transcripts of cell-specific structural genes, how markedly the phenotypic programmes of the definitive myoblast, the definitive chondroblast and the definitive fibroblast differ from those of their respective mother cells. Even more striking, however, is the fact that the data in Table 2 indicate that nothing is known of the luxury molecules which must be synthesised by the different kinds of precursors. The transition from penultimate to ultimate compartment, from final stem cell PMb and PCb to terminal cell Mb and Cb, involves not merely the synthesis of one or two luxury molecules, but the synthesis of co-ordinated sets of such molecules. It is likely that equally marked differences obtain in the metabolic options of the bipotent Ms cell as contrasted with its daughters, the postulated bipotent PMbFb and PCbFb cells.

UNIPOTENT CELLS AND LINEAGES WITH FINITE NUMBERS OF GENERATIONS

There are lineages of unipotent cells. Such lineages may have evolved to 'count' and limit the number of descendants derived from a single progenitor. Such lineages function in mosaic developmental systems (Nöthiger, 1972; Okazaki, 1975; Freeman, 1976) and particularly in forms such as rotifers and nematodes where the mature organism consists of a constant number of cells. These lineages also occur among 'maturing' white and red blood cells. For example, in forms as diverse as man and chickens there are approximately six successive generations of erythroblasts, each generation synthesising characteristic amounts of haemoglobin (Hb) and having characteristic morphological features (Lajtha, 1970; Lajtha & Schofield, 1974; Weintraub, Campbell, Holtzer & Holtzer, 1971; Weintraub, 1975; Holtzer *et al.*, 1975*b*). This type of self-limited lineage in which the first-generation erythroblast divides symmetrically, producing two second-generation erythroblasts, which divide symmetrically,

that become available to each of the definitive cells involve a co-ordinated set of molecules, not just an option to synthesise one or two. A more detailed analysis of the options of chondroblasts and fibroblasts will be found in Schiltz, Mayne & Holtzer (1973) and Mayne, Schiltz & Holtzer (1973). The designation 'c' heavy and light chains and 'c' light meromyosin (LMM) refers to the possibility that all cells synthesise a constitutive type of myosin (Holtzer *et al.*, 1972, 1975*b*).

and so on until the sixth generation, may be more common and more important biologically than currently is realised.

The most compelling evidence that passage through a quantal cell cycle directly affects transcription comes from comparing the production of Hb-transcripts in erythrogenic haematocytoblasts with that of their daughters, the erythroblasts. Groudine, Holtzer, Scherrer & Therwath (1974), using the reverse transcriptase technique, prepared a labelled cDNA against Hb mRNA and assayed the erythrogenic haematocytoblasts and their descendent erythroblasts for Hb-transcripts. They concluded that, within the limits of reliability of such hybridisation tests, the mother of the first generation erythroblast, the erythrogenic haematocytoblast, was not more capable of transcribing its Hb-genes than fibroblasts, blastodisc cells, muscle cells or nerve cells. According to their calculations, if the erythrogenic haematocytoblast transcribes its globin genes, it does so at a rate 10^4–10^5 less frequently than functional erythroblasts (see Harrison, this volume, for further discussion of this issue).

Evidence that the erythrogenic haematocytoblast itself will not initiate transcription of its Hb-genes comes from experiments using FudR, cytosine arabinoside and colcimid. These agents block cell division by preventing DNA synthesis or mitosis. The erythrogenic haematocytoblasts exposed to these inhibitors did not initiate the synthesis of Hb as determined microspectrophotometrically, using benzidine to detect the accumulating haem molecules (Campbell, Weintraub & Holtzer, 1974 and unpublished data). The incorporation of BudR into the DNA of replicating erythrogenic haematocytoblasts yields the same results (Weintraub, Campbell & Holtzer, 1972).

These experiments with embryonic chick red blood cells indicate that such terms as 'undifferentiated', or 'uncommitted', or 'pluripotent' stem cell are no longer useful in working out the molecular events responsible for erythrogenesis on a cellular level. The data strongly suggest the physical existence of a unipotent erythrogenic haematocytoblast. This erythrogenic cell is past the point where it can yield white blood cells; this cell has not acquired the option to transcribe its globin genes and though it 'matures', it will never itself transcribe its globin genes. This is an option that the erythrogenic haematocytoblast will generate and transmit to its daughters, the first generation erythroblasts, as a consequence of passing through a particular quantal cell cycle.

In this context it is worth noting that if Friend leukaemic cells are to acquire the option to synthesise Hb following exposure to such non-specific inducing agents as DMSO, butyric acid, etc., they too must pass through a quantal cell cycle (McClintock & Papaconstantinou, 1974; Levy, Terada, Rifkind & Marks, 1975).

STEM CELLS, TUMOUR CELLS AND SUPPRESSED PHENOTYPIC PROGRAMMES

The distinction between stem cells and other dividing cells is based on the former having the option of self-renewal, as well as yielding daughters with different phenotypic options. By this definition, several types of neoplastic cells must be viewed as stem cells. The embryonal carcinoma cells studied by Stevens (1967), Kleinsmith & Pierce (1964) and Martin (1975) are in all likelihood transformed, differentiated morula or blastula cells that may (1) continue to replicate as transformed, but differentiated, morula or blastula cells, or (2) be channelled by normal quantal cell cycles into normal cell lineages. Some recent experiments using a temperature-sensitive mutant of the Rous sarcoma virus (ts-RSV) support this view, and further underscore the relationship between cell lineages, stem cells and the ability of cells to undergo many proliferative cell cycles without expressing, or losing their commitment to, their unique phenotypic programmes.

Myogenic, chondrogenic and melanogenic cells have been transformed with a ts-RSV mutant. At permissive temperature these three types of cells replicate and assume a nondescript fibroblastic morphology. By available biochemical criteria they are as 'undifferentiated' as any blastoderm or anaplastic tumour cell. That they are not uncommitted, however, but are functioning as stem cells is readily demonstrated by shifting to non-permissive temperature.

Approximately 24 hours after ts-transformed myogenic cells are shifted to non-permissive temperature many (1) withdraw from the cell cycle, (2) initiate the synthesis of the myoblast-unique myosin heavy and light chains, (3) assemble normal interdigitating thick and thin filaments, (4) fuse to form multinucleated myotubes, (5) spontaneously contract (Holtzer *et al.*, 1975*a*; Cohen *et al.*, 1977).

Essentially similar results have now been obtained with ts-transformed chondrogenic and ts-transformed melanogenic cells. Fig. 4(*a*) and (*b*) illustrates the differences in morphology between normal and transformed chondrogenic cells grown at permissive

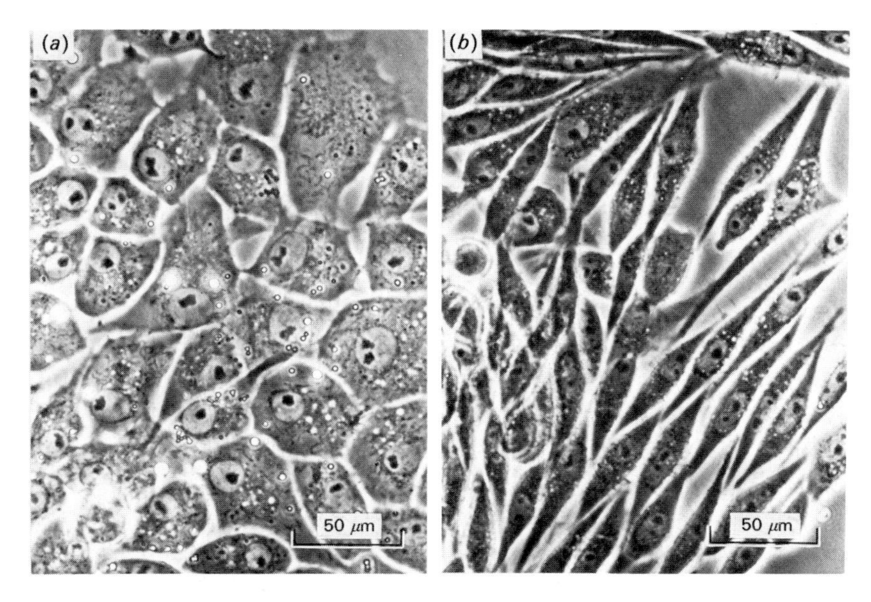

Fig. 4. (*a*) Phase micrograph of living normal chondroblasts. The cells display a polygonal morphology, secrete a metachromatic matrix, and synthesise a chondroblast-unique Type IV sulphated proteoglycan. The cells do not overlap. (*b*) At permissive temperature the transformed chondroblasts are bipolar, fail to secrete metachromatic matrix, and do not synthesise Type IV sulphated proteoglycan. The cells do overlap. If these cells are shifted to non-permissive temperature they display the properties shown and described in (*a*).

temperature. The replicating cells illustrated in Fig. 4(*b*) do not synthesise the chondroblast-unique Type IV sulphated proteoglycan. If, however, these ts-transformed cells are shifted to non-permissive temperature, they revert to the morphology of normal chondroblasts and reinitiate the synthesis of Type IV sulphated proteoglycan (Roby, Boettiger, Pacifici & Holtzer, 1976; Pacifici *et al.*, 1977). Fig. 5(*a*) and (*b*) illustrates normal and ts-transformed retinal pigment cells. The replicating ts-transformed retinal pigment cells have lost their epithelioid arrangement, do not assemble melanosomes, and do not synthesise melanin. However, following a shift to non-permissive temperature, these cells reacquire their epithelioid morphology, assemble melanosomes and reinitiate the synthesis of melanin (Roby *et al.*, 1976; Boettiger *et al.*, 1977).

In these three cases the transformed cells behaved as 'normal' cells when they were shifted to non-permissive temperature. The myogenic cells permanently withdrew from the cell cycle, and so

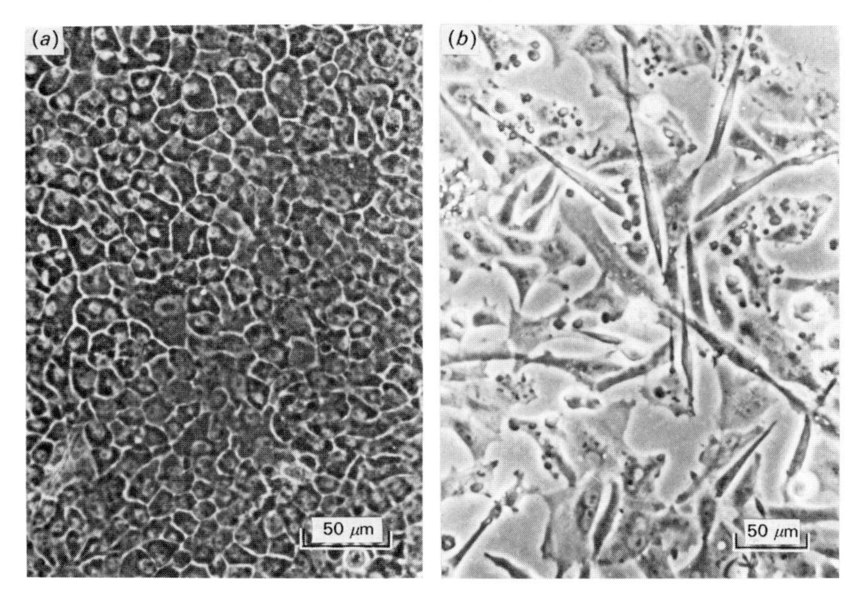

Fig. 5. (*a*) Phase micrograph of living normal retinal melanoblasts. The cells arrange themselves into tightly packed epithelioid colonies. The cells do not display amoeboid processes and virtually all cells are dense with melanin granules. (*b*) At permissive temperature the transformed retinal melanoblasts vary greatly in morphology, do not adhere to one another, display a variety of amoeboid processes, and are lacking in melanin granules. If these cells are shifted to non-permissive temperature they assume the characteristics shown in (*a*).

can no longer even be classified as 'neoplastic'. Clearly, a unique set of phenotypic options can be transmitted through a long series of proliferative cell cycles within that lineage. Equally clearly, the exogenous conditions that induce expression of the covert phenotypic programme must not be confused with the conditions which initially establish such programmes (Holtzer, 1963, 1968, 1970). Regrettably, most recent efforts to analyse cell differentiation on a molecular level fail to distinguish between these two very different classes of events (e.g., Rutter, Pictet & Morris, 1973; Slavkin & Greulich, 1975).

The transformation of myogenic, chondrogenic and retinal melanogenic cells by RSV is itself an illustration of reprogramming a cell's phenotypic options. For integrated virus to be transcribed the host cell must synthesise DNA (Boettiger, 1974). The mechanisms which permit viral DNA transcription only after a round of host DNA synthesis, may be analogous to those which permit 'new' predetermined regions of the genome to be transcribed in a cell whose mother has undergone a quantal cell cycle.

The Friend erythroleukaemic cell is another illustration of a phenotypic programme retained intact over many proliferative cell cycles (Scher, Preisler & Friend, 1973; Levy *et al.*, 1975). When induced to synthesize Hb, the Hb-positive cells cease dividing after a few cell cycles – as do their normal counterparts. Normal and viral-infected cells which enter the terminal compartments of the erythrogenic lineage appear to be programmed not only to synthesise Hb, spectrin, glycophorin, carbonic anhydrase, etc., but also to cease synthesising DNA (Campbell, Weintraub, Mayall & Holtzer, 1971; Holtzer *et al.*, 1975*b*). This endogenous control built into the programme of erythroblasts probably accounts for the observation that there are no Hb-synthesising neoplastic cells. By the same token, as stressed elsewhere (Holtzer *et al.*, 1975*b*), myoblasts cannot be transformed into tumour cells: the cell of origin of rhabdomyosarcoma is almost certain to be a neoplastic presumptive myoblast, or perhaps PMbFb cell.

The ts-transformed cells and the Friend leukaemic cells function as stem cells. In some manner the virus keeps these cells undergoing proliferative cell cycles, thus keeping them in the penultimate or perhaps even antepenultimate compartment of their respective lineages. A shift of 4 deg C in temperature, on the one hand, or molecules such as DMSO or butyric acid, on the other, can re-establish a more normal equilibrium between replicating precursor and terminal daughter.

Blocking the emergence, or suppressing the expression, of an entire phenotypic programme is not a unique property of the viral transforming factor. Similar effects have been obtained by exposing replicating cells to BUdR. Myogenic, chondrogenic and melanogenic cells that have incorporated BUdR into their DNA are also 'reversibly' blocked from expressing their terminal programmes (Stockdale *et al.*, 1964; Okazaki & Holtzer, 1965; Holtzer & Abbott, 1968; Bishoff & Holtzer, 1970; Coleman, Coleman, Kankel & Werner, 1970; Abbott, Mayne & Holtzer, 1972; Levitt & Dorfman, 1973).

That cells in early compartments of the myogenic and chondrogenic lineages differ fundamentally from their progeny in later compartments is shown in experiments using PMA (phorbol-12-myristate-13-acetate) and BUdR (Cohen *et al.*, 1977; Abbott,

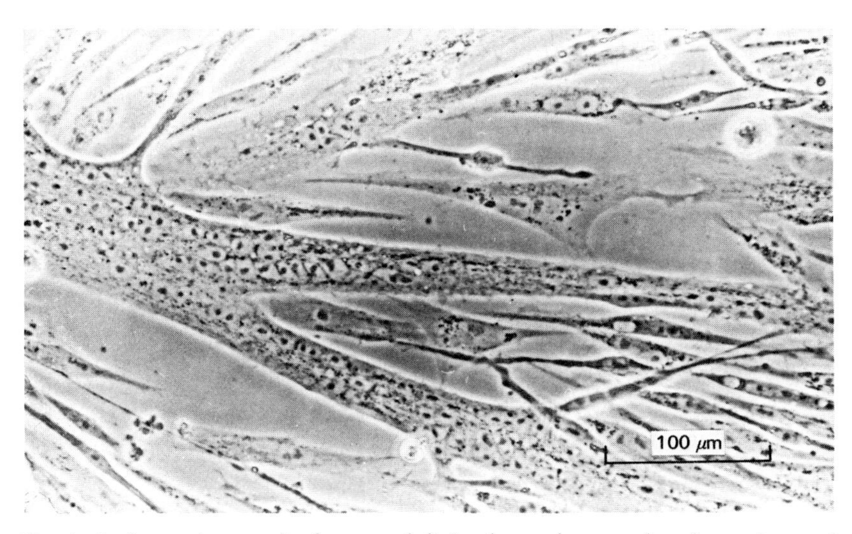

Fig. 6. A phase micrograph of a control, living late 3-day muscle culture. Approximately 50% of the total number of nuclei are in the multinucleated myotubes.

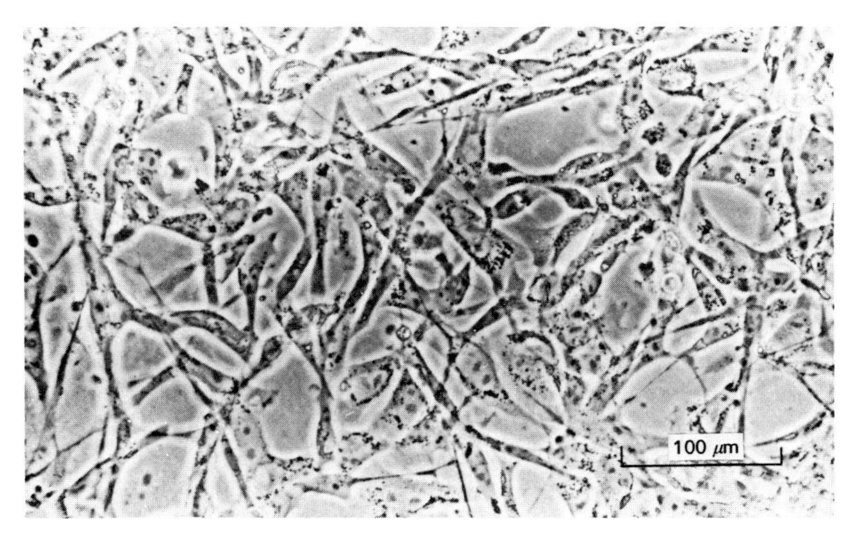

Fig. 7. A sister-culture to that shown in Fig. 6, but reared continuously in PMA. As described in Cohen *et al.* (1977) the tumour-promoting agent has two effects: (1) it enhances the frequency of DNA synthesis, and (2) it blocks fusion. If the PMA-containing medium is replaced with normal medium, within 24–48 hours many of the myogenic cells will permanently withdraw from the cell cycle and fuse to form multinucleated myotubes as shown in Fig. 6.

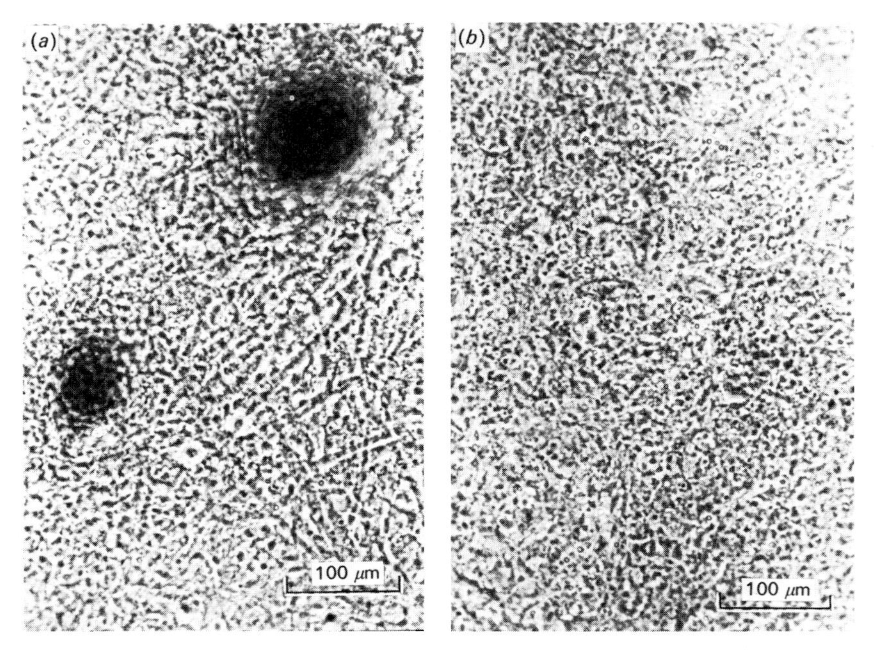

Fig. 8 (*a*) Micrograph of a fixed and stained 5-day-old culture prepared from a suspension of cells from 4-day-old limb buds. The culture has been stained with toluidine blue. Two distinct metachromatic cartilage nodules are readily seen. Observe the numerous criss-crossing multinucleate myotubes in the background. (*b*) A sister-culture to that shown in (*a*), but reared continuously in medium containing PMA. The tumour-promoting agent has not interfered with cell replication, but has blocked the emergence of terminal chondroblasts and terminal myoblasts. This suppression is irreversible.

Mayne & Holtzer, 1972). PMA and BUdR reversibly block cells in the ultimate and penultimate compartments (Figs. 6, 7 and 9), but *irreversibly* block terminal differentiation when acting on early limb bud or somite cells (Figs. 8*a* and *b*). Clearly the intracellular effect of these exogenous molecules varies depending upon the phenotypic properties of the cells in the antepenultimate and penultimate compartments respectively. It will be of interest to compare (1) PMA- and BUdR-suppressed limb bud cells with PMA- and BUdR-suppressed myogenic and chondrogenic cells, and (2) PMA- and BUdR-suppressed limb bud cells with PMA- and BUdR-suppressed neurogenic cells.

These experiments with the ts-RSV mutant, BUdR, and now with PMA, demonstrate that diverse exogenous molecules can induce stem cells which normally would have passed through a terminal

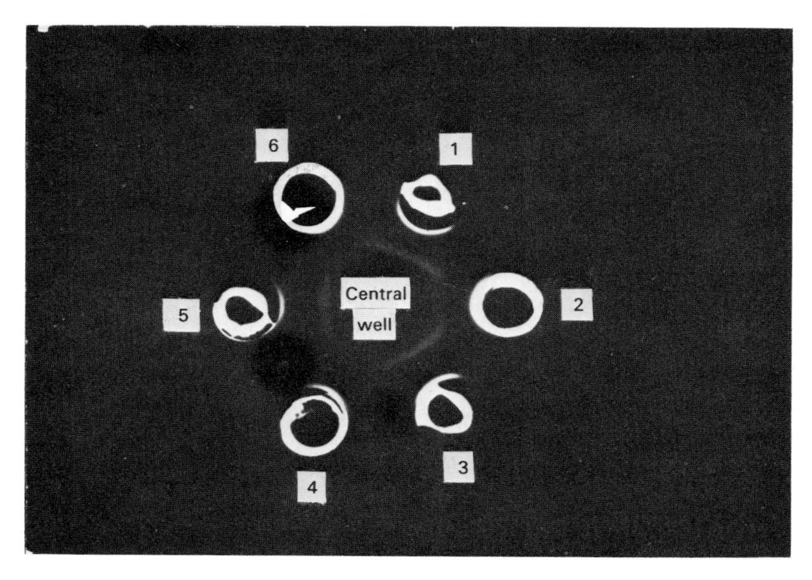

Fig. 9. Antibody against myosin heavy chains from mature muscle is in the central well. This myosin antibody does not form precipitin bands with myosin heavy chains from (*a*) presumptive myoblasts, (*b*) ts-transformed myogenic cells grown at permissive temperature, (*c*) fibroblasts, (*d*) mesenchyme cells, (*e*) smooth muscle cells, (*f*) many kinds of non-myogenic cells. It reacts exclusively with myosin from (*a*) post-mitotic, mononucleated myoblasts, (*b*) multinucleated myotubes, and (*c*) mature muscle. Wells 1, 3 and 5 contain 4 mg ml^{-1} actomyosin extracted from 4-day normal muscle cultures. Wells 2, 4 and 6 contain 5 mg ml^{-1} actomyosin from PMA-treated 4-day muscle cultures. Experiments using fluorescein-labelled antibodies to mature muscle myosin and mature muscle light meromyosin yield results consistent with the agar diffusion tests (Holtzer, Marshall & Finck, 1957; Okazaki & Holtzer, 1966; Holtzer *et al.*, 1975*b*; Fellini & Holtzer, 1976).

quantal cell cycle to undergo, instead, many proliferative cell cycles. The interpolation of these proliferative cell cycles does not alter the fidelity of the transmitted phenotypic programme. Such fidelity of transmission requires an exceptionally stable physical basis. The stability and the inheritance of the covert differentiated state of precursor cells is impressive and must be largely independent of exogenous molecules.

Lastly, these experiments with the ts-RSV mutant, BudR and PMA are consistent with the notion of an intracellular 'master-switch' common to all cells and which controls the activation of that cell's set of phenotypic options (Holtzer, Weintraub, Mayne & Mochan, 1972; Dienstman & Holtzer, 1975; Holtzer *et al.*, 1975*a*, *b*). This master-switch does not control the synthesis of the individual luxury molecules which belong to that set of phenotypic options possessed by a particular cell, rather it allows the set of options, as a unit, to be activated. Setting the master-switch 'on' is prerequisite to the initiation of the transcription of that cluster of structural genes which comprises the cell's set of phenotypic options. This master-switch can be flipped 'off' or 'on', either directly or indirectly, by very different kinds of molecules. This accounts for the lack of specificity of so many 'inducing' molecules which have been reported in the literature.

This work was supported in part by USPHS grants HL-18708, CA-18194, GM-20138, HL-15835, and the Muscular Dystrophy Association.

REFERENCES

Abbott, J., Mayne, R. & Holtzer, H. (1972). Inhibition of cartilage development in organ cultures by thymidine analogue 5-bromodeoxyuridine. *Developmental Biology*, **28**, 430–42.

Abbott, J., Schiltz, J. & Holtzer, H. (1974). Phenotypic complexity of myogenic clones. *Proceedings of the National Academy of Sciences, USA*, **71**, 1506–10.

Bergink, E. W., Wallace, R. A., Vandenberg, J. A., Bos, E. S., Gruber, M. & Geert, A. B. (1974). Estrogen induced synthesis of yolk proteins in roosters. *American Zoologist*, **14**, 1177–93.

Bischoff, R. & Holtzer, H. (1968). The effect of mitotic inhibitors on myogenesis. *Journal of Cell Biology*, **36**, 111–28.

Bischoff, R. & Holtzer, H. (1970). Inhibition of myoblast fusion after one round of DNA synthesis in 5-bromodeoxyuridine. *Journal of Cell Biology*, **44**, 134–50.

Boettiger, D. (1974). Reversion and induction of Rous sarcoma virus expression in virus-transformed baby hamster kidney cells. *Virology*, **62**, 522–9.

Boettiger, D., Roby, K., Pacifici, M., Brumbaugh, J. & Holtzer, H. (1977). Ts-Rous sarcoma virus transformed melanogenic cells. *Cell*, **11**, 881–90.

Borun, T. W. (1975). Histones, differentiation, and the cell cycle. In *Cell cycle and cell differentiation*, ed. J. Reinert & H. Holtzer, pp. 249–90. Berlin: Springer-Verlag.

Campbell, G. L., Weintraub, H. & Holtzer, H. (1974). Effect of FudR on Hb synthesis. *Journal of Cell Physiology*, **83**, 11–18.

Campbell, G. L., Weintraub, H., Mayall, B. & Holtzer, H. (1971). Primitive erythropoiesis in early chick embryogenesis. II. Correlation between hemoglobin synthesis and the mitotic history. *Journal of Cell Biology*, **50**, 669–81.

Caplan, A. I. & Koutroupas, S. (1973). The control of muscle and cartilage development in the chick limb: the role of differential vascularization. *Journal of Embryology and experimental Morphology*, **29**, 571–80.

Caplan, A. I. & Rosenberg, M. J. (1975). Interrelationship between poly(ADP-rib) synthesis, intracellular NAD levels and muscle or cartilage differentiation from mesodermal cells of embryonic chick limb. *Proceedings of the National Academy of Sciences, USA*, **72**, 1852–7.

Chacko, S., Abbott, J., Holtzer, S. & Holtzer, H. (1969). The loss of phenotypic traits by differentiated cells. IV. Behavior of the progeny of a single chondrocyte. *Journal of experimental Medicine*, **130**, 417–41.

Chi, J. C., Fellini, S. A. & Holtzer, H. (1975). Differences among myosins synthesized in non-myogenic cells, presumptive myoblasts, and myoblasts. *Proceedings of the National Academy of Sciences, USA*, **72**, 4999–5003.

Cohen, R., Pacifici, M., Rubinstein, N., Biehl, J. & Holtzer, H. (1977). Effects of a tumor promoter on myogenesis. *Nature, London*, **266**, 538–40.

Coleman, A., Coleman, J. R., Kankel, D. & Werner, I. (1970). The reversible control of animal cell differentiation by the thymidine analog5-bromodeoxyuridine. *Experimental Cell Research*, **59**, 319–28.

de la Haba, G., Cooper, G. & Elting, V. (1966). Hormonal requirements for myogenesis of striated muscle *in vitro*: insulin and somatotropin. *Proceedings of the National Academy of Sciences, USA*, **56**, 1719–23.

Dienstman, S. R., Biehl, J., Holtzer, H. & Holtzer, S. (1974). Myogenic and chondrogenic lineages in developing limb buds grown *in vitro*. *Developmental Biology*, **39**, 83–95.

Dienstman, S. R. & Holtzer, H. (1975). Myogenesis: a cell lineage interpretation. In *Cell cycle and cell differentiation*, ed. J. Reinert & H. Holtzer, pp. 1–25. Berlin: Springer-Verlag.

Fambrough, D. & Rash, J. (1972). Development of acetylcholine sensitivity during myogenesis. *Developmental Biology*, **26**, 55–68.

Fellini, S. & Holtzer, H. (1976). Localization of fluorescein-labelled anti-light meromyosin in myogenic cells. *Differentiation*, **6**, 71–4.

Flickinger, R. A. (1974). The effect of 5-bromodeoxyuridine on chick embryo limb bud mesenchyme in organ culture. *Cell Differentiation*, **4**, 295–304.

Freeman, G. (1976). The effects of altering the position of cleavage planes on the process of localization of developmental potential in Ctenophores. *Developmental Biology*, **51**, 332–7.

Gospodarowicz, D., Wiseman, J., Moran, S. & Lindstrom, J. (1976). The effect of fibroblast growth factor on the division and fusion of bovine myoblasts. *Journal of Cell Biology*, **70**, 395–405.

Groudine, M., Holtzer, H., Scherrer, K. & Therwath, A. (1974). Lineage-dependent transcription of globin genes. *Cell*, **3**, 243–7.

Hauschka, S. D. & Konigsberg, I. (1966). The influence of collagen on the development of muscle clones. *Proceedings of the National Academy of Sciences, USA*, **55**, 119–26.

Holtzer, H. (1961). Aspects of chondrogenesis and myogenesis. In *Synthesis of molecular and cellular structure. Proceedings of the 19th Growth Symposium*, ed. D. Rudnick, pp. 35–85. New York: Ronald Press.

Holtzer, H. (1963). In *Induktion und Morphogenese*, p. 217. Berlin: Springer-Verlag.

Holtzer, H. (1968). Induction of chrondrogenesis: a concept in quest of a mechanism. In *Epithelial–mesenchymal interactions*, ed. R. Fleischmajer & R. E. Billingham, pp. 152–64. Baltimore: Williams & Wilkins.

Holtzer, H. (1970). Myogenesis. In *Cell differentiation*, ed. O. Schjeide & J. DeVillis, pp. 476–503. New York: Van Nostrand-Reinhold.

Holtzer, H. & Abbott, J. (1968). Oscillations of the chondrogenic phenotype *in vitro*. In *Stability of the differentiated state*, ed. H. Ursprung. Berlin: Springer-Verlag.

Holtzer, H., Abbott, J. & Lash, J. W. (1957). On the formation of multinucleated myotubes. *Anatomical Record*, **131**, 567–71.

Holtzer, H. Abbott, J., Lash, J. W. & Holtzer, S. (1961). The loss of phenotypic traits by differentiated cells. I. Dedifferentiation of cartilage cells. *Proceedings of the National Academy of Sciences, USA*, **46**, 1533–42.

Holtzer, H., Biehl, J., Yoeh, G., Meganathan, R. & Kaji, A. (1975*a*). Effect of oncogenic virus on muscle differentiation. *Proceedings of the National Academy of Sciences, USA*, **72**, 4051–5.

Holtzer, H., Fellini, S., Rubinstein, N., Chi, J. & Strahs, K. (1976). Cells, myosins and 100 Å filaments. In *Cell motility*, ed. R. Goldman, T. Pollard & J. Rosenbaum. *Cold Spring Harbor Conferences on Cell Proliferation*, **3**, 823–40.

Holtzer, H. & Holtzer, S. (1976). Lineages, quantal cell cycles, and cell diversification. In *Progress in differentiation research*, ed. Muller-Berat. Amsterdam: North-Holland Publishing Co.

Holtzer, H., Marshall, J. & Finck, H. (1957). An analysis of myogenesis by the use of fluorescent antimyosin. *Journal of biophysical and biochemical Cytology*, **3**, 705.

Holtzer, H. & Mayne, R. (1973). Experimental morphogenesis: the induction of somitic chondrogenesis by embryonic spinal cord and notochord. In *Pathobiology of development*, ed. E. Perrin & M. Finegold, pp. 52–64. Baltimore: Williams & Wilkins.

Holtzer, H., Rubinstein, N., Fellini, S., Yeoh, G., Chi, J. D., Birnbaum, J. & Okayama, M. (1975*b*). Lineages, quantal cell cycles, and the generation of diversity. *Quarterly Review of Biophysics*, **8**, 523–57.

Holtzer, H., Sanger, J., Ishikawa, H. & Strahs, K. (1973). Selected topics in skeletal myogenesis. *Cold Spring Harbor Symposia on quantitative Biology*, **23**, 549–66.

Holtzer, H., Strahs, K., Biehl, J., Somlyo, A. P. & Ishikawa, H. (1975*c*). Thick and thin filaments in postmitotic, mononucleated myoblasts. *Science*, **188**, 943–5.

Holtzer, H., Weintraub, H., Mayne, R. & Mochan, B. (1972). The cell cycle cell lineages and cell differentiation. *Current Topics in Developmental Biology*, **1**, 229–56.

Kleinsmith, L. & Pierce, B. (1964). Multipotentiality of the single embryonal carcinoma cell. *Cancer Research*, **24**, 1544–52.

Konigsberg, I. R. (1971). Diffusion-mediated control of myoblast fusion. *Developmental Biology*, **26**, 133–62.

Lajtha, L. G. (1970). Stem cell kinetics. In *Regulation of hematopoiesis*, vol. 1, ed. A. S. Gordon, pp. 111–31. New York: Appleton-Century-Crofts.

Lajtha, L. G. & Schofield, R. (1974). On the problem of differentiation in haemopoiesis. *Differentiation*, **2**, 313–20.

Lash, J. W. (1968). Somitic mesenchyme and its response to cartilage induction.

In *Epithelial–mesenchymal interactions*, ed. R. Fleischmajer & R. Billingham, pp. 165–72. Baltimore: Williams & Wilkins.

Lash, J. W., Rosene, K., Minor, R., Daniel, J. C. & Kosher, R. A. (1973). Environmental enhancement of *in vitro* chondrogenesis. III. The influence of external potassium and chondrogenic differentiation. *Developmental Biology*, **35**, 370–8.

Leder, A. & Leder, P. (1975). Butyric acid, a potent inducer of erythroid differentiation in cultured erythroleukemic cells. *Cell*, **5**, 319–22.

Levitt, D. A. & Dorfman, A. (1973). Control of chrondrogenesis in limb bud cell cultures by bromodeoxyuridine. *Proceedings of the National Academy of Sciences, USA*, **70**, 2201–5.

Levy, J., Terada, M., Rifkind, R. & Marks, P. (1975). Inductions of erythroid differentiation by dimethyl sulfoxide in cells infected with Friend virus: relationship to the cell cycle. *Proceedings of the National Academy of Sciences, USA*, **72**, 28–32.

Linsenmeyer, T., Toole, B. & Trelstad, R. (1973). Temporal and spatial transitions in collagen types during embryonic chick limb development. *Developmental Biology*, **35**, 232–9.

Martin, G. R. (1975). Teratocarcinoma as a model system for the study of embryogenesis and neoplasia. *Cell*, **5**, 229–43.

Mauro, A. (1961). Satellite cell of skeletal muscle fibers. *Journal of Cell Biology*, **9**, 493.

Mauro, A., Shafiq, S. A. & Milhorat, A. T. (eds.) (1970). *Regeneration of striated muscle and myogenesis*. Amsterdam: Excerpta Medica.

Mayne, R., Schiltz, J. & Holtzer, H. (1973). Some overt and covert properties of chondrogenic cells. In *Biology of the fibroblast*, ed. E. Kulonen & J. Pikkarainen, pp. 61–78. New York & London: Academic Press.

McClintock, P. & Papaconstantinou, J. (1974). Regulation of hemoglobin synthesis in a murine erythroblastic leukemic cell: the requirement for replication to induce hemoglobin synthesis. *Proceedings of the National Academy of Sciences, USA*, **71**, 4551–5.

Medoff, J. & Zwilling, E. (1972). Appearance of myosin in chick limb bud. *Developmental Biology*, **16**, 118–26.

Miller, E. & Matoukas, J. (1974). Biosynthesis of collagen: the biochemist's view. *Federation Proceedings*, **33**, 1197–1204.

Nöthiger, R. (1972). The larval development of imaginal discs. In *The biology of imaginal discs*, ed. H. Ursprung & R. Nöthiger, pp. 1–34. Berlin: Springer-Verlag.

Okayama, M., Pacifici, M. & Holtzer, H. (1976). Differences among sulfated proteoglycans synthesized in non-chondrogenic cells, presumptive chondroblasts, and chondroblasts. *Proceedings of the National Academy of Sciences, USA*, **73**, 3224–8.

Okazaki, K. (1975). Spicule formation by isolated micromeres of the sea urchin embryo. *American Zoologist*, **15**, 567–81.

Okazaki, K. & Holtzer, H. (1965). An analysis of myogenesis *in vitro* using fluorescein-labelled antimyosin. *Journal of Histochemistry and Cytochemistry*, **13**, 726–39.

ORKIN, R. W., POLLARD, T. D. & HAY, E. (1973). SDS gel analysis of muscle proteins in embryonic cells. *Developmental Biology*, **35**, 388–94.

PACIFICI, M., BOETTIGER, D., ROBY, K. & HOLTZER, H. (1977). Ts-Rous sarcoma virus transformed chondrogenic cells. *Cell*, **11**, 891–9.

PIERCE, G. B., Jr (1967). Teratocarcinoma: model for developmental concept of cancer. *Current Topics in Developmental Biology*, **2**, 223–46.

ROBY, K., BOETTIGER, D., PACIFICI, M. & HOLTZER, H. (1976). Effects of Rous sarcoma virus on the synthetic programs of chondroblasts and retinal melanoblasts. *American Journal of Anatomy*, **14**, 401–5.

RUTTER, W., PICTET, R. & MORRIS, P. (1973). Toward molecular mechanism of developmental processes. *Annual Review of Biochemistry*, **42**, 601–46.

SAXÉN, L. & TOIVONEN, S. (1962). *Primary embryonic induction*. New York & London: Academic Press.

SCHER, W., PREISLER, H. D. & FRIEND, C. (1973). Hemoglobin synthesis in murine virus induced leukemic cells *in vitro*. III. Effects of 5-bromo-2-deoxyuridine dimethylforamide and dimethylsulfoxide. *Journal of Cell Physiology*, **81**, 63–74.

SCHILTZ, J., MAYNE, R. & HOLTZER, H. (1973). The synthesis of collagen and glycosaminoglycans by dedifferentiated chondroblasts in culture. *Differentiation*, **1**, 97–107.

SEARLS, R. & JANNERS, M. (1969). The stabilization of cartilage properties in cartilage-forming mesenchyme of embryonic chick limb. *Journal of experimental Zoology*, **170**, 365–76.

SELMAN, K. & KAFATOS, F. (1974). Transdifferentiation in the labial gland of silk moths: is DNA required for cellular metamorphosis? *Cell Differentiation*, **3**, 81–94.

SLAVKIN, H. & GREULICH, R. (1975). *Extracell matrix and gene expression*. New York & London: Academic Press.

SOLURSH, M. & REITER, R. S. (1975). Determination of limb bud chondrocytes during a transient block of the cell cycle. *Cell Differentiation*, **4**, 131–7.

SPEMANN, H. (1938). *Embryonic development and induction*. Yale University Press.

STEVENS, L. C. (1967). The biology of teratomas. *Advances in Morphology*, **6**, 1–31.

STOCKDALE, F. & HOLTZER, H. (1961). DNA synthesis and myogenesis. *Experimental Cell Research*, **24**, 508–20.

STOCKDALE, F.. OKAZAKI, K., NAMEROFF, M. & HOLTZER, H. (1964). 5-Bromodeoxymidine: effect on myogenesis *in vitro*. *Science*, **146**, 533–5.

TURNER, D. C., GMUR, R., SIEGRIST, M., BUCKHARDT, E. & EPPENBERGER, H. (1976). Differentiation in cultures derived from embryonic chicken muscle. I. Muscle-specific enzyme changes before fusion in EGTA-synchronized cultures. *Developmental Biology*, **48**, 258–83.

VON DER MARK, H., VON DER MARK, K. & GAY, S. (1976). Collagen synthesis during development of the chick embryo by immunofluorescence. *Developmental Biology*, **48**, 237–9.

WEINTRAUB, H. (1975). The organization of red cell differentiation. In *Cell cycle and cell differentiation*, ed. J. Reinert & H. Holtzer. Berlin: Springer-Verlag.

WEINTRAUB, H., CAMPBELL, G. L. & HOLTZER, H. (1972). Identification of a developmental program using bromodeoxyuridine. *Journal of molecular Biology*, **70**, 337–50.

Weintraub, H., Campbell, G. L., Holtzer, S. & Holtzer, H. (1971). Primitive erythropoiesis in early chick embryogenesis. I. Cell cycle kinetics and the control of cell division. *Journal of Cell Biology*, **50**, 652–88.

Weintraub, H. & Groudine, M. (1976). Chromosomal subunits in active genes have an altered conformation. *Science*, **193**, 848–56.

Weintraub, H. & Holtzer, H. (1972). Fine control of DNA synthesis in developing chick red blood cells. *Journal of molecular Biology*, **66**, 13–26.

White, N. K., Bonner, P. H., Nelson, D. R. & Hauschka, S. D. (1974). Clonal analysis of vertebrate myogenesis. IV. Medium-dependent classification of colony-forming units. *Developmental Biology*, **44**, 346–61.

Woodruff, R. & Telfer, W. (1973). Polarized intercellular bridges in ovarian follicles of the Cecropia moth. *Journal of Cell Biology*, **58**, 172–88.

Yeoh, G. C. T. & Holtzer, H. (1977). The effect of cell density, conditioned medium and cytosine arabinoside on myogenesis in primary and secondary cultures. *Experimental Cell Research*, **104**, 63–78.

Zalin, R. & Montague, W. (1975). Changes in cyclic AMP, adenylate cyclase and protein kinase levels during the development of embryonic chick skeletal muscle. *Experimental Cell Research*, **93**, 55–62.

Zwilling, E. (1968). Morphogenetic phases in development. *Developmental Biology*, Suppl. **2**, 184–90.

Cell position and cell lineage in pattern formation and regulation

L. WOLPERT

Department of Biology as Applied to Medicine,
The Middlesex Hospital Medical School, London W1P 6DB, UK

Pattern formation and cellular differentiation (or cytodifferentiation) are not the same. Pattern formation deals with the problem of the spatial organisation of cellular differentiation. Thus, from the point of view of cellular differentiation there might seem to be no differences occurring with respect to muscle and cartilage differentiation in the arm and leg of a vertebrate. But it is clear from the point of view of pattern formation that the spatial organisation in the arm and leg are significantly different. Pattern formation can be viewed as assigning cell states to cells in an ensemble such that they will differentiate so as to provide a reliable spatial pattern (Wolpert, 1969, 1971).

Homeostasis in relation to pattern formation is the regulation and regeneration of the pattern when the system is perturbed by removal, addition, or transposition of parts. Thus, while it is adult appendages, such as the limbs of urodeles or the heads of hydra, which usually come to mind in relation to regeneration, such cases can be considered part of general regulative mechanisms which are exhibited by early embryos as well. It is thus necessary to consider how patterns are set up if one is to understand how regulation and regeneration take place.

THE PRINCIPLE OF NON-EQUIVALENCE

In considering both the development and regeneration of pattern it is necessary to have a clear idea of the cell states involved. The principle of non-equivalence says that cells of the same differentiation class may have intrinsically different internal states (Lewis & Wolpert, 1976), and they will as a rule be non-equivalent if they give rise to structures differing in shape, pattern, or function. This means that cells are to be characterised not merely by their histological type but by other characteristics such as their position in the animal.

The requirement for emphasising such a principle arises from the natural tendency to classify cells that look alike as being similar. Thus, the histologist can classify some 200 cell types in man. The histologist's discrete classification of the cell types of the adult generates a classification of embryonic cells according to their potentialities in the sense that two cells are regarded as belonging to the same differentiation class if the range of terminally differentiated types to which they give rise is the same. For example, in terms of this definition the mesenchyme of the early bud of the vertebrate hindlimb belongs to the same differentiation class as that of the forelimb, since both give rise to the same classes of terminally differentiated cell types – muscle, cartilage, tendon, dermis, connective tissue. However, this is misleading, since the mesenchyme of forelimbs and hindlimbs is intrinsically different. As Saunders, Cairns & Gasseling (1957) showed, a small block of mesenchyme from the leg bud, from the prospective thigh region, grafted to the tip of an early wing bud developed into a toe, and even induced the overlying ectoderm to form scales instead of feathers. This experiment shows that the leg mesenchyme carries its leg label with it when grafted into the wing, even though it can respond to signals within the wing bud which can affect its proximo-distal character. There are numerous other examples of non-equivalence which can be drawn from both vertebrate and invertebrate systems (Lewis & Wolpert, 1976). There are of course other systems, such as liver, in which the cells may be equivalent.

The significance of non-equivalence in the present context is that there are more cell states than are revealed by terminally differentiated cell types. Moreover, these other cell states are a central feature of development in the development of pattern, which, it can be argued, is the central feature of development. Thus, discussions of development at the cellular level in terms of binary choices (Holtzer *et al.*, 1975) have to be viewed very critically, since they are almost always couched in terms of choices between overt terminal states of differentiation and thus take no account of the variety of non-equivalent cell states.

CELL LINEAGE AND POSITION IN PATTERN FORMATION

For pattern formation, the key question is how different states are assigned to cells in order both that the different terminal cell types develop and that non-equivalent states are specified. The relative importance of lineage and position in this process is an old problem and underlies the distinction between mosaic and regulative embryos. In the former, cytoplasmic localisation in the egg enables cells to be assigned different states in the egg during cleavage, and in principle no interaction between cells is required. In the case of regulative embryos the suggestion is that cells acquire their cell states as the result of cell-to-cell interactions, and are specified according to their relative positions in the embryo. Here, I wish to argue that apart from some special cases of cytoplasmic localisation, the specification of different cell states arises from position-dependent cell-to-cell interaction. Cytoplasmic localisation apart, there is no case known where two daughter cells differ from one another in the absence of position-dependent external influences in the way suggested by Holtzer *et al.* (1975, fig. 1) or by the model of Holliday & Pugh (1975) which proposed somatic segregation of gene activities.

This in no way excludes the possibility that cell division – quantal mitosis – may be necessary for certain changes in state, as suggested by Holtzer *et al.* (1975), or even generates new states, as in the progress zone model (Summerbell, Lewis & Wolpert, 1973), or provides a developmental clock (Holliday & Pugh, 1975). Nor does it exclude a cell lineage associated with the maturation of a determined cell type such as the erythroblast or myoblast as studied by Holtzer *et al.* (1975). The case I am making is that the particular lineage a cell may undergo is determined by its past history and environmental influences. As we have emphasised, signals between cells may be very simple since the specificity of cell state does not lie in the signal, but in the cellular response (Wolpert & Lewis, 1975). At any stage in development a cell has very few courses open to it – though not necessarily only two – and the choice may be guided by simple yes/no signals. Since the cell has a memory, that choice, once made, will in turn control what options become available for the next decision.

The clearest example of cytoplasmic localisation is the specification of germ cells by polar plasm in insects. Transplantation of the pole plasm to the anterior end of the egg can result in cells, which would normally give rise to head structures, forming germ cells (Illmensee & Mahowald, 1974). It would, in principle, be possible to make cytoplasmic localisation so fine-grained that each cell would be uniquely specified at the end of cleavage. There are, however, numerous experiments to show that this is not the case and that the fate of a cell is position-dependent. It is now clear that even in those eggs that were once thought to be mosaic, in the sense that cell specification was based on cytoplasmic localisation, pattern formation involves gradients and interactions within the cytoplasm (Sander, 1975).

It is very important to realise that a well defined lineage in a developmental system, such that the lineage of each cell can be followed, does not imply a lineage mechanism for pattern formation. Nor does it in any way exclude cell-to-cell interactions playing an important role. For example, mollusc development can be characterised by well defined lineages, but there is now good evidence for interactions (Cather, 1971). Again, until recently, it was thought that the development of the highly ordered cellular pattern in the ommatidia of the insect eye represented a lineage, but Shelton & Lawrence (1974) and Reedy, Hanson & Benzer (1976) have now shown that this is not the case. It is also now clear that Mintz's (1974) evidence for clonal determination early in development is not valid (Lewis, Summerbell & Wolpert, 1972; McLaren, 1972).

That the behaviour of a cell in the development of pattern is largely specified by its position within the embryo, in relation to other cells, is a very old one, going back to Driesch in the 1890s, and implies a central role for cell-to-cell interaction. There are numerous experiments both classical and recent, in the embryological literature to justify this view (see Cooke, 1975). For example, the animal half of an early sea urchin embryo will only give rise to a ciliated hollow ball of ectoderm. If a few cells from the vegetal pole – the micromeres – are added, then a more or less normal larva will develop, parts of the ectoderm now forming endoderm under the influence of the added micromeres (Hörstadius, 1973). There is undoubtedly a cell-to-cell interaction involved here and it is difficult to understand the objection of Holtzer *et al.* (1975) to this influence on cell differentiation. In their scheme, if exogenous influences play a role in epigenetic diversification, it is only by permitting the action of predetermined endogenously defined gene sequences. They liken such effects to those of hormones on already differentiated target cells. However, in the simple experiment just described, this is clearly not the case. Whether or not cell division is necessary for changes to be brought about is irrelevant here. The crucial point is that interactions alter the fate of the cells.

A recent experimental approach illustrates again the importance of cell position and interaction. Garcia-Bellido, Ripoll & Morata (1976) have carried out a clonal analysis of wing development in *Drosophila* using the *Minute* mutant. Cells which are heterozygous for this mutant grow more slowly than the wild-type, but nevertheless a normal fly develops. By appropriate genetic technique a clone of wild-type cells can be initiated in tissues the rest of which are heterozygous for *Minute*. This clone then grows much faster than the surrounding cells and can give rise to almost half the cells of the wing, which is, nevertheless, quite normal. If a clone had been initiated from the same cell but grew at the same rate as surrounding cells it would only have formed about one-tenth of the wing. We thus have two situations in which the progeny of a single cell can form either a small part or a large part of the wing, yet in both cases the pattern of cellular differentiation is normal. This result would seem to exclude any mechanism in which the progeny of the cell could be specified by the unequal distribution of cytoplasmic determinants or any other endogenous mechanism based on programmed cell lineage. The behaviour of the cells must be specified by their relative position in the wing by some sort of cellular interaction. It should also be noted that the recent studies on the development of compartments or polyclones (Crick & Lawrence, 1975) require that groups of cells be assigned different states early in development. A compartment is made up of the descendants of a small group of cells and is never a clone. The properties of a compartment are assigned not singly, but in groups of cells. This idea that determination occurs not in single cells but in groups of cells, has been argued very cogently by Nöthiger (1972) from other lineage studies, and clearly requires cell-to-cell interaction.

POSITIONAL INFORMATION

The basic idea of positional information is that pattern formation may result from cells first having their positions specified with respect to boundary regions, as in a co-ordinate system, and then the cells interpreting this positional information according to their genome and developmental history (Wolpert, 1969, 1971). Interpretation can lead to cytodifferentiation or other cellular activities such as cell division or cell movement. The idea is thus intimately related to the idea of non-equivalence, since cells of the same differentiation class may have intrinsically different internal states corresponding to their positional value. Viewed in these terms pattern regulation and regeneration essentially require the respecification of positional values when the system is perturbed, these values then being interpreted again.

Four main kinds of model have been suggested for specifying positional information (Fig. 1).

(i) In *diffusion reaction* schemes, position can be specified by the concentration of a morphogen whose profile falls off monotonically

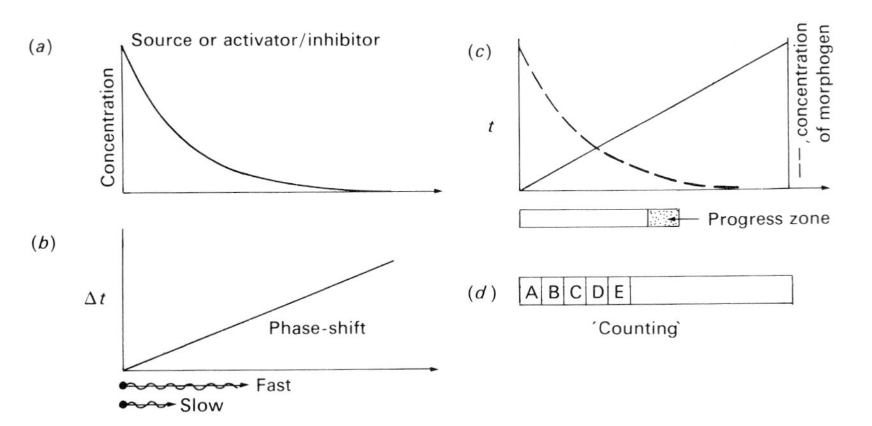

Fig. 1. Some mechanisms for specifying positional information. (*a*) The concentration of a morphogen will fall off with distance from a source, which could result from activator/inhibitor reaction. (*b*) In the phase-shift model, the time difference (Δt) between the arrival of two signals increases with distance from their origin. (*c*) The cells' positions can be specified by how long they remain in the progress zone, where all the cells are dividing. This could be done by measuring time (t), in terms of, for example, cell cycles, or the concentration of a morphogen could decrease the longer they stay in the zone. (*d*) Position could be specified by some sort of 'counting' mechanism from one end.

with distance from a boundary region. This can be achieved by localised source/sink models (Crick, 1970; Wolpert, 1971) or more complex schemes (Gierer & Meinhardt, 1972).

(ii) In the *phase-shift* model of Goodwin & Cohen (1969) position is specified by the difference in time between the arrival of two signals, one propagated faster than the other, from a boundary region.

(iii) In the *progress zone* model, position in a growing system is specified by how long cells spend in a particular region (Summerbell *et al.*, 1973).

(iv) In principle, a *counting* model could provide positional information, but a detailed model has not been worked out (see, for example, Wolpert & Gingell, 1969).

As noted above, the interactions and signal required for specification of positional information can be very simple. The complexity of observed patterns lies in the interpretation of positional information. The first step in interpretation can probably be regarded in terms of a threshold response to the concentration of some chemical compound, and we have shown how this might be realised and how a gradient in a morphogen could plausibly specify 30 discrete states in a field about 1 mm long (Lewis, Slack & Wolpert, 1977).

DEVELOPMENT OF THE VERTEBRATE LIMB

Some of the problems of pattern formation can be seen with respect to the development of the chick limb. We may consider the development of the chick limb in terms of a three-dimensional co-ordinate system and the extreme view would be that the patterns of cartilage, muscles and tendons are specified independently of one another but with respect to the same co-ordinates (Wolpert, Lewis & Summerbell, 1975), the pattern being the result of interpretation of positional values. There seem to be three different mechanisms for specifying position along the three axes (proximo-distal, anteroposterior and dorso-ventral).

For the proximo-distal axis the mechanism is based on how long the cells spend in the progress zone at the tip of the limb (Summerbell *et al.*, 1973). It is not known how time is measured but clearly cell cycles would provide one means. In this form, the change in positional value in the progress zone is in some ways analogous to a lineage mechanism and requires no cell-to-cell interaction other than the apical ectodermal ridge specifying the progress zone. The cells in the progress zone would take on progressively more distal positional values with each cycle. This can be contrasted with the antero-posterior axis for which we have suggested that there is a signal from a boundary region – the zone of polarising activity, ZPA (Saunders, 1972) – at the posterior margin of the bud (Tickle, Summerbell & Wolpert, 1975). Grafts of an additional ZPA to the anterior margin of the bud can result in a mirror image duplication of the limb. This pattern of duplication includes cartilage, muscle, and tendons (Shellswell & Wolpert, 1977). The results from grafting an additional ZPA to different positions along the antero-posterior axis are consistent with a model in which there is a diffusible morphogen produced by the ZPA, the concentration of which provides positional information along the antero-posterior axis. This mechanism involves respecification of cell state under the influence of the ZPA and does not involve cell lineage. The dorso-ventral axis is specified by the ectoderm and can be reversed by inverting the ectoderm (MacCabe, Errick & Saunders, 1974). In this case too, there is a positional signal specifying the pattern.

In terms of our model the cartilaginous elements, for example, are non-equivalent, and there is very good evidence that their growth characteristics are quite different. The wrist, for example, which is initially almost as long as the radius and ulna, hardly grows at all (Lewis, 1975, 1977; Summerbell, 1976). Our model also suggests that the final length of a cartilaginous element is largely determined by two factors: the length of the primordium after it has emerged from the progress zone, and its positional value which controls the programme of subsequent growth, both cell division and matrix secretion.

There is some disagreement as to what extent regulation can occur during early development, when pieces of the limb are removed, or added (see Sengel in Wolpert *et al.*, 1975; Kieny & Pautou, 1976). It is, however, clear that if cells which have left the progress zone are once again placed in it, they can acquire more distal characteristics. The growth of the cartilaginous elements, after they have been laid down, is autonomous. In a large number of experiments in which pieces are exchanged between buds of different ages, the lengths of the elements of the graft are quite unaffected by their host; their growth is donor-specific (Summerbell & Lewis, 1975). There are no long range influences controlling the growth of the parts of the limb after exit from the progress zone; neither hormones nor chalones are involved. Removal of complete transverse slices of tissues of varying thickness from young limb bud in the region of the presumptive forearm and humerus has been found, as would be expected, to give defects in the overall length of the elements. No evidence of overall growth regulation was found. However, if instead of a slice, a large block of mesenchyme is removed, the limbs have a much smaller defect than expected and the difference in behaviour in the two cases is not clearly understood (Summerbell, 1977).

Differentiation of tissue type

It is important to recognise that all these experiments relate to tissues rather than cells and we are ignorant of what is happening at the level of individual cells. We do not know whether positional information is used to specify which cells will become muscle, cartilage, tendon and connective tissue, or whether these are already established as suggested by Dienstman, Biehl, Holtzer & Holtzer (1974) and environmental influences merely favour the growth of one with respect to another, or determine the migration of such determined cells to specific positions. The time of the determination of the different mesenchymal cell types is not known and is controversial, and both Searls (1973) and Caplan & Rosenberg (1975) suggest that mesenchymal cells are labile until a relatively late stage (stage 25) and that up till then environmental influences can direct cells to alternative pathways of differentiation, such as muscle and cartilage.

Our own studies (Cioffi, 1975) on the early development of cartilage and soft tissue suggest that the mesenchyme is labile at early stage of development. Quail tissue, whose cells have a nucleolar marker, was taken from the prospective soft tissue and cartilage at the level of the humerus and grafted to ectopic sites within the limb bud. At stage 20 the tissues appear to be labile and develop according to the position in which they are placed, in agreement with Searls' (1967) observation. By stage 23, the central core of the limb was determined as cartilage – which is earlier than that found by Searls & Janners (1969). When the presumptive soft tissue at stage 23 is grafted so as to be completely surrounded by cartilage it differentiates as cartilage; however, if it is adjacent to soft tissue on at least one side, it differentiates as soft tissue. In none of the experiments was there evidence for cell migration or cell death and the results strongly suggest that the presumptive fate of tissues can be altered by local interactions. However suggestive this is, the behaviour of single cells remains unknown.

We (McLachlan, Bateman & Wolpert, 1976) have also shown that *in vivo* 3-acetylpyridine does not switch mesenchymal cells from muscle differentiation to cartilage differentiation as claimed by Caplan & Rosenberg (1975). It is also important to remember that only about 6 % of the cells of the early limb bud give rise to cartilage and another 6 % to muscle (Lewis, 1977).

Important new information relating to muscle cell differentiation comes from the studies of Chevalier, Kieny, Mauger & Sengel (1977). They have shown that if the somites from a quail are grafted in place of those adjacent to the presumptive limb bud at an early stage then all the muscle cells, and only the muscle cells – not the epimysium or tendons – will be quail. This suggests that the muscle cells have a different lineage to other mesenchymal cells in the limb and come from the somitic mesoderm and not the lateral plate mesoderm. However, this does not exclude the possibility that the respecification can occur. On the contrary, complete removal of the somitic mesoderm at an early stage resulted in the muscles still forming, in which case they must have developed from the lateral plate mesoderm. These results raise important issues which we are currently investigating.

PATTERN REGULATION

Pattern regulation or regeneration can occur by two different mechanisms. In morphallaxis the missing parts are formed by remodelling what remains in order to form a normal but smaller pattern, whereas in epimorphosis there is localised growth and most of the system remains unchanged. These mechanisms can be treated within the context of positional information (Fig. 2). In epimorphosis the positional values in the system remain constant over most of the field, but new values are generated at the cut surface by localised growth. In morphallaxis, on the other hand, positional values change over most of the field and growth is not involved. Morphallaxis is believed to occur in the regulation of early amphibian and sea urchin embryos, and also in the regeneration of hydra. Epimorphosis is more characteristic of later stages in development and the regeneration of adult structures such as limbs in both insects and amphibians.

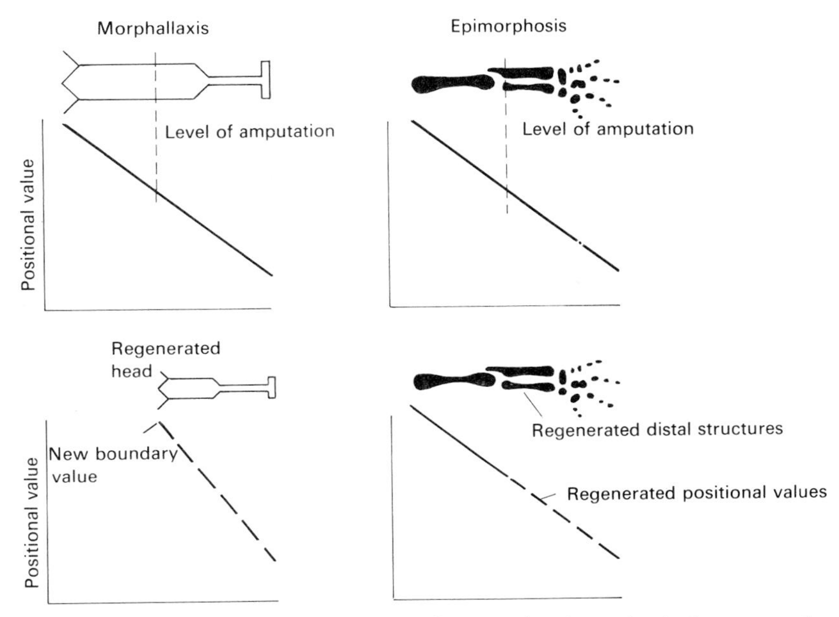

Fig. 2. The difference between morphallaxis and epimorphosis in terms of a gradient in positional information. In both hydra and the amphibian limb, positional values are represented by a simple linear gradient. If the distal (head end) half of hydra is removed, morphallaxis results in a new boundary region being established at the cut end and a steeper gradient is set up. The amphibian limb regenerates by epimorphosis and growth from the region of the amputation re-establishes the original gradient.

Morphallaxis

Hydra provides a good example of morphallaxis. A small fragment will generate a more or less normally proportioned hydra with a head comprising hypostome and tentacles and a basal disc at the foot end. Almost any region can form a foot end or a head end. Polarity is strictly maintained and whether a piece of tissue forms a head or a foot depends upon its relative position. An extensive series of grafting experiments has led to a quite simple model which accounts for head end regeneration (Fig. 3) (Wolpert, Hornbruch & Clarke, 1974). This is based upon two gradients, P and S. P is the positional value and S the positional signal. P is a rather stable parameter whereas S can be regarded as a diffusible morphogen made by a source at the head end. Regulation and regeneration in hydra can be viewed as re-establishing first the boundary values of these gradients and then the intermediate values. The rule for making the

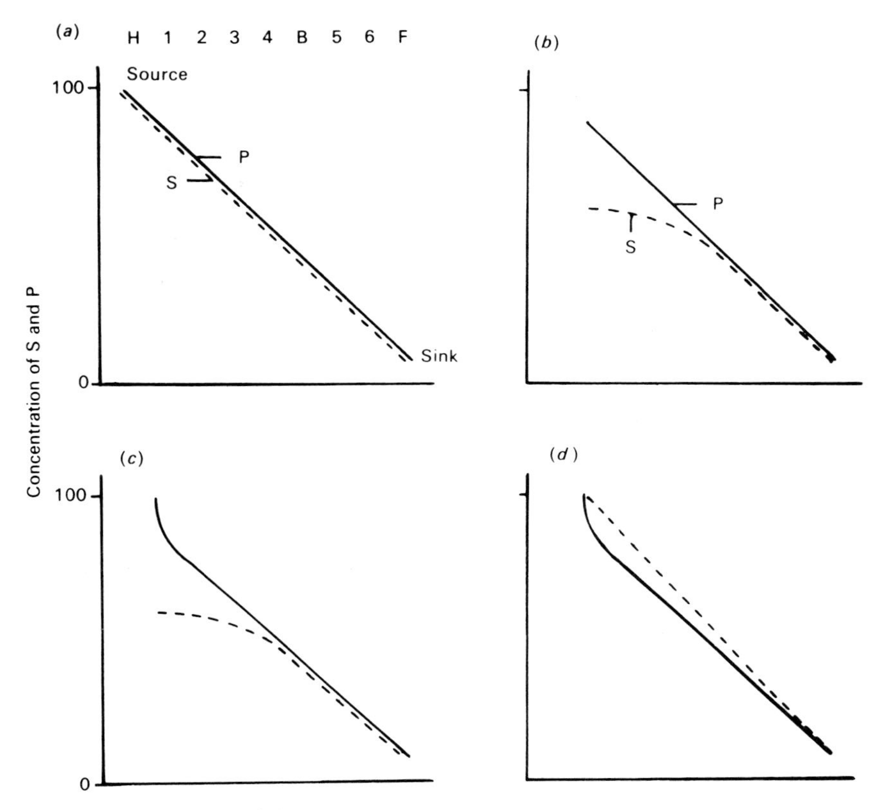

Fig. 3. A simple model to account for regeneration of the head end of hydra. (*a*) The positional value P is taken to be rather stable, whereas the positional signal S is assumed to be diffusible and made at the head end and destroyed at a sink at the foot end. (*b*) Removal of the head end results in a fall of S at the head end due to diffusion. (*c*) When S falls a threshold amount below P, P is synthesised to the boundary value of the head end. (*d*) At the same time S begins to be made and the head end boundary once again acts as a source of S. P returns to a linear gradient more slowly.

head end boundary is that S should fall a threshold amount below P, which leads to P increasing to the head end boundary value. Gierer & Meinhardt (1972) have proposed an analogous model but based upon defined molecular kinetics which does not rely on localised sources.

One of the important features of change in cell state in hydra is that cells can change their positional value up or down the gradient. This can arise from the tissue forming a head end or a foot end, or from being placed in a position such that the boundaries will alter the character of adjacent pieces. This is most dramatically illustrated by the complete reversal of polarity when a head is placed at the proximal end of a gastric region (Wilby & Webster, 1970). The time taken for reversal of polarity was long, up to 72 hours, in contrast to the time taken for a new head end to be determined which is only 4 hours. It seems that changes in P occur more rapidly at a boundary-forming region than away from a boundary. It should be emphasised that these processes occur in differentiated epithelio-muscular cells of the endoderm and ectoderm and have nothing to do with interstitial cells or nerves, since Marcum, Campbell & Romero (1977) have shown that hydra free of both interstitial and nerve cells behave normally with respect to regeneration and polarity reversal.

Cell division does not seem to be a requirement for cells forming the head end of hydra, for hydra exposed to 25 000 rads, which completely blocks cell division, are still able to regenerate a head and initiate new buds (Hicklin & Wolpert, 1973). Moreover it was shown by appropriate grafting procedures that the time for head end determination is only slightly longer than for normal animals. These results strongly suggest that cell division is not necessary for a change in cell state: however, it must be recognised that the irradiation blocks mitosis and not necessarily DNA synthesis. It is not known whether the reversal of polarity involves cell division.

Epimorphosis

The regeneration of the vertebrate limb is the classical example of epimorphosis. The mature limbs of urodeles can regenerate distal structures after amputation and the regenerate arises from a blastema formed from tissue localised at the region of the cut. The progress zone model provides a mechanism for setting up new positional values for the missing distal parts by equating the early blastema with the setting up of the progress zone (Smith, Lewis, Crawley & Wolpert, 1974). If we assume that during early blastema formation the mesodermal cells retain their positional values and these correspond to the level of amputation, the progress zone will then generate distal positional values as in the development of the chick limb. We have put forward a quantitative model which can provide a good description of the observed growth curves for newt limb regeneration, including Spallanzani's original observation that the total time for regeneration is almost independent of the level of amputation. The validity of our model has by no means been established. It successfully accounts for the reduplication of elements when a proximal blastema is grafted in place of a distal one (Iten & Bryant, 1975; Stocum, 1975) but does not account for the intercalary regeneration when a distal blastema is placed on a proximal stump.

In the cockroach leg, regeneration occurs along the proximo-distal axis (Bohn, 1970). In addition, if normally non-adjacent levels are placed together localised growth is stimulated and intermediate structures regenerated (Fig. 4*a*). This is a most important result since it shows that merely the apposition of non-adjacent positional values can stimulate cell division. Moreover, the intercalary re-

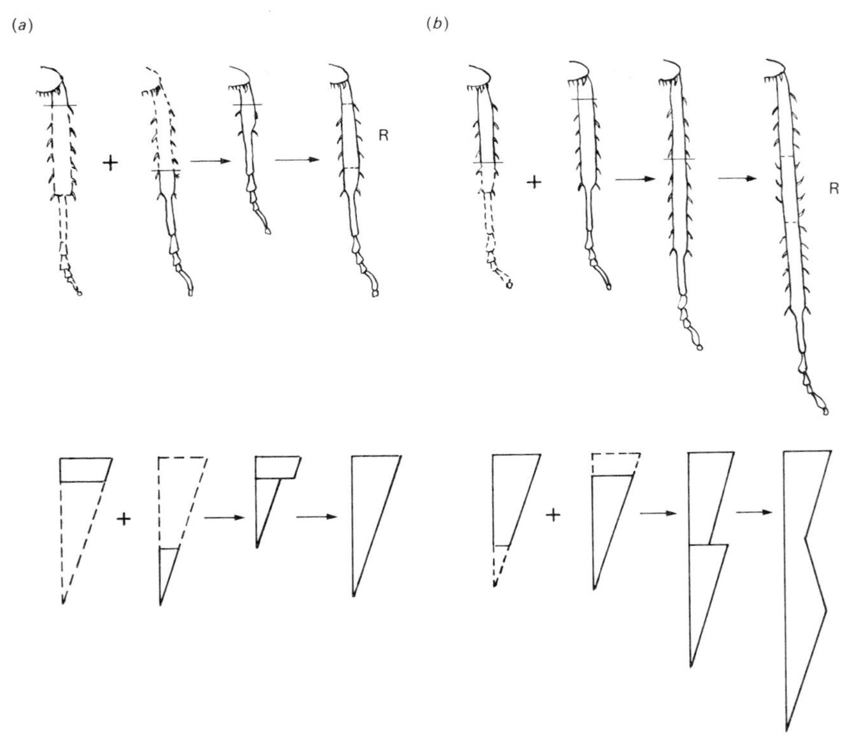

Fig. 4. (*a*) Intercalary regeneration, R, occurs when a proximal part of the tibia of a cockroach leg is grated onto a more distal part. This can be interpreted in terms of re-establishing the gradient in positional value. (*b*) When the same levels are grafted together but this time in a different proximo-distal sequence the same intercalary regeneration, R, takes place, making the tibia even longer. Notice that the bristles point down the gradient (see Bohn, 1970).

generate is formed from both host and graft which shows that tissues can move both up and down with respect to positional values (Bohn, 1976). Of particular importance is the observation that when non-adjacent values are placed next to each other an even longer segment can develop and the regenerate has reversed polarity (Fig. 4*b*). Intercalary regeneration can be regarded as an example of pattern homeostasis and such experiments exclude long range influences, stem cells, and lineage mediated change.

A model to account for this might involve, as in hydra, a stable positional value P and a positional signal S. When non-adjacent positional values are opposed and S can diffuse, it will fall on the high side and rise on the low side of the discontinuity. This could provide the stimulus for growth, and simple rules could generate new positional values.

Similar conclusions can be drawn from other studies on intercalary regeneration, particularly in relation to the polar co-ordinate model of French, Bryant & Bryant (1976). In this model one component of positional information is a value corresponding to a position on a circle and the second value is for a position on a radius. The radial value has in effect been considered above in relation to the proximo-distal axis of the cockroach leg. For the circular value similar rules apply. When normally non-adjacent positional values are confronted in a graft combination growth will occur at the junction until all the intermediate positional values have been intercalated, and for the circular values this is by the shortest route. Again there is evidence that the intercalated structures are derived from both sides of the graft.

Thus, intercalary homeostasis might be viewed in terms of growth continuing until the slope of the gradient in positional values is restored to its original value or until no non-adjacent positional values are present. Moreover Lawrence (1973) has provided evidence that in early insect development growth of segments is controlled by the steepness of a gradient. These concepts for explaining epimorphic regeneration and growth control are radically different from those usually used to account for tissue homeostasis which rely on ideas of chalones and stem cells. It is hard to resist the temptation to apply such ideas to the pattern of cell division in renewing populations, particularly those in the intestinal crypt, renewing epidermal populations, and epiphyseal growth plates. It is important in these systems to bear in mind that position and cell interaction may control growth.

SUMMARY AND CONCLUSIONS

Pattern formation involves assigning cells different states which can lead to varying cellular activities and terminal differentiation. The principle of non-equivalence emphasises that there are many more cell states than are represented by histological cell types. Claims that the changes in cell state are restricted to binary choices should therefore be treated with suspicion, since many cell states are not easily assayed at the level of the single cell. Cytoplasmic differences in the egg can result in daughter cells being assigned different fates, but this is the exception, not the rule. Changes in cell state are mediated by position-dependent cell-to-cell interactions. Whenever lineages diverge this is mediated by such interactions. The signals may be simple, the complexity observed in pattern formation arising from cellular response to a sequence of signals. Pattern formation may be viewed in terms of cells acquiring positional information and then interpreting it in terms of appropriate cellular activities including cellular differentiation. Specification of positional information may involve positional signals, but could also involve a lineage-type mechanism for generating positional values, as in the proximo-distal axis of the chick limb bud.

Pattern regulation can be regarded as the respecification of positional values. In morphallaxis, as in hydra, regeneration involves specification of a new boundary region without growth. Cells may change their positional value either up or down the gradient. In epimorphosis, new positional values are specified during localised growth and most of the system remains unchanged, as in the regeneration of the amphibian limb. Intercalary regeneration in the cockroach leg and insect imaginal discs shows that positional values may increase or decrease. Growth occurs when normally non-adjacent positional values are placed in apposition and continues until the original gradient is restored. It might be possible to apply such concepts to the pattern of cell division and differentiation in typical stem cell systems.

This work is supported by the Medical Research Council.

REFERENCES

Bohn, H. (1970). Interkalare Regeneration und Segmentale Gradienen bei ein Extremitäten von Leucophaea-Larven (Blattaria). I. Femur und Tibia. *Wilhelm Roux Archiv für Entwicklungsmechanik der Organismen*, **165**, 303–41.

Bohn, H. (1976). Regeneration of proximal tissues from a more distal amputation level in the insect leg (*Blaberus craniifer*, Blattaria). *Developmental Biology*, **53**, 285–93.

Caplan, A. & Rosenberg, M. J. (1975). Interrelationship between poly (ADP-Rib) synthesis, intracellular NAD levels, and muscle or cartilage differentiation from mesodermal cells of embryonic chick limb. *Proceedings of the National Academy of Sciences, USA*, **72**, 1852–7.

Cather, J. N. (1971). Cellular interactions in the regulation of development in annelids and molluscs. *Advances in Morphogenesis*, **9**, 67–126.

Chevalier, A., Kieny, M., Mauger, A. & Sengel, P. (1977). Developmental fate of the somitic mesoderm in the chick embryo. In *Vertebrate limb and somite morphogenesis*, ed. D. A. Ede, J. R. Hinchliffe & M. Balls. London: Cambridge University Press. (In press.)

Cioffi, M. (1975). Determination and stability during differentiation in the avian limb. PhD Thesis, University of London.

Cooke, J. (1975). The emergency and regulation of spatial organization in early animal development. *Annual Review of Biophysics and Bioengineering*, **4**, 185–217.

Crick, F. H. C. (1970). Diffusion in embryogenesis. *Nature, London*, **225**, 420–2.

Crick, F. H. C. & Lawrence, P. A. (1975). Compartments and polyclones in insect development. *Science*, **189**, 346–7.

Dienstman, S. R., Biehl, J., Holtzer, S. & Holtzer, H. (1974). Myogenic and chondrogenic lineages in developing limb buds grown *in vitro*. *Developmental Biology*, **39**, 83–95.

French, V., Bryant, P. J. & Bryant, S. V. (1976). Pattern regulation in epimorphic fields. *Science*, **193**, 969–81.

Garcia-Bellido, A., Ripoll, P. & Morata, G. (1976). Developmental compartmentalization in the dorsal mesothoracic disc of *Drosophila*. *Developmental Biology*, **48**, 132–47.

Gierer, A. & Meinhardt, H. (1972). A theory of biological pattern formation. *Kybernetik*, **12**, 30–9.

Goodwin, B. C. & Cohen, M. H. (1969). A phase shift model for the spatial and temporal organization of developing systems. *Journal of theoretical Biology*, **25**, 49–107.

Hicklin, J. B. & Wolpert, L. (1973). Positional information and pattern regulation in hydra: the effect of γ-irradition. *Journal of Embryology and experimental Morphology*, **30**, 741–52.

Holliday, R. & Pugh, J. E. (1975). DNA modification mechanisms and gene activity during development. *Science*, **187**, 226–32.

Holtzer, H., Rubinstein, N., Fellini, S., Yeoh, G., Chi, J., Birnbaum, J. & Okayama, M. (1975). Lineages, quantal cell cycles, and the generation of cell diversity. *Quarterly Review of Biophysics*, **8**, 523–57.

Hörstadius, S. (1973). *Experimental embryology of the echinoderms*. Oxford: Clarendon Press.

Illmensee, K. & Mahowald, A. P. (1974). Transplantation of posterior polar plasm in *Drosophila*. Induction of germ cells at the anterior pole of the egg. *Proceedings of the National Academy of Sciences, USA*, **71**, 1016–20.

Iten, L. E. & Byrant, S. V. (1975). The interaction between the blastema and stump in the establishment of the anterior-posterior and proximo-distal axis of the limb regenerate. *Developmental Biology*, **44**, 119–47.

Kieny, M. & Pautou, M.-P. (1976). Régulation des excédents dans le développement du burgeon de membre de l'embryon d'oiseau. Analyse expérimentale de combinaisons xénoplastiques caille/poulet. *Wilhelm Roux Archives*, **179**, 327–38.

Lawrence, P. A. (1973). The development of spatial patterns in the integument of insects. In *Developmental systems: Insects*, ed. S. J. Counce & C. H. Waddington, pp. 157–211. New York & London: Academic Press.

Lewis, J. H. (1975). Fate maps and the pattern of cell division. *Journal of Embryology and experimental Morphology*, **33**, 419–34.

Lewis, J. H. (1977). Growth and determination in the developing limb. In *Vertebrate limb and somite morphogenesis*, ed. D. A. Ede, J. R. Hinchliffe & M. Balls, pp. 215–28. London: Cambridge University Press.

Lewis, J. H., Slack, J. M. W. & Wolpert, L. (1977). Thresholds in development. *Journal of theoretical Biology*, **65**, 579–90.

Lewis, J. H., Summerbell, D. & Wolpert, L. (1972). Chimaeras and cell lineage in development. *Nature, London*, **239**, 276–9.

Lewis, J. H. & Wolpert, L. (1976). The principle of non-equivalence in development. *Journal of theoretical Biology*, **62**, 479–90.

Marcum, B. A., Campbell, R. D. & Romero, L. J. (1977). Polarity reversal in nerve-free hydra. *Science*, **197**, 771–3.

MacCabe, J. A., Errick, J. & Saunders, J. W. (1974). Ectodermal control of the dorso-ventral axis in the leg bud of chick embryo. *Developmental Biology*, **39**, 69–83.

McLachlan, J., Bateman, M. & Wolpert, L. (1976). Effect of 3-acetylpyridine on tissue differentiation of the embryonic chick limb. *Nature, London*, **264**, 267–9.

McLaren, A. (1972). Numerology of development. *Nature, London*, **239**, 274–6.

Mintz, B. (1974). Gene control of mammalian differentiation. *Annual Review of Genetics*, **8**, 411–70.

Nöthiger, R. (1972). The larval development of imaginal disks. In *The biology of imaginal discs*, ed. H. Ursprung & R. Nöthiger, pp. 1–34. Berlin: Springer-Verlag.

Reedy, D. F., Hanson, T. E. & Benzer, S. (1976). Development of the *Drosophila* retina, a neurocrystalline lattice. *Developmental Biology*, **53**, 217–40.

Sander, K. (1975). Pattern specification in the insect embryo. In *Cell patterning, Ciba Foundation Symposium*, **29**, pp. 241–55. Amsterdam: Associated Scientific Publishers.

Saunders, J. W. (1972). Developmental control of three dimensional polarity in the avian limb. *Annals of the New York Academy of Sciences*, **193**, 29–42.

Saunders, J. W., Cairns, J. M. & Gasseling, M. T. (1957). The role of the apical ridge of ectoderm in the differentiation of the morphological structure and inductive specificity of limb parts in the chick. *Journal of Morphology*, **101**, 57–88.

SEARLS, R. L. (1967). The role of cell migration in the development of the embryonic chick limb bud. *Journal of experimental Zoology*, **166**, 39–45.

SEARLS, R. L. (1973). Newer knowledge of chondrogenesis. *Clinical Orthopaedics*, **96**, 327–44.

SEARLS, R. L. & JANNERS, M. Y. (1969). The stabilization of cartilage properties in the cartilage-forming mesenchyme of the embryonic limb bud. *Journal of Experimental Zoology*, **170**, 365–76.

SHELLSWELL, G. & WOLPERT, L. (1977). The pattern of muscle and tendon development in the chick wing. In *Vertebrate limb and somite morphogenesis*, ed. D. A. Ede, J. R. Hinchliffe & M. Balls. London: Cambridge University Press. (In press.)

SHELTON, P. M. J. & LAWRENCE, P. A. (1974). Structure and development of ammatidia in *Oncopeltus fasciatus*. *Journal of Embryology and experimental Morphology*, **32**, 337–53.

SMITH, A. R., LEWIS, J. H., CRAWLEY, A. & WOLPERT, L. (1974). A quantitative study of blastemal growth and bone regression during limb regeneration in *Triturus cristatus*. *Journal of Embryology and experimental Morphology*, **32**, 375–90.

STOCUM, D. L. (1975). Regulation after proximal or distal transposition of limb regeneration blastemas and determination of the proximal boundary of the regenerate. *Developmental Biology*, **45**, 112–36.

SUMMERBELL, D. (1976). A descriptive study of the rate of elongation and differentiation of the skeleton of the developing chick wing. *Journal of Embryology and experimental Morphology*, **35**, 241–66.

SUMMERBELL, D. (1977). Regulation of deficiencies along the proximal-distal axis of the chick wing bud: a quantitative analysis. *Journal of Embryology and experimental Morphology*, **41**, 137–59.

SUMMERBELL, D. & LEWIS, J. H. (1975). Time, place, and positional value in the chick limb bud. *Journal of Embryology and experimental Morphology*, **33**, 621–43.

SUMMERBELL, D., LEWIS, J. H. & WOLPERT, L. (1973). Positional information in chick limb morphogenesis. *Nature, London*, **244**, 492–6.

TICKLE, C., SUMMERBELL, D. & WOLPERT, L. (1975). Positional signalling and specification of digits in chick limb morphogenesis. *Nature, London*, **254**, 199–202.

WILBY, O. K. & WEBSTER, G. (1970). Experimental studies on axial polarity in hydra. *Journal of Embryology and experimental Morphology*, **24**, 595–613.

WOLPERT, L. (1969). Positional information and the spatial pattern of cellular differentiation. *Journal of theoretical Biology*, **25**, 1–47.

WOLPERT, L. (1971). Positional information and pattern formation. *Current Topics in Developmental Biology*, **6**, 183–224.

WOLPERT, L. & GINGELL, D. (1969). The cell membrane and contact control. In *Homeostatic Regulators*, ed. G. E. W. Wolstenholme & J. Knight, pp. 241–59. *Ciba Foundation Symposium*. London: Churchill.

WOLPERT, L., HORNBRUCH, A. & CLARKE, M. R. B. (1974). Positional information and positional signalling along hydra. *American Zoologist*, **14**, 647–63.

WOLPERT, L. & LEWIS, J. H. (1975). Towards a theory of development. *Federation Proceedings*, **34**, 14–20.

WOLPERT, L., LEWIS, J. H. & SUMMERBELL, D. (1975). Morphogenesis of the vertebrate limb. In *Cell Patterning, Ciba Foundation Symposium*, **29**, pp. 95–119. Amsterdam: Associated Scientific Publishers.

Stem cells in early mammalian development

V. E. PAPAIOANNOU, J. ROSSANT AND
R. L. GARDNER
Department of Zoology, South Parks Road,
Oxford, OX1 3PS, UK

WHAT IS A STEM CELL?

It is difficult to trace the origin of the stem cell concept although one of the earliest references must be that of Regaud (1901) who described stem cells – 'cellules souches' – in his studies of spermatogenesis in the rat. The definition he used is still the one in general use today, namely that stem cells are unspecialised cells which have the ability to divide throughout life so as to produce more stem cells and to produce differentiated cells (Leblond, Clermont & Nadler, 1967). Stem cells are found in 'renewing populations' within the adult (Leblond, 1964) where they counter the loss of cells by destruction, migration or other processes. Examples of such populations are the haemopoietic cells in the lymphomyeloid system, intestinal crypt cells, and the cells of the seminiferous epithelium in the testis. Other adult cell populations can be divided into 'static populations', which never undergo cell division, e.g. neurons, or into 'expanding populations', in which apparently differentiated cells continue to divide very slowly throughout life, e.g. liver cells (Leblond, 1964).

It seems clear, therefore, that the concept of a stem cell can only be strictly applied to the adult where renewing cell populations are in equilibrium, such that the number of stem cells remains more or less constant. This means that on average half the progeny of stem cell divisions form new stem cells while half form differentiating cells. Proof is lacking that this occurs by a strict process of differential mitosis, each division of a stem cell giving rise to one stem cell and one differentiated cell (Rolshoven, 1951; see Leblond & Cheng, 1976), but the end result is to produce populations of stem cells and differentiated cells in dynamic equilibrium.

The concept of stem cell populations cannot be readily applied to the early embryo, because it is a developing system rather than one in equilibrium. Embryonic development involves continuous growth with cytological and morphological changes and consequently permanently self-renewing cell populations do not appear until later. In the past the fertilised ovum has often been referred to as a stem cell (e.g. Lamerton, 1976) because it eventually gives rise to all the stem cell populations of the adult body as well as all differentiated tissues. However, the fertilised ovum certainly cannot be considered a classical stem cell. Its function is not to divide and renew itself but to divide and differentiate. Eventually the ovum will, of course, renew itself by producing new germ cells but this is not a direct process of self-renewal in the usual sense applied to stem cells, as we will discuss later.

The idea that the fertilised egg is a stem cell has probably arisen because of confusion between the concept of a totipotent cell and a stem cell. We would define a *totipotent* cell as one which retains the full range of differentiative capacities of the fertilised egg, i.e., it can, given the right circumstances, form every differentiated tissue found in the embryo and in the adult. In contrast with the *unrestricted* potential of such cells, other cells may be *restricted* in their potential to a greater or lesser degree. Some cells retain a wide range of capabilities although not totipotent and these will be termed *pluripotent*, while other cells may be restricted to the production of one cell type only and these are called *unipotent*. All these terms refer only to the potential of a given cell and do not necessarily have any bearing on its possible stem cell properties. Stem cell populations may be totipotent, pluripotent or unipotent. On the other hand, a cell may be totipotent in that it can differentiate into every tissue of the adult without being a stem cell capable of self-renewal.

In this paper we intend to survey early mammalian embryogenesis looking for evidence of totipotent cells and stem cells, bearing in mind that not all totipotent cells are stem cells and also that any stem cell populations that do occur in early development are not in stable equilibrium as in the adult.

IS THERE A POPULATION OF TOTIPOTENT CELLS IN EARLY DEVELOPMENT AND ARE THEY STEM CELLS?

During early cleavage of the mouse egg, isolated blastomeres appear to be totipotent at least until the 8-cell stage. One cell from a 2-cell egg can actually produce a whole mouse (Tarkowski, 1959*a*, *b*; Hoppe & Whitten, 1972) and although isolated 4- and 8-cell blastomeres cannot apparently form complete embryos (Tarkowski, 1959*a*, *b*; Tarkowski & Wroblewska, 1967; Rossant, 1976*b*), this appears to be due to their small size and not to restriction in cell potency. Such isolated blastomeres can contribute to all tissues of adult mice as well as to earlier extraembryonic membranes when combined with enough 'carrier' blastomeres to restore the normal cell number (Kelly, 1975). However, approximately three cell divisions later, at the 64-cell blastocyst stage (around 3½ days in the mouse), two committed cell types are found – the inner cell mass (ICM) and the trophectoderm. Various experiments suggest that blastocyst formation involves a restriction in the potential of both cell types, trophectoderm giving rise to later trophoblast tissues – ectoplacental cone and giant cells – while the ICM has the potential to form all the foetus and most of its membranes, but not the trophoblast (Gardner, 1968, 1971, 1975; Gardner, Papaioannou & Barton, 1973; Rossant, 1975*a*, *b*, 1976*a*; Van Blerkom, Barton & Johnson, 1976). Thus, although ICM cells resemble the totipotent cleavage stage cells in several ways (Gardner, 1972; Rossant, 1975*a*), they are no longer a totipotent cell population. The exact time between the 8-cell and blastocyst stage when totipotency is lost is not yet known. The totipotent cells of early cleavage do not, therefore, represent a true stem cell population, renewing itself as well as yielding committed trophectoderm cells. Nobody has succeeded in promoting continued division of totipotent blastomeres by suppressing blastocoel formation, which always seems to occur at a set point in development, possibly related to the number of cell divisions undergone or to some other internal clock mechanism (see Rossant, 1977).

ARE THERE CELLS IN THE EARLY EMBRYO WHICH CAN BE CONSIDERED AS RESTRICTED POTENTIAL STEM CELLS?

By the blastocyst stage the first restrictions in the potency of cells have occurred, and later embryonic development appears to involve further progressive restriction of cell potency (Gardner & Rossant, 1976). However, it is possible that within these restricted populations there may be situations in which stem cell properties are apparent, at least for a limited period of time.

Two types of cell develop from the trophectoderm of the blastocyst: the diploid trophoblast cells which continue dividing and form the ectoplacental cone and extraembryonic ectoderm (Gardner & Papaioannou, 1975) and the trophoblast giant cells which do not divide but endoreduplicate their DNA to levels as high as 1024 times the haploid DNA value (Barlow & Sherman, 1972). The primary giant cells arise from the mural trophectoderm cells, beginning at 4½ days, while the secondary giant cells arise from the outside of the ectoplacental cone later in development and eventually come to surround the whole conceptus. Grafting the diploid trophoblast cells of the extraembryonic ectoderm to ectopic sites or explanting them *in vitro* has shown that these cells retain the potential to form giant cells at least until 8½ days gestation and possibly longer (Gardner & Papaioannou, 1975; Diwan & Stevens, 1976; Rossant & Ofer, 1977). This is in marked contrast to blastocyst formation in which morphological differentiation of two cell populations coincides with a restriction in their potential. Thus one can consider that the diploid trophoblast cells act as a unipotent stem cell population, producing more diploid, proliferating cells and also terminally differentiated giant cells. The length of time over which the diploid cells act as stem cells is not known, but toward the end of gestation the giant cells begin to degenerate and are not replaced. It should again be emphasised that since the embryo is not an equilibrium system, the stem cell population of the diploid trophoblast is not likely to remain constant for any prolonged period. It has not yet proved possible to maintain proliferation of the diploid trophoblast once it has been removed from the intact embryo – all cells initiate giant transformation (Rossant & Ofer, 1977; Rossant, 1977).

Within the ICM, the first event after blastocyst formation is the delamination of a monolayer of primitive endoderm on its blastocoelic surface. By 24 hours after initial blastocyst formation these cells can be readily distinguished morphologically from the rest of the ICM (Enders, 1971; Gardner & Papaioannou, 1975) which remains relatively undifferentiated and is known as the primitive ectoderm. Between the 3½ and 4½ day blastocyst stages there might possibly be a short period over which the ICM cells act as stem cells for primitive endoderm and continue to renew themselves. However, this must be of extremely short duration, if it exists at all, since by the time primitive endoderm is morphologically differentiated the primitive ectoderm has lost the potential to form more primitive endoderm (Gardner & Papaioannou, 1975; Hogan & Tilley, 1977; R. L. Gardner & J. Rossant, unpublished data).

The next stage in development is the formation of the egg cylinder and although no new cell types can be detected within the ICM derivatives between 4½ and 7 days, changes in cell relationships do occur because of the formation of the proamniotic cavity. It is at the time of proamniotic cavity formation that size regulation in chimaeras is thought to occur without extensive cell death. An initially double-sized blastocyst formed by aggregating two morulae together will have regulated to normal size by this stage (Buehr & McLaren, 1974). Also, in embryos formed from single 2-cell stage blastomeres (Tarkowski, 1959*a*, *b*) upwards, size regulation must occur but it is not certain exactly when. The phenomenon of size regulation suggests that there is a population of cells within the postimplantation embryo which can vary its rate of proliferation before undergoing further differentiation. Since there are no outward morphological signs of new differentiating cells appearing at the same time as the presence of this varying proliferative population, it is not clear whether it can be considered a true stem cell population.

In the normal embryo between 6½ and 7 days, there is a small area of the ectoderm, the proliferative zone, with a very high mitotic index, and a cell cycle time of about 2½ hours (Snow, 1976 and personal communication). Although neither this area nor the products of these rapid cell divisions are morphologically distinguishable, we have here a strong candidate for a stem cell population. The proliferative zone remains small (increasing in size from 60 to only 600 cells between 6½ and 7½ days) indicating that some of the products of division move out of the zone and assume slower cell cycle times in keeping with the rest of the embryonic ectoderm while others renew the rapidly cycling population. It is not clear what the differentiated products of this hypothetical stem cell population are.

Around 7½ days the primitive streak becomes apparent opposite the proliferative zone. Through the primitive streak arise the mesoderm cells, apparently pushed out from the ectoderm and coming to lie between the ectoderm and the endoderm. At the anterior end of the primitive streak a structure known as the head process is formed and this may be the region where the definitive endoderm of the foetus arises (see Rossant & Papaioannou, 1977). Methods of experimentally approaching the study of cell commitment in these tissues are restricted by the inaccessibility of the postimplantation embryo to direct manipulation and most of our information is derived from studying ectopic grafts of isolated tissues and various tissue combinations. These studies have various limitations (see Rossant, 1977) and interpretations of the results must be made with caution. However, one or two tentative conclusions can be drawn.

Isolated ectoderm from two-layered rat egg cylinders and ectoderm plus mesoderm from primitive streak stage embryos will form gut derivatives in ectopic sites (Levak-Svajger & Svajger, 1971, 1974) but slightly later, by the head-fold stage, isolated ectoderm will no longer form gut derivatives. However, endoderm plus mesoderm isolated from the head-fold embryo does form gut structures (Levak-Svajger & Svajger, 1974). This experiment suggests that the definitive endoderm is established by the head-fold stage and that the ectoderm has lost the potential to form more endoderm by that stage. The exact time of origin of the definitive endoderm is not certain but it is probably close to the head-fold stage, since endoderm isolated from earlier mouse egg cylinders forms only primitive endoderm derivatives in ectopic sites (Diwan & Stevens, 1976). Thus, it seems unlikely that there can be a population of ectoderm cells acting as a stem cell population for definitive endoderm over any prolonged period of time and more likely that definitive endoderm formation is a two-way commitment event like primitive endoderm formation.

The formation of mesoderm from the ectoderm does not seem to be such a narrowly timed event and a stem cell population may be involved. Although mesoderm formation begins at the primitive streak stage, it is not known how long new mesoderm continues to be produced from the ectoderm. When isolated and grafted in ectopic sites, ectoderm can form mesodermal derivatives at least until the head-fold stage (Levak-Svajger & Svajger, 1974) and this potential is not confined to the ectoderm in the area of the primitive streak (Svajger & Levak-Svajger, 1976). This does suggest that there may be a stem cell population in the ectoderm which can produce mesoderm over a period of time. Could this stem cell population be related to the proliferative zone? Calculations of the proliferative capacity of this zone are compatible with this idea (M. H. L. Snow, personal communication). However, since the products of the zone cannot be traced it is equally possible that they form other tissues.

After the primitive streak stage, a circulatory system is established and organogenesis begins. From this time on one can observe the formation of systems which in the adult are characterised by the presence of stem cell populations. Other authors will be discussing such populations and we will not follow their development any further.

WHAT ABOUT THE GERM CELLS?

We have suggested in the previous sections that there is no totipotent cell line during early mammalian embryogenesis, development occurring by a series of steps in which potency is restricted. However, this discussion has avoided the question of the origin of the germ cells which are destined to express totipotency when male and female cells fuse to form the fertilised egg. Continuity of the germ line was formulated as a hypothesis by Weissmann (1892), who suggested that germ-cell determining factors or germ plasm is present in the gametes and in the fertilised egg. This germ plasm is transferred during embryogenesis to certain cells which are thus destined to form more germ cells. By this process the germ cell line would be continuously maintained from generation to generation as a segregated totipotent cell population. Evidence for germ plasm has been found in various vertebrate and invertebrate species (Beams & Kessel, 1974; Eddy, 1975), particularly anuran amphibians (see Blackler, 1970) and *Drosophila* (see Mahowald, 1970). It has been shown in these species that there is a certain region of the fertilised egg (vegetal pole cytoplasm in anurans and posterior pole cytoplasm in *Drosophila*) which if destroyed by ultraviolet light or other means results in sterile animals (e.g. Geigy, 1931; Smith, 1966). Fertility can be restored by transplanting such cytoplasm from another egg to the damaged region (Smith, 1966; Okada, Kleinman & Schneiderman, 1974). However, it has been suggested that the effect of ultraviolet light in anurans may be to derange the orderly processes by which germ cells are segregated rather than or as well as specifically destroying the germ plasm (Züst & Dixon, 1975; Beal & Dixon, 1975). In *Drosophila* any such objections have been overcome by the demonstration that functional germ cell precursors can be formed even when germ plasm is grafted to abnormal sites in intact eggs (Illmensee & Mahowald, 1974; Illmensee, Mahowald & Loomis, 1976). Evidence for such a precociously segregating germ plasm has not been forthcoming in mammals. Primordial germ cells are first seen in the endoderm of the yolk sac, near the base of the allantois (Chiquoine, 1954; Spiegelman & Bennett, 1973; Clark & Eddy, 1975), although they probably arise from the ectoderm or mesoderm (Ozdzenski, 1967; Gardner & Rossant, 1976). From the yolk sac they apparently migrate to the hindgut endoderm and thence to the gonadal ridges via the hindgut mesentery (Everett, 1943; Mintz, 1957). There is now good evidence to suggest that germ cells of the adult are derived from the initial population of primordial germ cells recognised in the early genital ridges (Peters, 1970). However, events in germ cell formation prior to this time are unclear. There is no clear morphological evidence of germ plasm in the early mammalian embryo, the only possible candidate being the 'nuage', which is dense fibrous material, recognisable in the electron microscope, present throughout much of the life history of the germ cells of mammals and other species (Eddy, 1974, 1975). However, no-one has yet traced this material back from the germ cells into the early embryo to see if it really has the characteristics of germ plasm and so the relationship between nuage and germ cell determination remains open in mammals and, indeed, in other animals as well (Eddy, 1975).

Apart from the lack of obvious germ plasm, the mammalian embryo also shows no evidence of an early segregating line of cells which can only produce germ cells and no somatic cells. Gardner & Lyon (unpublished data) have shown that single late 4½ day primitive ectoderm cells can produce both somatic and germ cell line chimaerism when injected into blastocysts, thus showing that a segregated germ *cell* line as such does not exist by this stage (Gardner & Rossant, 1976). However, this experiment does not preclude the existence of segregating germ *plasm* on the lines envisaged by Whitington & Dixon (1975) in *Xenopus*. They suggest that the reason why the total number of cells containing germ plasm increases very little between the 4-cell and tail-bud stage in *Xenopus* is that the germ plasm is only distributed to one daughter cell at each mitosis; later it moves to a perinuclear position and is distributed to both daughters of each mitosis and so the germ cell pool increases in size. If such a system were operating in the mouse embryo, one could explain how 4½ day ectoderm cells produce both somatic and germ cell segregants by assuming that the germ plasm is still being transferred to only one daughter cell at this stage.

Thus the presence of a segregating germ plasm in mammals cannot be excluded, but it is still possible to consider that the formation of germ cells is a purely *de novo* epigenetic event, occurring in much the same way as any other process of cell commitment. There is already evidence to suggest that primordial germ cells may arise epigenetically from somatic cells in urodele amphibians (Kocher-Becker & Tiedemann, 1971; Sutasurja & Nieuwkoop, 1974). The commitment of germ cells would occur at an unknown time between 4½ and 8½ days in the mouse. The stumbling block in most people's minds over such an interpretation is that it apparently requires formation of a totipotent cell line from a population of cells with restricted potential, namely the embryonic ectoderm. This problem can be overcome by considering germ cells as just another differentiated cell line and not as cells which are necessarily totipotent throughout their life history. Certainly mature sperm and eggs possess a high degree of morphological differentiation from each other and from somatic cells. One must postulate that one of the differentiated properties of germ cells is that they have the capacity to express totipotency under certain specific conditions. The first obvious condition is that of fertilisation, i.e. the fusion of two germ cells. Secondly, parthenogenetic activation of the egg may also cause the expression of totipotency, although mice have not yet been born from such eggs. The latest stage reported so far is 10½ days (Kaufman, Barton & Surani, 1977). Finally, conditions leading to the formation of teratocarcinomas may also promote expression of totipotency as will be discussed shortly.

Thus, the observations on germ cells in the early embryo can be used either to support the notion of a segregating germ plasm or of an epigenetic origin for the germ cells, and without suitable markers specific for the earlier progenitors of germ cells, this problem is extremely difficult to resolve. Whichever hypothesis is true, the renewal of germ cells from one generation to the next can hardly be considered as a simple stem cell renewal system.

HOW DO TERATOCARCINOMA STEM CELLS RELATE TO CELLS OF THE NORMAL EMBRYO?

In this consideration of the potential of embryonic cells during mammalian embryogenesis we have found no conclusive evidence of totipotent cells beyond the blastomeres of the 8-celled embryo. There are examples of possible restricted potency stem cell populations, but these are transitory in nature due to the non-equilibrium state of the growing embryo. We have also discussed two theories on the origin and nature of the germ cells, which are cells capable of expressing totipotency given certain conditions. It is perhaps relevant at this point to discuss the presence or absence of totipotent stem cells in the malignant derivatives of germ cells and normal embryos.

The formation of the adult mammal obviously requires an extremely orderly and complex sequence of developmental events. The explantation and grafting of embryos or parts of embryos to ectopic sites disrupts this sequence and can result in disorganised growth and the formation of tumours. These are characteristically composed of a variety of differentiated cell types and in some cases contain a proliferating 'stem cell' population (for reviews see Damjanov & Solter, 1974; Martin, 1975; Graham, 1977). Similar tumours occur spontaneously in the gonads of certain strains of mice. The general term for these gonadal or embryo derived tumours is teratoma, although current usage favours a subdivision into benign tumours called teratomas and malignant tumours called teratocarcinomas (Pierce, 1967; Stevens & Pierce, 1975). Malignancy is defined in this case by continuous growth, retransplantability in histocompatible hosts and, of particular interest for us, the presence of an undifferentiated stem cell population, the embryonal carcinoma cells, so called because of their resemblance to early embryonic cells. These embryonal carcinoma cells have characteristics of classical stem cell populations. They have the capacity for indefinite proliferation and also the capacity to produce cells for maturation or differentiation pathways. We pointed out earlier that embryonic stem cells differ from adult stem cells in that they function in a growing and developing organism whereas the adult stem cells are concerned with the maintenance of tissue homeostasis in the mature animal. The stem cells of teratocarcinomas are likewise contributing to growing and expanding tissues and although there may be some controls on their proliferation, these do not have the effect of maintaining a steady state. They are similar to stem cells of other tumours in that growth of the tumour occurs by stem cell renewal as well as differentiation of cells characteristic of the particular tumour (Pierce & Wallace, 1971; Wylie, Nakane & Pierce, 1973). The differentiated cells in this case are representatives of all three embryonic germ layers.

To understand the relationship between teratocarcinoma stem cells and embryonic cells one must first consider the various ways in which teratocarcinomas may arise. Spontaneously occurring testicular tumours are first seen in the foetal testes of 129 strain mice on the 15th day of gestation (Stevens, 1959) and are presumed to develop from the primordial germ cells (Stevens, 1967). Spontaneous ovarian tumours frequently occur in LT strain females starting at 30 days of age. These arise from parthenogenetically activated cleaving eggs. (Stevens & Varnum, 1974). These gonadal tumours are usually teratomas but occasionally are transplantable teratocarcinomas. Experimental foetectomy in rats has been found to lead to formation of teratomas by regions of the yolk sac trapped outside the uterus during operation (Sobis & Vandeputte, 1975). Furthermore, it is claimed that such growths are not due to primordial germ cells located in this tissue because they also occur in busulphan-treated pregnant rats subjected to this operation (Merchant, 1975; Sobis & Vandeputte, 1976). Teratocarcinomas can be experimentally produced by transplantation of pre- or post-implantation embryos to ectopic sites in isogeneic hosts (Stevens, 1968, 1970; Solter, Škreb & Damjanov, 1970). Tumours are formed in this way with increasing frequency from the early cleavage stages through the 8th day of gestation. From the 9th day, transplantation of whole embryos results only in teratoma formation and it is not until the 11th to 13th day that teratocarcinomas can once again be produced, this time by the explantation and grafting of male genital ridges to ectopic sites (Stevens, 1964). The reason teratocarcinomas can be derived from some embryonic stages and not others is not understood. The primary teratocarcinomas formed in these diverse ways, however, are indistinguishable from one another although their characteristics on transplantation *in vivo* or *in vitro* may diverge.

It is germane to our discussion of potency in embryonic cells to examine the potency of the stem cells of teratocarcinomas. These stem cells, the embryonal carcinoma cells, can either be studied *in vivo* as progressively growing transplantable tumours or *in vitro* as tissue culture cell lines. The multipotentiality of individual embryonal carcinoma cells has been demonstrated by in-vivo clones from a spontaneous testicular tumour (Kleinsmith & Pierce, 1964). A wide range of embryonic and adult cells was found in tumours from these clones, including trophoblast giant cells, visceral and parietal yolk sac, muscle, bone, cartilage, epithelium, astrocytes, and numerous other examples of cell types derivative of all three germ layers. Differentiation of embryonal carcinoma cells that have been maintained *in vitro* is often more limited. Transplantation of clones to histocompatible hosts confirms their pluripotency however (Kahan & Ephrussi, 1970), and under certain conditions clones of embryonal carcinoma cells can form a wide range of differentiated cell types comparable to in-vivo tumours (Evans & Martin, 1975). Foetal erythrocytes and cells with morphological characteristics of trophoblastic giant cells have been reported in clonal organ cultures of embryonal carcinoma cells (Nicolas, Avner, Gaillard & Jakob, 1976; Cudennec & Nicolas, 1977). It is clear that the culture conditions are important for obtaining differentiation of a range of cell types *in vitro*. Specific limitations of particular cell lines may thus be due to environmental factors, and not to inherent differences in cell potential. Another possibility is that genetic changes have occurred during culture that limit the potential of some cell lines. Karyotypic changes do in fact commonly occur *in vitro* (Guénet, Jakob, Nicolas & Jacob, 1974; McBurney, 1976); embryonal carcinoma cell lines with normal karyotypes are rare.

The absence *in vitro* and *in vivo* of highly organised structures such as lung or liver and certain highly differentiated cell types such as sperm is perhaps not surprising due to the disorganised state of the tissue in a tumour or in a culture dish. Tissues that require complex inductive interactions, hormonal influences or particular structural configurations are unlikely to form in spite of competence of the cells. A definitive test for the totipotency of embryonal carcinoma stem cells is to put them back into an embryo and to see what cell types are formed from the progeny of the embryonal carcinoma cells under the influence of normal embryonic cells in a uterine environment. It can be seen from the resulting chimaeric mice that when embryonal carcinoma cells are injected into blastocysts they can make a functional contribution to various adult tissues (Brinster, 1974; Mintz & Illmensee, 1975; Papaioannou, McBurney, Gardner & Evans, 1975; Illmensee & Mintz, 1976). Clonal in-vitro embryonal carcinoma cells can contribute to trophoblast as well as adult tissues of all three germ layers following injection into blastocysts (Papaioannou *et al.*, 1975). Lacking so far, however, is evidence of colonisation of the germ line by in-vitro clones, but again, this may be due largely to the known karyotypic abnormalities of these cell lines. However, injection of cells from in-vivo tumours has demonstrated the totipotency of embryonal carcinoma cells, since in addition to a range of adult tissues, functional sperm were formed (Mintz & Illmensee, 1975). Although totipotency has only been conclusively demonstrated for a single embryonal carcinoma tumour, it is reasonable to suggest that given optimal conditions totipotency may be a general feature of embryonal carcinoma cells.

Assuming the totipotency of these stem cells, what is their relationship to normal embryonic cells and what can they tell us about the existence of totipotent stem cells in the embryo? The first problem is that of the origin of the embryonal carcinoma cells. Ultrastructurally the embryonal carcinoma cells are remarkably similar to primordial germ cells from 15–18 day foetal testes (Pierce & Beals, 1964) and to embryonic ectoderm cells of the presomite embryo (Solter, Damjanov & Škreb, 1970). All of these cells contain the enzyme alkaline phosphatase and have an undifferentiated appearance with many free ribosomes and little endoplasmic reticulum, unlike the differentiated embryonic endoderm. No evidence of a transition or transformation point can be found between embryonic and tumour cells (Damjanov & Solter, 1975). As mentioned earlier the spontaneous testicular tumours as well as the tumours formed following genital ridge transplantation are believed to be derived from the primordial germ cells that reach the genital ridges between the 9th and 11th days of gestation. Likewise the ovarian tumours arise from germ cells, but later in development and only after parthenogenetic activation of ova. Development of these activated ova proceeds through blastocyst formation to the stage of endoderm and ectoderm formation before the tissues become disorganised and a tumour is obvious (Stevens, 1975).

Whereas testicular tumours appear to arise directly from primordial germ cells, it has been argued that this may not be the case for other teratocarcinomas (Stevens, 1968, 1970). Teratocarcinomas derived from ectopic transplantation of presomite embryos are believed to arise from the embryonic ectoderm, largely because of ultrastructural and histochemical similarities (Damjanov & Solter, 1975) but also because of the increasing efficiency of teratocarcinoma production from the 1st to 8th days, which is coincident with the embryonic ectoderm formation. Direct equivalence of embryonic ectoderm and embryonal carcinoma is, however, difficult to imagine since the available evidence discussed earlier leads us to believe that the embryonic ectoderm has lost its totipotent character at the time of delamination of the primitive endoderm. One would have to postulate a differentiation of the embryonic ectoderm to an earlier totipotent cell type. However, we have not completely ruled out the possibility of a continuous, totipotent germ cell line. It would be in accord with the presumed germ cell origin of testicular teratocarcinomas if embryonal carcinoma cells were derived from this cell line.

If, on the other hand, one favours epigenetic factors as being important in the determination of the germ line, one could postulate the following explanation. We have suggested the possibility that it is a differentiated property of mature germ cells to be able to express totipotency, but only after they have been activated by the event of fertilisation. The primordial germ cells and their precursors might also possess this property once they have been committed to the germ line pathway by epigenetic factors. A restoration of totipotency by some external stimulus similar to that observed at fertilisation might then account for the formation of embryonal carcinoma cells. In the case of ovarian teratocarcinomas this stimulus would take the form of parthenogenetic activation, creating the abnormal situation of a cleaving egg still within the ovary. The observation that an embryonic-like structure with ectoderm and endoderm is formed before any uncontrolled growth (Stevens & Varnum, 1974) indicates that germ cell precursors could be determined first and embryonal carcinoma cells arise secondarily as the result of the stimulus of disorganised growth. A similar case could be made for the formation of teratocarcinomas from ectopically transplanted presomite embryos. The low incidence of teratocarcinoma formation from early cleavage stages could be a reflection of poor conditions for growth and differentiation of embryonic cells (Damjanov & Solter, 1974) including germ cell precursors. The increased incidence with age of the embryo would thus be due to the increase in the probability of reaching the stage of germ line commitment before disturbed conditions provide the stimulus for restoration of totipotency. This line of reasoning would lead us to expect that teratocarcinomas could be produced at any stage after the determination of germ cell precursors, so that the absence of teratocarcinoma formation during the 9th–11th days of gestation requires an explanation. It may be that this is an experimental artefact dependent on the relationship between commitment and cell number. If germ cell commitment is an epigenetic event, one would expect proportionally more potential germ cells than other cells in early stages. If we postulate that a commitment event occurs around the 8th–9th day, these cells would undoubtedly make up a small proportion of the whole embryo. Also the inherent organisation of the embryo may have reached such a state that ectopic transplantation does not easily disrupt cellular relationships to the extent necessary to provide a sufficient stimulus for restoration of totipotency in primordial germ cells. The transplantation of later genital ridges overcomes these restrictions in that the proportion of germ cells to non-germ cells is increased by the removal of the rest of the embryo and, since only a small part of the foetus is transplanted, it is easy to imagine that the disruption of cellular relationships is severe.

The malignant derivatives of normal embryos do form stem cell populations with many classical stem cell characteristics. Totipotency is probably a feature of these cells which makes it likely that they are malignant derivatives either of totipotent cells or cells that have the capacity to become totipotent (Pierce & Beals, 1964). In our discussion of normal embryogenesis we have suggested that the only potentially totipotent cells beyond early cleavage stages are the germ cells. Two ideas on their origin were considered, that of a continuous totipotent cell line and that of an epigenetic control of germ cell commitment with the capacity to become totipotent being one of their differentiated characteristics. We have suggested that all embryonal carcinoma stem cells could be derived from germ cells or their precursors. The stem cells could arise directly from totipotent germ cells, primordial germ cells, or their precursors as required by the germ line theory or they could arise epigenetically from the primordial germ cells or their precursors which become totipotent following an external stimulus.

We would like to thank J. K. Heath and M. H. L. Snow for valuable discussion. The authors' own work was supported by the Medical Research Council. V. E. Papaioannou was also supported by the Cancer Research Campaign. J. Rossant is a Beit Memorial Junior Research Fellow.

REFERENCES

Barlow, P. W. & Sherman, M. I. (1972). Biochemistry of differentiation of mouse trophoblast: studies on polyploidy. *Journal of Embryology and experimental Morphology*, **27**, 447–65.

Beal, C. M. & Dixon, K. E. (1975). Effect of U.V. on cleavage of *Xenopus laevis*. *Journal of experimental Zoology*, **192**, 277–83.

Beams, H. W. & Kessel, R. G. (1974). The problem of germ cell determinants. *International Review of Cytology*, **39**, 413–80.

Blackler, A. W. (1970). The integrity of the reproductive cell line in the Amphibia. *Current topics in developmental Biology*, **5**, 71–87.

Brinster, R. L. (1974). The effect of cells transferred into the mouse blastocyst on subsequent development. *Journal of experimental Medicine*, **140**, 1049–56.

Buehr, M. & McLaren, A. (1974). Size regulation in chimaeric mouse embryos. *Journal of Embryology and experimental Morphology*, **31**, 229–34.

Chiquoine, A. D. (1954). The identification, origin and migration of the primordial germ cells in the mouse embryo. *Anatomical Record*, **118**, 135–46.

Clark, J. M. & Eddy, E. M. (1975). Fine structural observations on the origin and associations of primordial germ cells of the mouse. *Developmental Biology*, **47**, 136–55.

Cudennec, C. A. & Nicolas, J. F. (1977). Blood formation in a clonal cell line of mouse teratocarcinoma. *Journal of Embryology and experimental Morphology*, **38**, 203–10.

Damjanov, I. & Solter, D. (1974). Experimental teratoma. *Current Topics in Pathology*, **59**, 69–130.

Damjanov, I. & Solter, D. (1975). Ultrastructure of murine teratocarcinomas. In *Teratomas and differentiation*, ed. M. I. Sherman & D. Solter, pp. 209–20. New York & London: Academic Press.

Diwan, S. B. & Stevens, L. C. (1976). Development of teratomas from the ectoderm of mouse egg cylinders. *Journal of the National Cancer Institute*, **57**, 937–9.

Eddy, E. M. (1974). Fine structural observations on the form and distribution of nuage in germ cells of the rat. *Anatomical Record*, **178**, 731–58.

Eddy, E. M. (1975). Germ plasm and the differentiation of the germ cell line. *International Review of Cytology*, **43**, 229–80.

ENDERS, A. C. (1971). The fine structure of the blastocyst. In *Biology of the blastocyst*, ed. R. J. Blandau, pp. 71–94. Chicago: University of Chicago Press.

EVANS, M. J. & MARTIN, G. R. (1975). The differentiation of clonal teratocarcinoma cell cultures *in vitro*. In *Teratomas and differentiation*, ed. M. I. Sherman & D. Solter, pp. 237–50. New York & London: Academic Press.

EVERETT, N. B. (1943). Observational and experimental evidence relating to the origin and differentiation of the definitive germ cells in mice. *Journal of experimental Zoology*, **92**, 49–91.

GARDNER, R. L. (1968). Mouse chimaeras obtained by the injection of cells into the blastocyst. *Nature, London*, **220**, 596–7.

GARDNER, R. L. (1971). Manipulations on the blastocyst. In *Advances in the Biosciences*, **6**, 279–96.

GARDNER, R. L. (1972). An investigation of inner cell mass and trophoblast tissue following their isolation from the mouse blastocyst. *Journal of Embryology and experimental Morphology*, **28**, 279–312.

GARDNER, R. L. (1975). Analysis of determination and differentiation in the early mammalian embryo using intra- and inter-specific chimaeras. In *The developmental biology of reproduction, 33rd Symposium of the Society for Developmental Biology*, ed. C. L. Markert, pp. 207–38. New York & London: Academic Press.

GARDNER, R. L. & PAPAIOANNOU, V. E. (1975). Differentiation in the trophectoderm and inner cell mass. In *The early development of mammals, 2nd Symposium of the British Society for Developmental Biology*, ed. M. Balls & A. E. Wild, pp. 107–32. London: Cambridge University Press.

GARDNER, R. L., PAPAIOANNOU, V. E. & BARTON, S. C. (1973). Origin of the ectoplacental cone and secondary giant cells in mouse blastocysts reconstituted from isolated trophoblast and inner cell mass. *Journal of Embryology and experimental Morphology*, **30**, 561–72.

GARDNER, R. L. & ROSSANT, J. (1976). Determination during embryogenesis. In *Embryogenesis in mammals, Ciba Foundation Symposium*, pp. 5–25. Amsterdam: Associated Scientific Publishers.

GEIGY, R. (1931). Action de l'ultra-violet sur le pole germinal dans l'oeuf de *Drosophila melanogaster. Revue suisse de Zoologie*, **38**, 187–288.

GRAHAM, C. F. (1977). Teratocarcinoma cells and normal mouse embryogenesis. In *Early mammalian embryogenesis, MIT monograph*, ed. M. I. Sherman. (In press.)

GUÉNET, J. L., JAKOB, H., NICOLAS, J. F. & JACOB, F. (1974). Tératocarcinome de la souris: étude cytogénètique de cellules à potentialités multiples. *Annales de microbiologie*, **125**, 135–50.

HOGAN, B. & TILLY, R. (1977). *In vitro* culture and differentiation of normal mouse blastocysts. *Nature, London*, **265**, 626–9.

HOPPE, P. C. & WHITTEN, W. K. (1972). Does X-inactivation occur during mitosis of first cleavage? *Nature, London*, **239**, 520.

ILLMENSEE, K. & MAHOWALD, A. P. (1974). Transplantation of posterior pole plasm in *Drosophila*. Induction of germ cells at the anterior pole of the egg. *Proceedings of the National Academy of Sciences, USA*, **71**, 1016–20.

ILLMENSEE, K., MAHOWALD, A. P. & LOOMIS, M. R. (1976). The ontogeny of germ plasm during oogenesis in *Drosophila. Developmental Biology*, **49**, 40–65.

Illmensee, K. & Mintz, B. (1976). Totipotency and normal differentiation of single teratocarcinoma cells cloned by injection into blastocysts. *Proceedings of the National Academy of Sciences, USA*, **73**, 549–53.

Kahan, B. W. & Ephrussi, B. (1970). Developmental potentialities of clonal *in vitro* cultures of mouse testicular teratomas. *Journal of the National Cancer Institute*, **44**, 1015–36.

Kaufman, M. H., Barton, S. C. & Surani, M. A. H. (1977). Normal post-implantation development of mouse parthenogenetic embryos to forelimb bud stage. *Nature, London*, **265**, 53–5.

Kelly, S. J. (1975). Studies of the potency of early cleavage blastomeres of the mouse. In *The early development of mammals, 2nd Symposium of the British Society for Developmental Biology*, ed. M. Balls & A. E. Wild, pp. 97–106. London: Cambridge University Press.

Kleinsmith, L. J. & Pierce, G. B. (1964). Multipotentiality of single embryonal carcinoma cells. *Cancer Research*, **24**, 1544–51.

Kocher-Becker, U. & Tiedemann, H. (1971). Induction of mesodermal and endodermal structures and primordial germ cells in *Triturus* ectoderm by a vegetalizing factor from chick embryos. *Nature, London*, **233**, 65–6.

Lamerton, L. F. (1976). Chairman's opening address at the Symposium on Stem Cells: a tribute to C. P. Leblond. In *Stem cells of renewing cell populations*, ed. A. B. Cairnie, P. K. Lala & D. G. Osmond, pp. 1–4. New York & London: Academic Press.

Leblond, C. P. (1964). Classification of cell populations on the basis of their proliferative behaviour. *National Cancer Institute Monographs*, **14**, 119–50.

Leblond, C. P. & Cheng, H. (1976). Identification of stem cells in the small intestine of the mouse. In *Stem cells of renewing cell populations*, ed. A. B. Cairnie, P. K. Lala & D. G. Osmond, pp. 7–31. New York & London: Academic Press.

Leblond, C. P., Clermont, Y. & Nadler, N. J. (1967). The pattern of stem cell renewal in three epithelia (esophagus, intestine and testis). *Proceedings of the Canadian Cancer Research Conference*, **7**, 3–30.

Levak-Svajger, B. & Svajger, A. (1971). Differentiation of endodermal tissues in homografts of primitive ectoderm from two-layered rat embryonic shields. *Experientia*, **27**, 683–4.

Levak-Svajger, B. & Svajger, A. (1974). Investigation on the origin of the definitive endoderm in the rat embryo. *Journal of Embryology and experimental Morphology*, **32**, 445–59.

Mahowald, A. P. (1970). Origin and continuity of polar granules. In *Results and problems in cell differentiation*, vol. 2; *Origin and continuity of cell organelles*, ed. J. Reinert & H. Ursprung, pp. 158–69. Berlin: Springer-Verlag.

Martin, G. R. (1975). Teratocarcinomas as a model system for the study of embryogenesis and neoplasia. *Cell*, **5**, 229–43.

McBurney, M. W. (1976). Clonal lines of teratocarcinoma cells *in vitro*: differentiation and cytogenetic characteristics. *Journal of Cell Physiology*, **89**, 441–5.

Merchant, H. (1975). Rat gonadal and ovarian organogenesis with and without germ cells. An ultrastructural study. *Developmental Biology*, **44**, 1–22.

MINTZ, B. (1957). Germ cell origin and history in the mouse: genetic and histochemical evidence. *Anatomical Record*, **127**, 333–6.

MINTZ, B. & ILLMENSEE, K. (1975). Normal genetically mosaic mice produced from malignant teratocarcinoma cells. *Proceedings of the National Academy of Sciences, USA*, **72**, 3585–9.

NICOLAS, J. F., AVNER, P., GAILLARD, J. & JAKOB, H. (1976). Cell lines derived from teratocarcinomas. *Cancer Research*, **36**, 4224–31.

OKADA, M., KLEINMAN, A. & SCHNEIDERMAN, H. (1974). Restoration of fertility in sterilized *Drosophila* eggs by transplantation of polar cytoplasm. *Developmental Biology*, **37**, 43–54.

OZDZENSKI, W. (1967). Observations on the origin of the primordial germ cells in the mouse. *Zoologica poloniae*, **17**, 367–79.

PAPAIOANNOU, V. E., MCBURNEY, M. W., GARDNER, R. L. & EVANS, M. J. (1975). Fate of teratocarcinoma cells injected into early mouse embryos. *Nature, London*, **258**, 70–3.

PETERS, H. (1970). Migration of gonocytes into the mammalian gonad and their differentiation. *Philosophical Transactions of the Royal Society of London*, **259B**, 91–101.

PIERCE, G. B. (1967). Teratocarcinoma: model for a developmental concept of cancer. *Current Topics in developmental Biology*, **2**, 223–46.

PIERCE, G. B. & BEALS, T. F. (1964). The ultrastructure of primordial germinal cells of the fetal testes and of embryonal carcinoma cells of mice. *Cancer Research*, **24**, 1553–67.

PIERCE, G. B. & WALLACE, C. (1971). Differentiation of malignant to benign cells. *Cancer Research*, **31**, 127–34.

REGAUD, C. (1901). Etudes sur la structure des tubes séminifères et sur la spermatogenèse chez les mammifères. *Archives d'Anatomie microscopiques et de Morphologie expérimentale*, **4**, 101–56, 231–80.

ROLSHOVEN, E. (1951). Ueber die Reifungsteilungen bei der Spermatogenese mit einer Kritik des bisherigen Begriffes der Zellteilungen. *Verhandlungen der anatomischen Gesellschaft*, **49**, 189–97.

ROSSANT, J. (1975*a*). Investigation of the determinative state of the mouse inner cell mass. I. Aggregation of isolated inner cell masses with morulae. *Journal of Embryology and experimental Morphology*, **33**, 979–90.

ROSSANT, J. (1975*b*). Investigation of the determinative state of the mouse inner cell mass. II. The fate of isolated inner cell masses transferred to the oviduct. *Journal of Embroylogy and experimental Morphology*, **33**, 991–1001.

ROSSANT, J. (1976*a*). Investigation of inner cell mass determination by aggregation of isolated rat inner cell masses with mouse morulae. *Journal of Embryology and experimental Morphology*, **36**, 163–74.

ROSSANT, J. (1976*b*). Postimplantation development of blastomeres isolated from 4- and 8-cell mouse eggs. *Journal of Embryology and experimental Morphology*, **36**, 283–90.

ROSSANT, J. (1977). Cell commitment in early rodent development. In *Development in mammals*, ed. M. H. Johnson, pp. 119–50. Amsterdam: Elsevier/North-Holland.

Rossant, J. & Ofer, L. (1977). Properties of extraembryonic ectoderm isolated from postimplantation mouse embryos. *Journal of Embryology and experimental Morphology*, **39**, 183–94.

Rossant, J. & Papaioannou, V. E. (1977). Biology of embryogenesis. In *Concepts in Mammalian embryogenesis*, ed. M. I. Sherman, pp. 1–36, Cambridge, Mass.: MIT Press.

Smith, L. D. (1966). The role of a 'germinal plasm' in the formation of primordial germ cells in *Rana pipiens*. *Developmental Biology*, **14**, 330–47.

Snow, M. H. L. (1976). Embryo growth during the immediate postimplantation period. In *Embryogenesis in mammals, Ciba Foundation Symposium*, pp. 53–70. Amsterdam: Associated Scientific Publishers.

Sobis, H. & Vandeputte, M. (1975). Sequential morphological study of teratomas derived from the displaced yolk sac. *Developmental Biology*, **45**, 276–90.

Sobis, H. & Vandeputte, M. (1976). Yolk sac-derived rat teratomas are not of germ cell origin. *Developmental Biology*, **51**, 320–3.

Solter, D., Damjanov, I. & Škreb, N. (1970). Ultrastructure of mouse egg-cylinder. *Zeitschrift für Anatomie und Entwicklungsgeschichte*, **132**, 291–8.

Solter, D., Škreb, N. & Damjanov, I. (1970). Extrauterine growth of mouse egg-cylinder results in malignant teratoma. *Nature, London*, **227**, 503–4.

Spiegelmann, M. & Bennett, D. (1973). A light- and electron-microscopic study of primordial germ cells in the early mouse embryo. *Journal of Embryology and experimental Morphology*, **30**, 97–118.

Stevens, L. C. (1959). Embryology of testicular teratomas in strain 129 mice. *Journal of the National Cancer Institute*, **23**, 1249–95.

Stevens, L. C. (1964). Experimental production of testicular teratomas in mice. *Proceedings of the National Academy of Sciences, USA*, **52**, 654–61.

Stevens, L. C. (1967). Origin of testicular teratomas from primordial germ cells in mice. *Journal of the National Cancer Institute*, **38**, 549–52.

Stevens, L. C. (1968). The development of teratomas from intertesticular grafts of tubal mouse eggs. *Journal of Embryology and experimental Morphology*, **20**, 329–41.

Stevens, L. C. (1970). The development of transplantable teratocarcinomas from intratesticular grafts of pre- and post-implantation mouse embryos. *Developmental Biology*, **21**, 364–82.

Stevens, L. C. (1975). Comparative development of normal and parthenogenetic mouse embryos, early testicular and ovarian teratomas, and embryoid bodies. In *Teratomas and differentiation*, ed. M. I. Sherman & D. Solter, pp. 17–32. New York & London: Academic Press.

Stevens, L. C. & Pierce, G. B. (1975). Teratomas: definitions and terminology. In *Teratomas and differentiation*, ed. M. I. Sherman & D. Solter, pp. 13–14. New York & London: Academic Press.

Stevens, L. C. & Varnum, D. S. (1974). The development of teratomas from parthenogenetically activated ovarian mouse eggs. *Developmental Biology*, **37**, 369–80.

Sutasurja, L. A. & Nieuwkoop, P. D. (1974). The induction of primordial germ cells in the urodeles. *Wilhelm Roux Archiv für Entwicklungsmechanik der Organismen*, **175**, 199–20.

Svajger, A. & Levak-Svajger, B. (1976). Differentiation in renal homografts of isolated parts of rat embryonic ectoderm. *Experientia* **32**, 378–80.
Tarkowski, A. K. (1959*a*). Experiments on the development of isolated blastomeres of mouse eggs. *Nature, London*, **184**, 1286–7.
Tarkowski, A. K. (1959*b*). Experimental studies on regulation in the development of isolated blastomeres of mouse eggs. *Acta theriologica*, **3**, 191–267.
Tarkowski, A. K. & Wroblewska, J. (1967). Development of blastomeres of mouse eggs isolated at the four- and eight-cell stage. *Journal of Embryology and experimental Morphology*, **18**, 155–80.
Van Blerkom, J., Barton, S. C. & Johnson, M. H. (1976). Molecular differentiation in the preimplantation mouse embryo. *Nature, London*, **259**, 319–21.
Weissmann, A. (1892). *Das Keiplasm. Eine Theorie der Vererburg*. Jena: Fischer.
Whitington, P. McD. & Dixon, K. E. (1975). Quantitative studies of germ plasm and germ cells during early embryogenesis of *Xenopus laevis*. *Journal of Embryology and experimental Morphology*, **33**, 57–74.
Wylie, C. V., Nakane, P. K. & Pierce, G. B. (1973). Degree of differentiation in nonproliferating cells of mammary carcinoma. *Differentiation*, **1**, 11–20.
Züst, B. & Dixon, K. E. (1975). The effect of U.V. irradiation of the vegetal pole of *Xenopus laevis* on the presumptive primordial germ cells. *Journal of Embryology and experimental Morphology*, **34**, 209–20.

Stem cells and tissue homeostasis in insect development

R. NÖTHIGER, T. SCHÜPBACH, J. SZABAD*
AND E. WIESCHAUS

Zoological Institute, University of Zurich, Zurich, Switzerland,
and *Biocentrum Szeged, Szeged, Hungary

'Stem cells' and 'tissue homeostasis' represent two biological principles that are encountered throughout the realm of multicellular organisms. It is easy to see that both are very useful to the organism: *stem cells* guarantee a constant supply of cells that are continuously used up, such as the red blood cells, or are wasted in enormous numbers, such as the gametes; *tissue homeostasis* renders the organism independent, at least to some degree, of environmental disturbances.

Both principles prompt a number of interesting questions some of which can be attacked experimentally. A relatively modest goal that may be set is the elucidation of the cellular *parameters*, or the *dynamics*, governing the two phenomena. An animal that is especially suited for a combined developmental and genetic analysis at the cellular level is *Drosophila*. Genetic engineering is highly developed and we have available a large number of mutations, some of which can be used to mark cells and others to alter the developmental fate of specific groups of cells. Because of these advantages and because we have been working with *Drosophila* for several years we shall restrict our discussion to this animal.

Our paper will be subdivided into two parts. The first part will deal with tissue homeostasis, or more specifically *developmental homeostasis* (Lerner, 1954), using the imaginal discs as a model system; in the second part we will discuss the female germ line as an example of stem cells.

DEVELOPMENTAL HOMEOSTASIS IN IMAGINAL DISCS

The system

The imaginal discs are groups of cells that are found in the larvae of holometabolous insects. The imaginal cells are set aside early in the embryo and are of no significance for the larva. Starting from some 10–30 cells their number increases exponentially throughout the larval period, at the end of which it has reached some 10 000–50 000 cells per disc. When metamorphosis is triggered by ecdysone and the larval organisation breaks down, the discs take over and form the epidermis of the adult insect, the imago, according to an exact predetermined plan. During embryonic and larval development, each disc has been assigned a very specific developmental programme, i.e. has become *determined* to form a precisely fixed part of the imago (see Fig. 1 in Nöthiger, 1972). Not only a disc as a whole, but also parts and perhaps even individual cells within it have been programmed to form a particular region or structure of the adult (Schubiger, 1968).

One might expect, and in fact it was the established view for many years, that such a precisely determined system has to follow a very rigidly prescribed pathway of development and consequently should be very sensitive towards any disturbances. However, recent experiments have revealed a great capacity of the discs to regain their developmental balance after severe experimental interference (review by Bryant, 1975). We will now analyse this developmental flexibility in terms of the number of cells that are engaged in the process, their growth rate and their developmental potential.

Techniques

A particularly elegant and powerful means for studying cell behaviour is genetic mosaics. In *Drosophila* such an analysis is especially easy due to the availability of marker mutations that alter the colour or morphology of bristles (Fig. 1) or trichomes on the body surface of the fly. Because these structures are derivatives of single imaginal cells, the patterns of mosaicism in the differentiated structures allow us to study development at the cellular level.

Genetic mosaics may be produced in different ways, e.g. (*a*) by *mixing together cells from genetically different discs*; (*b*) by *loss of a chromosome*, mainly an X-chromosome, during an early cleavage division of a female zygote; or (*c*) by *mitotic recombination* resulting in a *clone* of genetically labelled cells. All three techniques have recently been described (Nöthiger, 1972; Hotta & Benzer, 1973; Garcia-Bellido & Nöthiger, 1976), and we do not want to repeat this here, except for a brief presentation of mitotic recombination which is the most powerful of them (Fig. 1). The essence of the technique is that it is possible to produce a *clone of genetically labelled* (homozygous) *cells* in a growing population of heterozygous phenotypically wild-type cells. Mitotic recombination is an event that can be forced to occur at any time during larval development in about 1 % of the dividing cells by an X-ray dose of 1000 r. Further details are given in the legend to Fig. 1.

Developmental homeostasis

An imaginal disc is a closed system in that, once the founder cells have been chosen, the primordium increases in size by proliferation of these founder cells. The divisions occur at a constant rate throughout larval life until a fixed number of cells is reached, characteristic for each disc.

In 1975, Morata & Ripoll made the important discovery that the mitotic rate of a clone, or of a population of cells, is determined by its genotype and is largely independent of neighbouring cells with a different genotype and mitotic rate. These authors were able to induce within a disc a clone which due to its genotype exhibited a

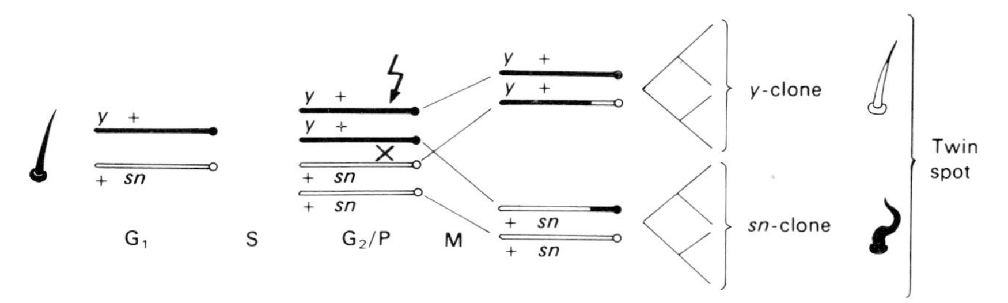

Fig. 1. Schematic representation of mitotic recombination taking place in a mitotically dividing diploid cell. G_1, S, G_2, P, M, phases of the cell cycle; *y*, *sn*, recessive mutations. After mitotic recombination (×) two daughter cells are produced homozygous for *y*/*y* or *sn*/*sn*, respectively, giving rise to light-coloured or gnarled bristles. Since the mutations express themselves autonomously in individual cells, two adjacently growing clones of genetically marked bristles are formed resulting in a so-called twin spot. The large majority of cells did not undergo mitotic recombination, and so the twin spot is an island surrounded by dark-coloured normal-shaped bristles.

much higher growth rate than the rest of the cell population. This resulted in giant clones which could cover large parts of the fly (Fig. 2*b*). Despite the potential of the fast-growing clones to flood large areas of a disc, they strictly respected certain borders which they did not cross (Fig. 2, A/P). These borders define a cell group that was called by Garcia-Bellido (1975) a 'compartment'. The cells of such a compartment are committed to a particular pathway of development. But these cells within a compartment are apparently equivalent and can replace each other: the adult part derived from a disc, e.g. a wing or a leg, is perfectly normal in size and shape although a huge fraction of it has been formed by the clonal descendants of a single cell (Fig. 2*b*). This cell and its clone would normally contribute much less to the inventory of the disc (compare Fig. 2*a* with 2*b*).

The pattern of structures – bristles, trichomes, sensilla – on the fly's body surface is highly constant. Nevertheless, the cell lineage relationships within a compartment are largely random and certainly not determined in any fixed way. The fast-growing clone that may sometimes fill almost the entire compartment clearly shows that a cell differentiates according to its *position* within a compartment, and not according to its genealogy or clonal descent. This forces us to assume that some sort of a co-ordinate system is established within

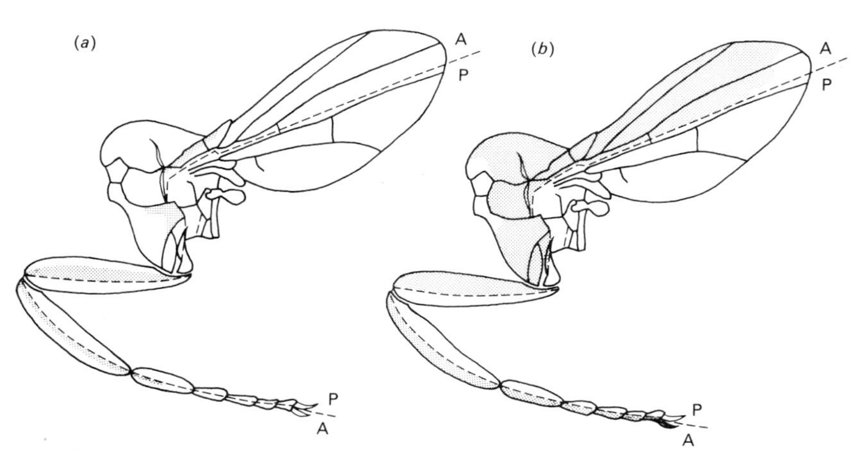

Fig. 2. Size and range of clones (stippled areas) induced at 3 hours of embryonic development. (*a*) Clone with normal growth rate. (*b*) Giant clone with faster growth rate. (Modified from Wieschaus (1977) and Steiner (1976).) Both types of clones follow a straight border (dashed line) that separates an anterior (A) from a posterior (P) compartment in both wing and leg.

each compartment independently of the clonal relationships of the constituent cells. This autonomous system of positional information (Wolpert, 1969) forms the basis for the developmental homeostasis made apparent by the fast-growing clones.

By comparing the sizes of normal and fast-growing clones we can estimate the degree of developmental flexibility of the system. Thus, a cell whose clonal descendants would normally form about one-twentieth of a compartment may now, when homozygous for the allele that allows for a faster growth rate, contribute as much as nine-tenths or even more of the structures of this compartment (Fig. 2; Steiner, 1976; Dübendorfer, 1977).

The production of a fast-growing clone represents transplantation of a single cell with a special genetic endowment. Another way of interfering even more drastically with the system consists in killing a large fraction of the cells within a disc. This is achieved by heavy doses of X-rays. The first careful study of X-ray damage and repair in imaginal discs was done by Schweizer (1972). He observed that genital discs irradiated with 4000 r gave abnormal genitalia when larvae of 92 hours (pupariation begins at 120 hours) were treated; but perfectly normal genitalia appeared when the treatment was given at 68 hours. Was this difference due to differential X-ray sensitivity at the two stages, or to the fact that the surviving cells of the younger discs had still enough time left before metamorphosis to compensate for the damage? Schweizer first showed that discs from 92-hour larvae are also capable of producing normal genitalia if they are transplanted into 68-hour larvae. A clonal analysis then demonstrated elegantly and unambiguously that the X-ray damage was repaired by additional divisions of the surviving cells: the size of the clones was dependent on the dose, i.e. on the degree of cell death. The larger clones found with higher doses indicate that the fewer surviving cells had to undergo more cell divisions in order to form a normal structure. We see again that the new cells produced by supernumerary divisions differentiate according to an exact plan which assigns specific roles to the cells at specific positions.

Haynie & Bryant (1977) have estimated that a dose of 1000 r may kill as many as 40% of the cells of a disc; and the death toll may go up to 80% for a dose of 4000 r. Using a temperature-sensitive cell lethal, Simpson & Schneiderman (1976) estimated that 80% of the cells of a wing disc were killed after a temperature shock. In both situations, however, the system could still recover and form a normal wing. The developmental homeostatis of our system is impressive: neither a fast-growing clone nor killing a large fraction of the cells can prevent it from finally forming a normal structure.

We wonder how a cellular system first monitors and then repairs the damage so that in most cases the resulting structure is normal. We have to assume either that the co-ordinate system of positional information remains undisturbed by the damage and the cells 'simply' fill in the free places; or alternatively, that the co-ordinate system is rebuilt, presumably through communication among the surviving cells. In order to distinguish between these two alternatives we can perform an experiment in which the tissue architecture is completely disrupted so that the co-ordinate system must have been destroyed. Imaginal discs are treated with trypsin and mechanically dissociated until no tissue pieces bigger than 10–15 cells can be found in the suspension. After reaggregation the clump of cells is cultured for 8–10 days in adult host flies where cell divisions can take place. It is then subjected to metamorphosis by transplantation into larval hosts. The result is quite informative and clear-cut: the implants differentiated parts of legs or wings with a normal pattern (Fig. 3). Whereas probably most of the dissociated cells died, the few survivors divided and were capable of reconstructing at least part of the co-ordinate system. Haynie & Bryant (1976) and Strub (1977) have demonstrated that cells from different parts of imaginal discs, when artificially juxtaposed, apparently interact and have the tendency and the capacity to reconstruct the co-ordinate system. Based on results with regenerating cockroach and vertebrate limbs and with fragments of imaginal discs, French, Bryant & Bryant (1976) have proposed the so-called 'polar co-ordinate model'. This formal model describes and predicts adequately the regulative behaviour of fragments and dissociated cells of imaginal discs. However, its physical basis and the mechanisms by which the growing cell population builds the co-ordinate system, as well as the mechanisms by which the genes register and interpret the co-ordinates, are still unknown.

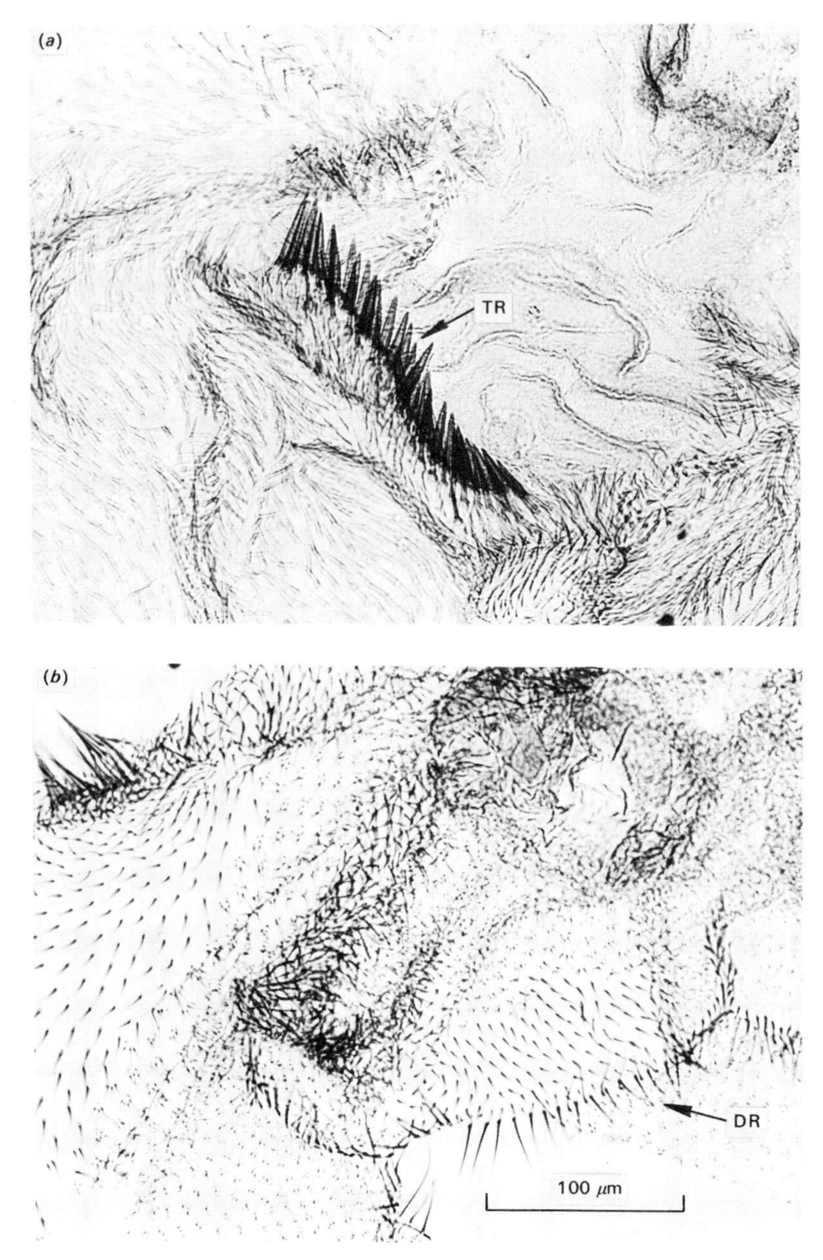

Fig. 3. Reconstruction of normal patterns from dissociated cells of the wing disc (after Garcia-Bellido & Nöthiger, 1976). (*a*) TR, triple row of bristles along anterior wing margin. (*b*) DR, double row of bristles along posterior wing margin.

STEM CELLS IN THE FEMALE GERM LINE

Sensu stricto we know of only one well-established case of stem cells in *Drosophila*, namely the gonial cells in the gonads of both sexes. Here we discuss the female germ line which we have been analysing in the past year using genetic mosaics.

A *Drosophila* female possesses two ovaries. Each ovary consists of some 17 smaller units, the *ovarioles* (Fig. 4). An ovariole is essentially a tube in which some six developing *cysts*, each containing an oocyte, can be seen arranged in linear order. New cysts are formed at the apex of an ovariole while the oldest cysts are found at the base from where the oocytes leave the ovary to be fertilised and deposited. A female lays some 1200 or more eggs during her lifespan.

Histological observations and the occurrence of clusters of induced mutations suggest that an ovariole possesses a small group of stem cells which divide and branch off prospective oocytes (King, 1970, p. 55). These stem cells of the germ line derive from the so-called pole cells, the first cells to be formed in an embryo. Unpublished data of our laboratory indicate that the entire population of stem cells in both ovaries descends from about five to 10 pole cells.

Our work has been directed at the following questions:

(i) How many stem cells are there per ovariole in the adult ovary, and how many founder cells were initially included in an ovariole when it was formed?

(ii) What are the dynamics and the pattern of stem cell divisions and egg production?

Two types of genetic mosaics were produced in order to analyse these questions. For the first type we made use of an unstable ring-X chromosome carrying the wild-type allele of a histochemical marker gene, *maroon-like* (mal^+). Larvae and flies homozygous for the mutant *mal* allele lack aldehyde oxidase (ALDOX). The mutation expresses itself autonomously in cells of mosaic animals. These were obtained by loss of the unstable ring-X from a triplo-X zygote with attached-Xs that were homozygous for *mal*. Since flies, but also cells, with three as well as with two X-chromosomes are of female sex we were able to produce mosaic females that contained triplo-X wild-type cells and diplo-X mutant cells.

When ovaries of such females were histochemically stained for

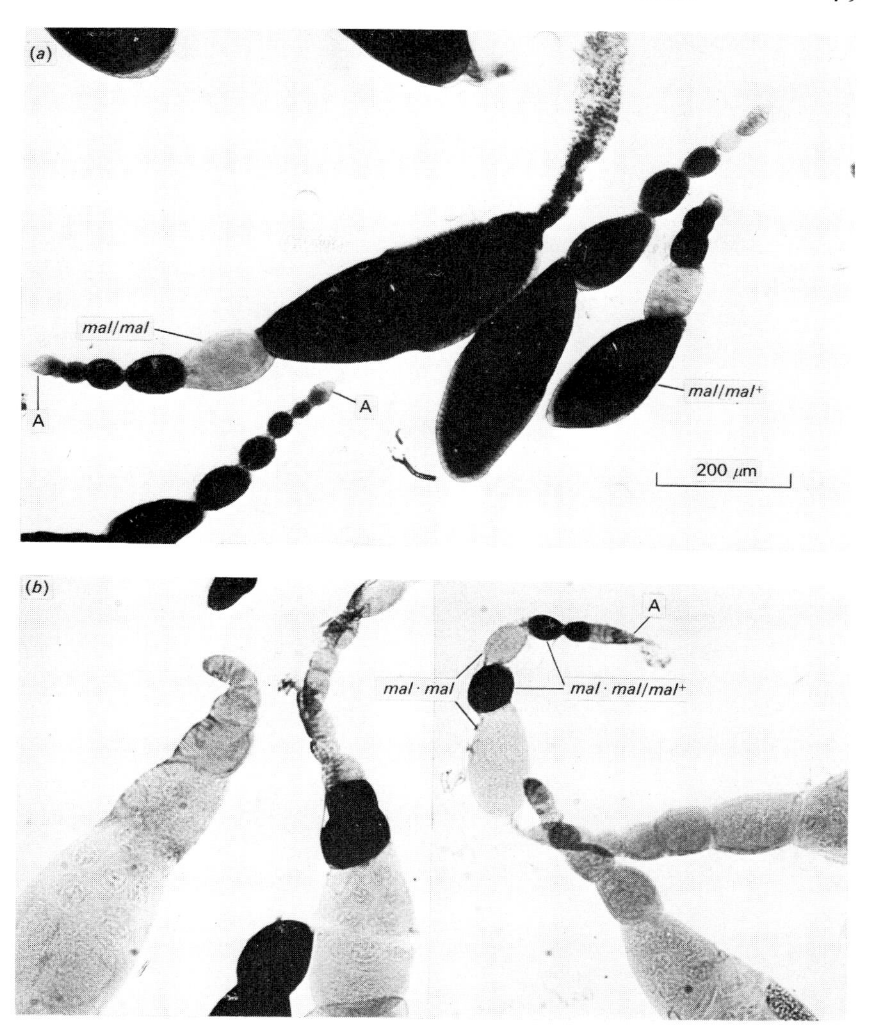

Fig. 4. Mosaic ovarioles with *mal* (light, unstained) and *mal*⁺ (black, stained) cysts after histochemical staining for ALDOX. *mal*/*mal* cysts produced by mitotic recombination (*a*) or by loss of a ring-X from triplo-X zygotes (*b*). A, apex containing germarium with stem cells.

ALDOX a mosaic pattern of stained (triplo-X, with ALDOX-activity) and unstained (diplo-X, without ALDOX activity) cysts were found (Fig. 4). Mosaic ovaries were quite frequent and, as Fig. 4(*b*) shows, even single ovarioles could also be mosaic. This already yields an important result, namely that an ovariole is not of clonal

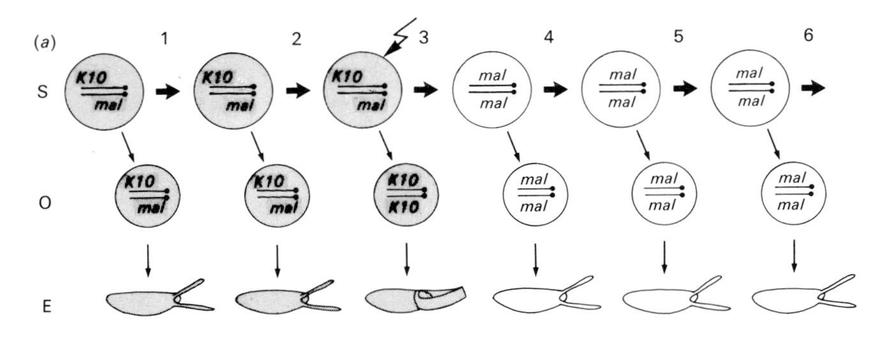

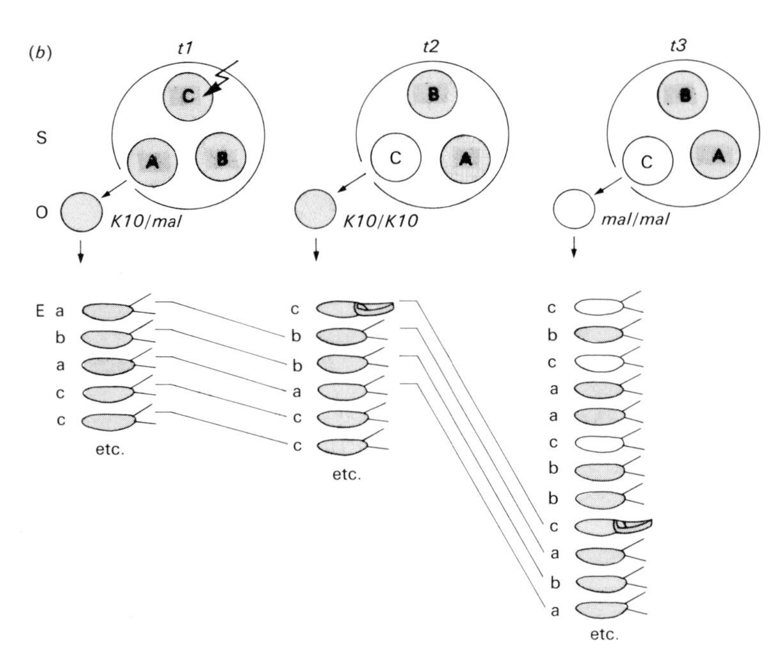

Fig. 5. Mitotic recombination in the female germ line. Adult females, heterozygous for *fs(1)K10* and *mal*, were X-irradiated in order to induce mitotic recombination (see Fig. 1). (*a*) After each division of a stem cell (S) one daughter cell remains in the stem line (thick arrows) whereas the other becomes a prospective oocyte (O) differentiating into an egg (E). In the absence of mitotic recombination (divisions 1 and 2), all the eggs produced by *K10/mal* cells are of normal morphology and possess ALDOX activity (stippled). After mitotic recombination in a stem cell (division 3), two daughter cells arise, one of them homozygous for *K10*, the other for *mal*. If the *mal/mal* cell remains in the stem line, only one *K10* egg will be produced. The *mal/mal* stem cell, however, will yield a series of *mal* eggs (divisions 4, 5, 6, etc.) that can be histochemically recognised by the absence of ALDOX

origin: in fact its stem cells may have a separate cell lineage from the time of the first cleavage division when the ring-X was lost.

The second type of mosaic was obtained by mitotic recombination in animals heterozygous in trans position for a female sterile mutation, *fs(1)K10*, and *mal*. Eggs produced from homozygous *fs(1)K10* oogonia are of abnormal shape (Fig. 5) and will not develop, even when fertilised with normal sperm. Homozygous *mal* flies have a maroon-like eye colour and lack ALDOX when derived from homozygous *mal* oogonia (Marsh & Wieschaus, 1977), but exhibit a wild-type eye colour when derived from heterozygous mal/mal^+ oogonia, due to a maternal effect of the mal^+ allele.

The results of the experiment are illustrated in Fig. 5. Females of the genotype $fs(1)K10\ mal^+/fs^+\ mal$ were irradiated with 1000 r and mated to males $fs^+\ mal/Y$. In the absence of mitotic recombination such females will lay normal eggs from which offspring with wild-type eye colour will develop. However, mitotic recombination proximal to *mal* in a diploid stem cell produces two daughter cells, one of them homozygous for *fs(1)K10*, the other for *mal* (see Fig. 1, replace *y* by *fs(1)K10*, and *sn* by *mal*!). The former will give rise to abnormally shaped eggs, the latter to flies with *mal* eyes. Thus, after mitotic recombination has occurred in a single stem cell, a female contains a mosaic ovary with heterozygous and homozygous germ line cells.

By examining the progeny of such mosaic females we can gain insights into the cellular dynamics of stem cell divisions and oogenesis. When 5-day-old adult females were irradiated almost 50% of them laid one or several abnormally shaped *K10* eggs, and most of them also yielded *mal*-eyed progeny. This result clearly shows

---

activity (unstippled). (*b*) Diagrammatic illustration of the dynamics of divisions in an ovariole containing three stem cells A, B and C. t_1 represents the time of irradiation (zig-zag arrow) of an adult female, resulting in mitotic recombination in stem cell C. Until t_1, only normal eggs have been produced. The letters a, b and c designate the origin of the eggs from stem cell A, B and C, respectively. At time t_2, cell C which has just undergone mitotic recombination gives rise to one homozygous *K10/K10* cell destined to form an abnormally shaped *K10* egg, and one *mal/mal* cell which remains in the stem line. At time t_3, the mosaic ovariole has been functioning for some time and has yielded three *mal* eggs (unstippled) interspersed at irregular intervals between the normal eggs derived from stem cells A and B. The *K10* egg formed at time t_2 has moved downward on the drawing which represents the temporal sequence of eggs produced by this particular mosaic ovariole. Note that this female possesses some 33 more ovarioles which produce normal eggs.

that mitoses take place in the germarium of the ovaries of adult flies. Thus, the *Drosophila* ovary differs from the mammalian ovary where all the eggs to be produced are ready and waiting in the early stages of meiosis.

Further analysis of the data reveals even more details. The mosaic females fall into two classes: about half of them produced only a single *K10* egg while the other half yielded several *K10* eggs deposited at irregular intervals throughout the remainder of the female's life. The single *K10* egg was generally followed by several *mal*-eyed progeny (Fig. 5*b*) whereas the series of *K10* eggs was usually accompanied by a single *mal* offspring. This result demonstrates that the stem cells in the germarium undergo asymmetric divisions: one of the daughter cells is shifted into the pathway of oogenesis, the other remains in the population of stem cells and forms the source of future eggs.

By studying the egg production of mosaic females we can also estimate the number of stem cells in the two ovaries. If we assume that each stem cell produces the same number of eggs, then the reciprocal of the fraction of *K10* eggs deposited by those mosaic females which laid several *K10* eggs equals the number of stem cells. On average, such a mosaic female produced one *K10* egg per hundred eggs, which means that there are some hundred stem cells in the two ovaries. Since we count 17 ovarioles per ovary, each ovariole must contain some three stem cells.

We now turn to the dynamics of the process of oogenesis. We have already seen that the divisions in the germarium are asymmetric, i.e. they are *differential divisions*. We can estimate the *length of a cell cycle* by measuring the average time interval between two *K10* eggs laid by the same mosaic female. The length of a cell cycle is, of course, dependent on many variables, but at 22 °C and under normal food conditions takes about 60 hours. The time elapsed between irradiation, i.e. induction of mitotic recombination, and the first *K10* egg being laid is about 11 days. This then is the time required for the whole process of oogenesis. Histological staining for ALDOX allows us to follow in further detail the kinetics of oogenesis. The time elapsed between the event of mitotic recombination and the first homozymous *mal* cyst to appear in the ovariole is some 7 days, and 3–4 more days are needed until the oocyte is ready for fertilisation.

A further interesting aspect is the pattern of 'firing' in an ovariole. If the three cells per ovariole release an oocyte in regular sequence then we would observe a corresponding pattern of *mal* and *mal*$^+$ cysts in an ovariole. Ovaries of mosaic females that were either produced by loss of the unstable ring chromosome or by mitotic recombination were histochemically stained for ALDOX. Among some 500 mosaic ovarioles examined we could detect no regularity in the staining pattern (Fig. 4). In view of the small number of stem cells per ovariole this result suggests that the stem cells 'fire' at irregular intervals.

CONCLUDING REMARKS

Drosophila is a small animal, and a biochemical approach to problems of development at the cellular level is very difficult. However, this handicap is more than compensated by the great potential of this fly for genetic analysis.

Our paper has shown what the simple techniques of producing genetic mosaics can achieve. We are able to characterise the dynamics of developmental processes in terms of number of cells, growth rate and change of developmental programmes. The genetically marked clones reveal the developmental potential of individual cells and their clonal descendants. In *imaginal discs* they made apparent an impressive *developmental homeostasis* and uncovered the cellular parameters underlying the phenomenon. This analysis has led to the important insight that the fate of a cell depends on its position more than on its clonal descent. In an analysis of the *female germ line* the genetic mosaics established that the ovary contains a dividing population of some hundred *stem cells*, and that these undergo asymmetric divisions in an irregular temporal pattern.

The use of more mutations and the application of more sophisticated genetical engineering will undoubtedly advance our understanding of the mechanics of development.

We would like to thank the Swiss National Science Foundation and the Julius Klaus-Stiftung for financial support of our research. One of us (J.Sz.) held a postdoctoral fellowship of the UNDP/UNESCO.

REFERENCES

BRYANT, P. J. (1975). Regeneration and duplication in imaginal discs. In *Cell patterning, Ciba Foundation Symposium 29*, pp. 71–93. Amsterdam: Associated Scientific Publishers.

DÜBENDORFER, K. (1977). Die Entwicklung der männlichen und weiblichen Genital-Imaginalscheibe von *Drosophila melanogaster*: eine klonale Analyse. PhD Thesis, University of Zurich.

FRENCH, V., BRYANT, P. J. & BRYANT, S. V. (1976). Pattern regulation in epimorphic fields. *Science*, **193**, 969–81.

GARCIA-BELLIDO, A. (1975). Genetic control of wing disc development in *Drosophila*. In *Cell patterning, Ciba Foundation Symposium, 29*, pp. 161–82. Amsterdam: Associated Scientific Publishers.

GARCIA-BELLIDO, A. & NÖTHIGER, R. (1976). Maintenance of determination by cells of imaginal discs of *Drosophila* after dissociation and culture *in vivo*. *Wilhelm Roux's Archives*, **180**, 189–206.

HAYNIE, J. L. & BRYANT, P. J. (1976). Intercalary regeneration in imaginal wing disk of *Drosophila melanogaster*. *Nature, London*, **259**, 659–62.

HAYNIE, J. L. & BRYANT, P. J. (1977). Clonal analysis of cell death and regeneration in irradiated imaginal wing discs of *Drosophila melanogaster*. *Wilhelm Roux's Archives*, **183**, 85–100.

HOTTA, Y. & BENZER, S. (1973). Mapping of behaviour in *Drosophila* mosaics. In *Genetic mechanisms of development*, ed. F. Ruddle, pp. 129–67. New York & London: Academic Press.

KING, R. C. (1970). *Ovarian development in* Drosophila melanogaster. New York & London: Academic Press.

LERNER, J. M. (1954). *Genetic homeostasis*. London: Oliver & Boyd.

MARSH, J. L. & WIESCHAUS, E. (1977). Germ line dependence of the *maroon-like* maternal effect in *Drosophila*. *Developmental Biology*, **60**, 396–403.

MORATA, G. & RIPOLL, P. (1975). Minutes: mutants autonomously affecting cell division rate in *Drosophila*. *Developmental Biology*, **42**, 211–21.

NÖTHIGER, R. (1972). The larval development of imaginal discs. In *The biology of imaginal discs*, ed. H. Ursprung & R. Nöthiger, pp. 1–34. Berlin: Springer-Verlag.

SCHUBIGER, G. (1968). Anlageplan, Determinationszustand und Transdeterminationsleistungen der männlichen Vorderbeinscheibe von *Drosophila melanogaster*. *Wilhelm Roux's Archives*, **160**, 9–40.

SCHWEIZER, P. (1972). Wirkung von Röntgenstrahlen auf die Entwicklung der männlichen Genitalprimordien von *Drosophila melanogaster* und Untersuchung von Erholungsvorgängen durch Zellklon-Analyse. *Biophysik*, **8**, 158–88.

SIMPSON, P. & SCHNEIDERMAN, H. A. (1976). A temperature sensitive mutation that reduces mitotic rate in *Drosophila melanogaster*. *Wilhelm Roux's Archives*, **179**, 215–36.

STEINER, E. (1976). Establishment of compartments in the developing leg imaginal discs of *Drosophila melanogaster*. *Wilhelm Roux's Archives*, **180**, 31–46.

STRUB, S. (1977). Developmental potentials of the cells of the male foreleg disc of *Drosophila*. II. Regulative behaviour of dissociated fragments. *Wilhelm Roux's Archives*, **182**, 75–92.

WIESCHAUS, E. (1977). The cell lineage relationships in the *Drosophila* embryo. In *Genetic mosaics and cell differentiation*, ed. W. Gehring. Berlin: Springer-Verlag, in press.

WOLPERT, L. (1969). Positional information and the spatial pattern of cellular differentiation. *Journal of theoretical Biology*, **25**, 1–47.

The concept of the stem cell in the context of plant growth and development

P. W. BARLOW

Agricultural Research Council,
Letcome Laboratory,
Wantage OX12 9JT, UK

INTRODUCTION

Plants, unlike most animals, grow in size throughout their life. The new cells that account for this growth are produced in discrete regions called meristems. Not all meristems are permanently active: for example, the meristematic zones of a developing leaf, or the internode of a young stem, produce only as many cells as are necessary for the formation of the tissue in question, whereupon the meristematic cells cease dividing and commence to differentiate as did their derivatives before them. These *impermanent* meristems can be considered to be residua of mitotic activity that have become displaced from some other meristem to occupy a position in the organism which does not permit more than a limited number of divisions. The meristems that will concern us here are those that are more or less continually active throughout the plant's life, bearing in mind the rhythms of activity and inactivity that may be imposed by the seasons of each successive year; I shall refer to them as *permanent* meristems.

The locations of permanent meristems favour their continual activity. They are at the tips of shoots and roots – i.e. the apical meristems that bring about the lengthening of these organs (the shoot apex in addition produces leaf primordia), and below the surface of stems and roots – i.e. the two 'cambium' meristems, one of which gives rise to new water- and nutrient-conducting tissues, the xylem and the phloem, and the other to the periderm. In these locations the meristems can produce cells for differentiation without at the same time becoming subject to the full physical weight or pressures that their derivatives would exert on them if they were located, say, in the heart of the organism. Moreover, some of the tissues derived from the meristems protect their parent meristem against externally-inflicted damage; for example, the shoot meristem is enclosed by the youngest leaves, and the root meristem and the vascular cambium are protected by the root cap and bark, respectively.

A permanent meristem that continually emits cells that differentiate and yet remains constant in size and function is a self-maintaining system. As such, it resembles many of the meristematic zones in animals; for example, those of the skin, intestinal epithelium, haematopoietic tissue. In these animal systems the property of self-maintenance (or self-renewal) is linked with the presence of stem cells. In a multicellular tissue unit the stem cells are believed to be relatively few in number, and *to be the source from which all the other cells, both meristematic and differentiated, that constitute the unit are derived.* With this general description of a stem cell we may now ask whether such cells exist in the permanent meristems of plants. But before proceeding, it must be said that the term 'stem cell' is not one that is commonly used by investigators of plant meristems; in the past many other terms (such as 'initials', 'promeristem', 'continuing meristematic residue', 'founder cells') have been employed to identify the cells from which tissues originate, and argument has surrounded the usage of these terms (Newman, 1961, 1965; Sussex & Steeves, 1967). I do not wish to enter this dispute; however, I shall continue to use the term 'stem cell', not only in order to conform with the other contributors to this volume, but also because 'stem cell' encompasses many of the attributes that the other terms mentioned have sought to convey with respect to the role that the corresponding cell groups play in plant development.

THE CONCEPT OF STEM CELLS IN RELATION TO PLANT MERISTEMS

In both the cambial meristems and in the apical meristems of roots and shoots there is either one cell, or a group of cells, that can be identified as the source from which all other cells are derived (Clowes, 1961). The identification of these cells can often be made simply on the basis of the spatial arrangement of the cells constructing the meristem; this is facilitated by the fact that the cells of plants, unlike those of animals, have rigid walls whose structure binds the cells together and, it is generally believed, prevents them from gliding over each other. Moreover, in certain, rather exceptional plants the cells that are the most probable primary source of all other cells can be directly seen. These are cases where the apex is chimaerical, being composed of two recognisable classes of cells: e.g. the shoot apices of *Epilobium* (Bartels, 1961) and *Ligustrum* (Stewart & Dermen, 1970). Such studies define the minimum set of stem cells.

To some extent the emanation of cell lineages from a minimum set of stem cells can be traced in certain tissue units of animals, not so much on the basis of cell geometry, though this is possible to a limited degree, but by observations of the kinetic properties of the cells in the tissue unit. Because one of the properties of stem cells in general is that they divide only occasionally, whilst the meristematic cells derived from them maintain a higher rate of division, the probable location of stem cells can be inferred from a gradient of mitotic activity (e.g. Al-Dewachi, Wright, Appleton & Watson, 1975). Within the apical meristem of both shoots and roots there is also a zone where the rates of entry into mitosis and S phase are conspicuously lower than in surrounding cells. Clowes (1956) was the first to demonstrate this experimentally in roots of *Zea mays* (sweet corn) and he called the zone of low mitotic activity the 'quiescent centre'. Since that time quiescent centres have been demonstrated to be a common, if not universal, feature of roots. In shoots an infrequently-dividing group of cells is located at the summit of the apex, defining what is often called the 'central zone' (Buvat, 1955; Lyndon, 1976) and, just as in the case of the quiescent centre (Clowes, 1954), this zone includes the cells from which the whole tissue must be derived (e.g. Paolillo & Gifford, 1961). It is important to note that the minimum number of cells essential for the construction of shoot or root tissue may often be less than the number of cells that composes, respectively, either the central zone or the quiescent centre. However, for the purposes of this discussion these two groups, each made up of cells with common cytophysiological properties and which include the minimum number of cells required for tissue construction, will be considered to be the functional stem-cell group of shoots and roots.

No especially slowly-dividing cells have been described within cambium meristems, but the cells of this meristem divide at a slow rate in any case (a cell cycle of 10 days was estimated by Wilson (1964) for the vascular cambium of *Pinus strobus* in the non-dormant condition), so any gradient of mitotic activity would be hard to detect. Nevertheless, in the vascular cambium of gymnosperms a single layer of cells (the initial cell layer) which is the origin of other dividing vascular cambium cells and their differentiating derivatives has been described (Sanio, 1873; Bannan, 1955; Newman, 1956); however, such a single layer of stem cells has not been recognised in the vascular cambium of *Acer pseudoplatanus* (sycamore), an angiosperm (Catesson, 1974).

A group of cells with the positional features of stem cells thus appears to be a rather general feature of meristems. Further, the candidates for a stem-cell role (i.e. the quiescent centre and central zone) behave similarly to stem cells in animal systems with respect to their rates of division in relation to the meristematic unit of which they are a part. Let us turn now to consider in more detail other properties that have been attributed to stem cells (see, for instance, Potten's (1975) description of the intestinal epithelial system) and see how far the cells of the quiescent centre and central zone correspond in their behaviour. Stem cells show or possess: (1) a slower rate of division and residence in an indeterminate state of the cell cycle (often called G_0 phase); (2) the capability of repopulating the meristem with which they are associated should the meristem be damaged; and (3) pluripotency. Stem cells have also been postulated to preserve an unblemished copy of the genome (Cairns, 1975). In the remainder of this section I shall consider each of these attributes in turn and then discuss features of stem cells as they are carried through the developmental sequences of the plant organism.

Nuclear cycle and structure of stem cells

Quiescent centre of roots. In many species the number of cells within the quiescent centre is sufficiently great to make possible a detailed study of the cell and nuclear division cycle. Recording the fraction of labelled mitoses in the quiescent centre at times after feeding roots for a brief period with [^{3}H]thymidine reveals that here the duration of both the G_2 and S phases is longer than elsewhere in the meristem (Clowes, 1965; Thompson & Clowes, 1968; Barlow, 1973; Barlow & Macdonald, 1973). With this technique of nuclear cycle analysis it is difficult to get an accurate measure of the duration of the G_1 phase, but it is undoubtedly long too (Clowes, 1971).

It is quite probable that what is called G_1 phase actually consists of two portions, one being a state of determinate duration in which the nucleus is making active progress towards S phase, the other being a state of indeterminate duration. These two states represent, respectively, part of the B-phase and the A-state (often called G_0 phase) of the cell cycle; A-state and B-phase are terms introduced by Smith & Martin (1973) to describe the structure of the intermitotic interval. It has been said that in the quiescent centre there are cells that are not in cycle and others that are cycling but which show considerable variation in the duration of the intermitotic period, of which the G_1 phase is probably the most variable component (Clowes, 1971, 1976). The hypothesis of Smith & Martin (1973) regarding the cell cycle allows this to be interpreted as meaning that cells in the quiescent centre have a more variable transition rate (k_{trans}) from the A-state to the B-phase; in addition, k_{trans} is lower in the quiescent centre than it is in cells elsewhere in the meristem. The fact that the G_2 and S phases are prolonged suggests that the pre-S portion of the B-phase may also be slower. The variation of k_{trans} in the quiescent centre also suggests that after mitosis the time to the next mitosis is independent of the duration of the previous interphase. This is attested by the fact that the 'fast' cycling cells do not displace the 'slow' or 'non-cycling' cells, as the quiescent centre maintains a roughly constant shape and size during root growth. It is quite possible that the apparently low metabolic rate of cells within the quiescent centre is intimately linked with the low and variable k_{trans}, as the initiator of the B-phase (the governor of k_{trans}) may itself be a product of the cells here. Only when a critical concentration is reached is the B-phase triggered (Smith & Martin, 1973); random fluctuation of the amount of B-phase initiator would foster the independence of successive cell cycles.

Measurement of the relative DNA content of nuclei in cells of the quiescent centre shows that the great majority of nuclei have a 2C value (Barlow, 1974). This is consistent with there being a control point in interphase that governs the rate of entry into S phase; other work of my own (as yet unpublished) shows that once S phase is initiated it goes to completion without interruption.

Central zone of shoots. More species have been investigated with attention to the duration of the cell cycle in various regions of the shoot apex (reviewed by Lyndon, 1976) than is the case for the root apex. This is surprising considering the greater difficulties of working with shoot tips. On the other hand, rather less is known about the temporal structure of their cell cycles. Measurements of DNA contents show that the great majority of nuclei of cells in the central zone have a 2C value, an unknown fraction of which may be in A-state. Thus the central zone and quiescent centre have similar cytological properties. Nuclei with 4C DNA contents are found in both these regions and it cannot be excluded that some of these may mark cells in the A-state also.

Another similarity is that within these two regions there are nuclei, with a 2C DNA content, whose chromatin structure is more dispersed than it is in the 2C nuclei of neighbouring, more mitotically active cells (for the quiescent centre of *Zea* roots see Fig. 1; for the central zone of shoots of *Tradescantia* and *Pisum* see Booker & Dwivedi (1973) and Nougarède (1977) respectively). The cause of this difference in chromatin structure is incompletely understood (though changes in histone composition may be implicated (e.g. Nougarède & Rondet, 1977) and likewise its significance, as far as understanding its role in nuclear physiology is concerned. However, the appearance of the chromatin in the central zone of *Tradescantia* buds is influenced by the level of nitrogen available to the plant (Yun & Naylor, 1973); this implicates nutritional or hormonal factors in the control of the activities of central zone nuclei.

Regenerative potential

One aspect of the A-state, particularly where it is encountered in a stem-cell population, is that cells in such a state can be recruited into the mitotic cycle should other cells of the tissue unit be destroyed. This implies that there is a transfer of information from the other meristematic cells, or maybe even the differentiated cells, to the stem cells which causes the latter cells to be dominated by the activities of the former. This feedback control presumably normally operates to determine the value of k_{trans} referred to in the previous section.

Exposing meristematic cells to ionising radiations is one way of impairing their proliferation and thereby investigating the potentialities of the stem cells. When roots of bean (*Vicia faba*) and corn (*Zea mays*) are irradiated it is found that while the rate of cell division in the meristem falls over a period of days, more cell divisions occur in the quiescent centre (Clowes, 1963, 1964). In irradiated apices of *Vicia* small groups, or islands, of mitotically

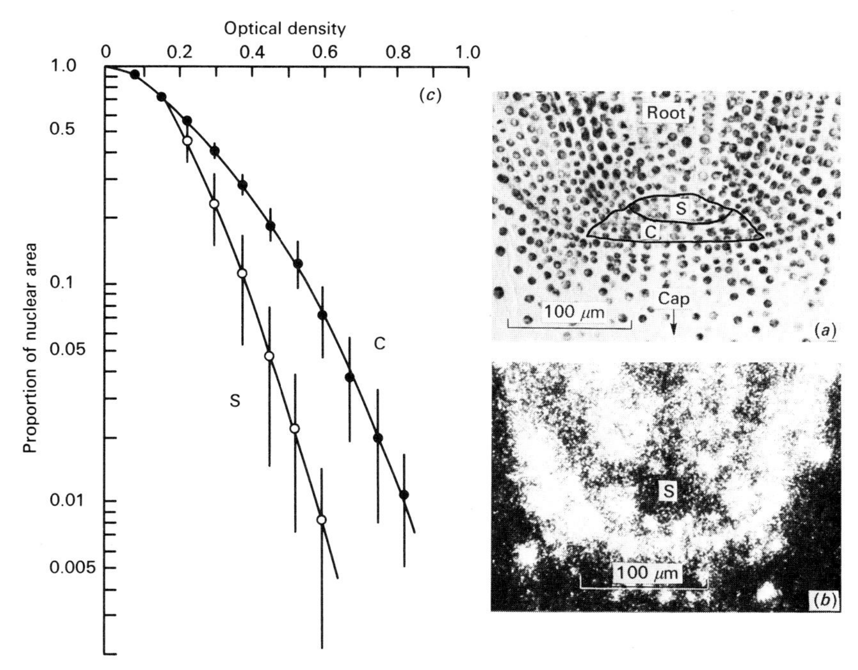

Fig. 1. (*a*) Median longitudinal section of a 6 day old primary root of *Zea mays*. The quiescent centre is outlined and its stelar and cortical components are labelled S and C respectively. The section was stained by Feulgen's reaction. The quiescent centre lies at the apex of the root proper and is surmounted by a cap. The cap is about 400 μm long and the arrow points towards its tip. In *Zea* there is quite a sharp demarcation between root and cap (not emphasised in this photograph). (*b*) Autoradiograph of a median longitudinal section of a primary root of *Zea mays* labelled for 22 hours with [^{14}C]thymidine. Note that the stelar portion of the quiescent centre (S) has remained unlabelled while nearly all the nuclei in every other region of the meristem are labelled. Photograph taken with dark-field illumination. (*c*) Optical density recordings of Feulgen stained nuclei with a 2C DNA content lying in the cortical (C) and stelar (S) portions of the quiescent centre of a 6 day old primary root of *Zea mays*. The data are the mean proportions of the total nuclear area seen in optical section that lie above different optical density thresholds. The error bars are the 95% confidence intervals of the mean. The readings show that the nuclei in the cortical portion of the quiescent centre have more dense chromatin than the chromatin of nuclei in the stelar portion. These latter nuclei are the slowest to enter S phase, as judged by autoradiography (see Fig. 1*b*).

active cells are seen located in the region formerly occupied by the quiescent centre and in other regions as well. These meristematic islands eventually give rise to a complete new root meristem (Clowes, 1959). In roots of *Zea* the regeneration that follows irradiation is from the quiescent centre zone only (Clowes, 1963, 1964). Accompanying the restoration of a new meristem is the formation of a new quiescent centre. Regenerating roots of *Vicia* sometimes become forked (Clowes, 1959), indicating that some of the meristematic islands are capable of independently organising new tissue. This observation parallels that made by Cairnie & Millen (1975) who found that the intestinal crypts of irradiated mice sometimes split during the course of the regeneration of the destroyed tissue. In these instances a group of viable cells of the original tissue unit (not necessarily the stem cells) splits up, resulting in new groups that can lead independent existences.

It is regrettable that irradiation as a probe of meristem behaviour has not been exploited to the extent it has been in animal systems, for it is clearly useful in uncovering otherwise hidden aspects of the cell cycle and cell interactions in a tissue unit. Irradiation of root apices causes a reduction in the length of the meristem by prematurely terminating the mitotic cycle in the most basal meristematic cells (von Wangenheim, 1975). It would be most interesting to know if the degree of stimulation of division in the quiescent centre by X-rays, for instance, is related to the number of remaining meristematic cells (a feedback control), or whether irradiation has some direct enhancing effect on the cycle of quiescent centre cells. The latter possibility is suggested by the observations of Clowes (1970) and Thompson (quoted by Clowes, 1970) that cells apparently enter mitosis more frequently in the quiescent centre of *Allium sativum* (onion) and *Zea mays* a few minutes after receiving an acute dose of X-rays; however, these observations do not preclude the possibility that there may be a feedback control that responds rapidly to perturbations.

Cell divisions in the quiescent centre of *Zea* are also stimulated by removal of part, or all, of the root cap (Clowes, 1972; Barlow, 1973, 1974). The descendants of these formerly quiescent cells then divide and differentiate in such a way as to reproduce the lost tissue. It seems as though the quiescent centre is poised for division should anything impair or curtail division in the rest of the meristem. We may conclude that its potential for root tissue regeneration is high.

The situation is less clear-cut with regard to the regenerative potential of the central zone of shoot apices. This may be because the appropriate experiments have not yet been done. Radiation work with shoot apices seems to have been more concerned with its potential for producing mutations and consequently high doses (many kilorads) of X-rays have often been employed. Nevertheless, on the present evidence the central zone does not seem to play a significant role in shoot regeneration after irradiation. The cells of this zone are not differentially resistant to X-rays, as are cells of the quiescent centre, and when shoot regeneration does occur it is the cells at the flank of the meristem that normally reorganise a new apex (Sekiguchi, Yamakawa & Yamaguchi, 1971; Iqbal, 1972; Langenauer & Davis, 1973). However, mechanical damage to the apex followed by its transplantation to a culture medium *in vitro* does stimulate cells of the central zone to synthesise nuclear DNA and divide (Sussex & Rosenthal, 1973; Davis & Steeves, 1977) and it may be supposed that some of their descendants contribute to the repopulation of the new meristem. Following this activation of the mitotic cycle the central zone reverts to its mitotically inactive state and it continues to be carried forward at the summit of the apex.

Although much detail is lacking for a definite conclusion, it does seem as though the central zone plays a less active role in shoot regeneration than does the quiescent centre for roots; this may be due to the sluggishness with which its cells can be provoked to division coupled with the relatively slow rates of cell division in shoots. This in turn may depend on the stage of development reached by the plant; for instance, the age of the central zone cells or changes in its internal microenvironment may both play a part in the way the cells respond. It is possible that prolonged quiescence may render the block to cycle activity much harder to overcome, in the same way that quiescence seems to become 'deeper' in nuclei of cultured mammalian cells the longer they remain in stationary phase (Rossini, Lin & Baserga, 1976). A corollary to this observation is the likelihood that the degree of quiescence may be related in some way to the structure of the chromatin (cf. Yun & Naylor, 1973; Rossini *et al.*, 1976) (see also Fig. 1).

As already intimated, there are examples of regeneration in which the stem cells are not the only source of the new tissue. A thorough investigation of the radiation response (as just one example of a disturbing influence) of cells within a tissue unit would help to clarify exactly what the potentialities of the stem cells are for regeneration *vis-à-vis* the potentialities of other proliferating cells. In addition, there are cases where regeneration occurs in the absence of stem cells (e.g. after their removal) and in the course of which new stem cells are re-established (Feldman, 1976; Arzee, Schwartz & Cohen, 1977; P. W. Barlow, unpublished data). These examples show that some, possibly all, meristematic cells are potential stem cells (cf. Cairnie & Millen, 1975); in other words, the developmental sequence stem cell → derived meristematic cell → differentiated cell is, in certain situations, reversible.

Pluripotency

The fact that in a regenerating tissue unit the stem cells may be largely responsible for the reconstruction of a new meristem and differentiated derivatives already suggests their pluripotent nature. However, it is possible to imagine a scheme of tissue development where, once a permanent stem-cell population has been laid down, the particular cell types that differentiate are the derivatives of stem cells each of which are imprinted with different developmental potentialities. One test of stem-cell equipotentiality, and thus of their pluripotency, is to devise ways of isolating the stem cells from each other within the tissue and to see the range of cell types that can then be derived from any one of them. It is already known from nuclear transplantation work in animals and single-cell cultures in plants that some, if not all, nuclei and cells are totipotent in that complete organisms can be derived from them; but the point at issue is whether stem cells are pluripotent within the environment of the organism rather than in the isolation of the culture chamber. Of course, pluripotency is a logical deduction if it could be shown that there is only one stem cell for a given tissue, because all the other cells must be derived from it. There are some plants (e.g. certain ferns) where there is a single mitotically active cell at the apex of root and shoot; while this apical cell itself remains uncommitted with regard to any path of differentiation, its derivatives acquire particular differentiated states that are presumably determined by the positions they ultimately come to occupy in the body of the plant. A further (though indirect) indication of pluripotency is that when the apical cell is destroyed further growth of the apex ceases (Wardlaw, 1949). Destruction of tissue around the apical cell and this cell's subsequent divisions (Kadej, 1960) also attest to its high regenerative potential. As far as I am aware, there are no reports of the successful culture and growth of an isolated apical cell; these would be testing conditions to reveal this cell's potential. However, Wardlaw (1957) was able to induce the shoot apical cell of *Dryopteris austriaca* to form a multicellular scale by application of certain chemicals to the apex. This result indicates that the apical cell is capable of direct transformation to a new cell type and this is probably caused by a significant disturbance to the internal milieu of the apex which normally maintains the pattern of growth and development therein.

The quiescent centre of *Zea* roots can be isolated as a complete unit and placed in culture (Feldman & Torrey, 1976). Here it proceeds to organise a new root without any noticeable disorganised callus stage intervening. While this regeneration of root tissue from an isolated quiescent centre is suggestive of the pluripotency of its individual cells, it does not prove it; for the organisation already inherent in this population may simply serve as a template for perpetuating the path of development that this group of cells and its derivatives normally take (cf. Barlow, 1976).

Actually, it is probable that some classes of plant stem cells are virtually totipotent. For example, the apical cell of certain fern roots from time to time may divide in uncharacteristic planes and the derivatives then proceed to organise a shoot (Rostowzew, 1890; Peterson, 1970). There are other examples of the transformation of roots into shoots in higher plants which have been reviewed by Champagnat (1971). Examples of the transformation of the shoot apex into a root are less well known. Beijerinck (1887) described such a transformation of shoots growing on root segments of *Rumex* and this was later confirmed by Priestley & Swingle (1929). All these cases are of interest for they indicate that in certain circumstances a fundamental switch can occur in the organisational activity of a certain portion of the meristem, so that cells that formerly organised one type of meristem now organise another type, But whether this involves a switch in the activities of the stem cells themselves is not known. One may surmise that there is little to distinguish in the properties of these cells in either shoot or root apices: they are totipotent; it is their derivatives that are guided into particular developmental pathways by environmental factors.

Immutability. According to Cairns (1975) most spontaneous mutations probably arise as a result of errors of DNA replication. The occurrence of a mutation in a stem cell would eventually lead to mutated derivative cells. To minimise the hazard resulting from stem-cell mutation Cairns (1975) proposed that the pattern of segregation of semi-conservatively replicated DNA is such that the older of the two parental DNA strands is retained in the stem cell, while the daughter cell (that leaves the stem-cell compartment) receives the younger DNA strand. Thus, from the time a stem cell comes into existence it retains the same strand of DNA throughout its life, and any erroneous replicas of this DNA strand are exported in the emigrating daughter cells. In organisms with many chromosomes, this 'retentive' type of DNA segregation in stem cells would be facilitated if there were the end-to-end linking together of the DNA strands that run the length of each chromosome, i.e. if all the chromosomes were an expression of just one or two circles of DNA (cf. chapter 11 of DuPraw, 1970). In fungi and plants end-to-end chromosome associations have been observed (McGinnis, 1956; Wagenaar, 1969) and these may be indicative of the circular continuity of DNA (see DuPraw (1970) for other examples).

If stem cells of plants operate a mechanism of DNA strand retention, then one place to look for it would be in the quiescent centre. After labelling a stem-cell nucleus for one round of DNA replication with, say, [^{14}C]thymidine, all radioactivity should be segregated from the nucleus of the cell that remains in the stem-cell location at the second mitosis. Therefore, assuming the multicellular quiescent centre as a whole to be a stem-cell population, one would expect to see within it the beginnings of a general clearing-out of labelled cells after two cell generations and at the same time to see the daughter nuclei of cells in anaphase and telophase carrying unequal amounts of label. Furthermore, these mitoses should be orientated so that the nucleus of the most apically positioned cell is unlabelled (i.e. there is bias towards retention of the parental DNA strands within the quiescent centre) and the cell with the labelled nucleus is the basally positioned one (i.e. is towards being displaced from the quiescent centre).

Accordingly, young primary roots of *Zea mays* have been labelled with methyl-[^{14}C]thymidine for 22 hours and the seedlings grown-on in potting compost at 27 °C. The primary root apices were harvested 3, 6 and 10 days later, and autoradiographs prepared of longitudinal sections. At these times, given an *average* intermitotic period of about 150–200 hours, one might expect many of the cells to be in their second division following their labelling with thymidine. The results of the experiment were inconclusive in their support of the hypothesis, for although label disappeared from nuclei in all regions of the meristem as a result of its dilution over the course of successive divisions, it persisted over the much more slowly dividing nuclei of the quiescent centre. Counts of silver grains over anaphase and telophase daughter nuclei in the quiescent centre showed most of them to be equally labelled (Fig. 2). A few such pairs of nuclei were unequally labelled; could they be showing the postulated mode of DNA segregation?

Another type of experiment aimed at testing strand retention would be to label both DNA strands and then to look for persistence

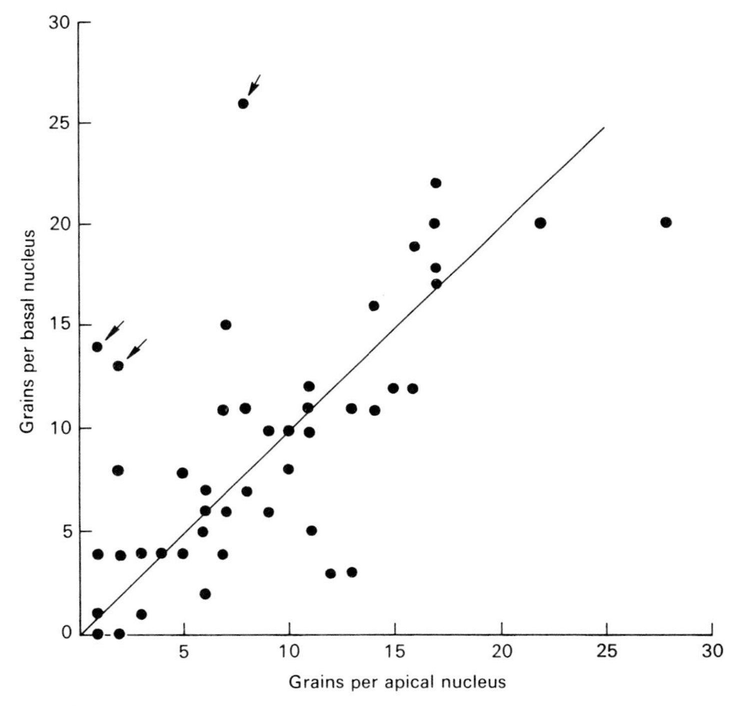

Fig. 2. The number of autoradiographic silver grains over each daughter nucleus in cells at either anaphase or telophase in the quiescent centre of primary roots of *Zea mays* fixed on the third, sixth and tenth day after labelling with [^{14}C]thymidine. The data from each day have been pooled. While the sorting-out of cells in a multicellular, hemispherical group, such as the quiescent centre, is undoubtedly complex, inequality of the distribution of label was looked for: hence the nuclei were classified as either basal or apical. This refers to their position in the dividing cell with respect to the axis of the cell-file in which it lies. A basal nucleus is nearest the end of the cell directed towards the base of the root, an apical nucleus is towards the apex. Points that denote daughter nuclei with a markedly unequal distribution of grains are arrowed. The line is drawn at an angle of 45° to indicate where the grain counts would lie if both daughter nuclei were covered by an equal number of grains.

of label within the stem-cell nuclei after a number of rounds of DNA replication and mitoses. Such an experiment has been performed by Rosenberger & Kessel (1968) using hyphae of *Aspergillus nidulans*; the hyphae are actually multinucleate, but the nuclei at the hyphal tip may be analogous to stem-cell nuclei. Conidia, each with a single labelled nucleus, were germinated; after the first two or three synchronous nuclear divisions a pair of strongly-labelled nuclei tended to persist at the distal tip of the young hyphae, while nuclei in the more proximal portion of the hyphae (the conidial end) were weakly labelled. The distribution of labelled nuclei within hyphae was not followed beyond the octonucleate stage, but nevertheless Rosenberger & Kessel concluded that 'the mechanism which disperses newly formed nuclei in the growing hypha can distinguish between nuclei containing DNA strands of different age'.

Mechanisms are not hard to envisage whereby the segregation of DNA strands to daughter nuclei is polarised to favour the persistence of one of the strands in a stem cell or hyphal tip, in contradistinction to a random pattern of segregation. For instance, in both fungal and higher plant protoplasm the chromatin is attached to the nuclear membrane which, in turn, is continuous with cytoplasmic membranes and thence to the external, wall-associated, plasma membrane (Bracker, Grove, Heintz & Morré, 1971; Morré & Mollenhauer, 1974). A polarised segregation of DNA might depend on a polarised segregation of membranes following mitosis. In addition, a shifting plane of division, such as that that causes the tetrahedral apical cell in roots and shoots of pteridophytes to cut off new cells from each of its faces in a regular sequence, may also help to segregate its DNA strands in a particular way. Alternation of the planes of stem-cell division may be a necessary requirement (Cairns, 1975) if plant chromosomes show the same spatial pattern of chromatid splitting as do animal cells (Herreros & Giannelli, 1967).

THE PERMANENCE OF STEM CELLS

Given the continuing existence and growth of a tissue or organ, the question to be discussed now is whether its stem cells always function as such. Or may it be that from time to time the stem-cell group either differentiates, or is displaced, resulting in other, previously non-stem cells assuming this function?

Where the stem cell is a single cell, as at the summit of an apex, it can probably never be displaced, but it may differentiate (Sossountzov, 1972). If its differentiation is incompatible with cell division, as has been suggested to obtain in the apical cell of certain pteridophytes (Buvat & Liard, 1953; Sossountzov, 1972), then either the stem-cell function is taken over by some other cell, or cells, or the apex becomes determinate in its growth. In many plants cell differentiation is accompanied by polyploidy of the nucleus; in many pteridophytes the apical cell becomes polyploid but is still able to divide, as shown by the fact that polyploid tracts of tissue are found apparently to originate from it (D'Amato, 1975). Chiang & Gifford (1971) believe that as roots of pteridophytes age so divisions in the apical cell become rarer; and as somatic polyploidisation reflects a breakdown of the usually rather tight coupling between DNA synthesis and mitosis, apical cells would seem eventually to relinquish their stem-cell role. However, this does not mean that they then become completely devoid of any influence or function in apical growth.

Where the stem-cell group is composed of a number of cells, permanency of this role is not assured for any of its cells either, but not necessarily as a consequence of cell differentiation. Observations of Catesson (1953) on the shoot apex of the woodrush *Luzula pedomontana* illustrate here another aspect of stem-cell impermanency. She plotted the distribution of mitoses in the apex during successive plastochrons (a plastochron is the time interval between the inception of one leaf and the next) and found the central zone, where mitoses are infrequent, to cant from one side of the apex to the other. Although a narrow zone was demonstrated in which mitoses are always infrequent, the point is that other cells of the central zone, which we have taken to possess characteristics of a stem-cell population, periodically lose these characteristics. Furthermore, direct observations of marker particles placed on shoot apices of *Lupinus* led Soma & Ball (1964) to conclude that no cell occupies a permanent position in the apex, though it may be of significance that some particles (and so by implication some cells) remained at least 42 days at the extreme tip. Other results obtained by Soma & Ball (based on a method that involved the killing of cells) suggested that a cell, or its descendants, can be moved up to the summit of the apex from a lower position and thereby displace the cells that formerly resided at the summit. It seems likely, therefore, that the cytophysiological properties of cells in the central zone of the shoot apex are not inherited from one division to the next, but are a response to the position the cells occupy in relation to the internal milieu of the apex (see also the discussion by Newman, 1961, 1965). Positional responses can arise as a result of gradients of metabolites. In permanent meristems such gradients may determine the pattern of mitotic activity found therein (Barlow, 1976). It is the displacement of cells through a fixed gradient, as well as the shifting of the gradient itself with respect to the cells, that can lead to the changing population of stem cells.

It is important to recall, however, that the stability of certain types of chimaerical shoot apices does attest to the permanence, both in a positional and functional sense, of a small number of cells within what we have taken to be the functional stem-cell group of the shoot apex (Bartels, 1961; Stewart & Dermen, 1970). Such cells, earlier referred to as the minimum set of stem cells, may be those that reside in the permanently quiescent portion of the central zone of *Luzula* (see Catesson, 1953) and at the apparently immobile summit of *Lupinus* (see table 3 of Soma & Ball, 1964).

There may be no permanent stem cell in some cambium meristems, the role of initiating the radial lineages of xylem and phloem falling on one cell then on another in response to shifting internal changes (see Catesson, 1974). But where a stem cell is recognisable, as it appears to be in *Pinus* and in other gymnosperms, it may not be at the same level as the stem cell in neighbouring radial cell rows (Newman, 1956). This implies that the positional specification of the stem cell is determined within the row of its derivatives and is independent of the cells that constitute other rows.

On the other hand, in root apices, particularly in those with a 'closed' construction where there is an eye-catching cell wall boundary between the cap and the rest of the root, the position of the quiescent centre is probably quite constant. This may be due, in part, to its maintenance by the cap cells and the constraints to cell growth mediated by the boundary wall between the two populations (Barlow, 1973, 1974). Where the boundary wall is less prominent, as in roots with an 'open' construction, the size and position of the quiescent centre may be more able to fluctuate. However, there is no evidence that it does.

Although we find that permanency of behaviour is not necessarily an attribute of the cells within a stem-cell group, I do not think that this is sufficient reason to reject the concept of stem cells as it applies to plants. For one thing, the permanence of shoot chimaeras argues against this. Secondly, stem cells must surely be considered in terms of their function and behaviour at any particular time, even though these parameters may be determined by a temporally shifting positional control. In addition, the positional specifications that determine stem-cell properties may be reinforced by the constraints imposed upon the cells by the geometry of the tissue unit (Barlow, 1974, 1976; Potten, 1974).

DEVELOPMENTAL CHANGES AND AGEING OF STEM CELLS

The developmental climax of stem cells

Throughout their life the stem cells of cambium and roots propagate the same types of cells. As mentioned in the previous section, this is probably due to the continuing (homoeostatic) stability of the cellular environment. For much of a plant's life the same is true of the central zone of the shoot. However, in many species environmental changes induce the shoot apex to form a flower or inflorescence; the molecular mechanisms which underlie this induction are incompletely known. In this circumstance mitoses in the central zone and elsewhere in the apex become more frequent as the average duration of the cell cycle shortens (Lyndon, 1976). Flowering brings apical growth to an end, so it seems as though the flowering stimulus causes the apex to become determinate with respect to the number of divisions remaining to the cells there. Stem cells in the shoot apex cease to exist: the morphogenesis of a flower may be considered to be their developmental climax.

It could be imagined that the formation of a flower is the culmination of a developmental sequence inherent in the shoot apex. The relative quiescence of the cells of the central zone of the vegetative apex may be responsible for the holding in check of this developmental potential. The increased tempo of mitosis that accompanies flower induction could be a necessary step towards the climax of shoot development, with the inductive stimulus causing the breakdown of the homoeostatic condition that normally maintains the vegetative state. The cells of the central zone therefore play a stem-cell role for only as long as their milieu dictates.

The floral apex is not the only determinate state of a shoot meristem. In some species lateral shoot apices may develop as tendrils (e.g. in vines), in other species as thorns (e.g. in gorse) (Tucker, 1968); these developmental courses are also in part mediated by stimuli from the outside environment. It may also be recalled that leaves develop from a determinate meristem that was originally a portion of the shoot apical meristem.

Ageing of stem cells

The potential for growth of animal cell populations both *in vivo* and *in vitro* has been found to decline in relation to the number of cell doublings already achieved (reviewed by Hayflick, 1973). At the cellular level the decline in vitality is considered to be an expression of 'ageing'. If the primary cause of ageing is truly a function of the number of cell divisions completed, then stem cells, like any other cell, must take a step towards their death every time they divide. Indeed, stem cells would be most vulnerable to a division-determined death for they are the cells with potentially the greatest number of divisions ahead of them. However, since stem cells often divide at a slower rate than other cells they may have a correspondingly greater opportunity to repair certain types of damage, in addition to having a longer lifespan, in chronometric terms, than more rapidly dividing cells. Further, the probability of stem-cell death may be considerably less than the probability of the whole organism being afflicted by some fatal disease or accident. But, on the other hand, ageing of the stem cells could lead to their derivatives inheriting their senescent property; should there be many derivatives, and these in turn divide many times, then the senescent property would be amplified, and again this may cause the death of the organism before the death of the stem cell.

Whether or not there is a limit to the number of divisions that plant cells can accomplish before dying is not known. There is a remarkable range of lifespans in the plant kingdom (Molisch, 1929). This could be due to differences in susceptibility, with time, to fatal diseases, structural or metabolic failures of the whole plant, as well as to inherent differences in ageing rates, the latter being coupled with the mitotic rate and a division-limited lifespan of the stem cells. As an example of stem-cell longevity we may take the baobab tree *Adansonia digitata*: this tree is recorded as living about 5000 years (Molisch, 1929); if the stem cells of the shoot or the cambium divide just once a year (a reasonable minimum value) this represents 5000 cell divisions. This remarkable figure of course does not mean that plant stem cells aspire to immortality, only that we must have patience in establishing their rate of ageing.

In the foregoing paragraphs the term 'ageing' has been used in a way common among physiologists, i.e. being the antithesis of pristine and allied to senescence and its attendant pathological associations. But we may view both ageing and development, the latter in the sense of the unfolding of potentialities inherent in the hereditary material, as equivalent and to be functions of a phenomenon that may not inappropriately be called 'experience'. Organisms, tissues, and cells and their nuclei experience changes in the environment around them. Qualitatively different experiences can evoke in sensitive cells (dividing cells) different responses, each being the discrete actualisation of a potentiality resident within them. We may postulate that development concludes when all the cells' potentialities have been actualised; this working-out of the developmental programme is ageing, while the events that follow the completion of the programme are manifestations of senescence. Therefore, the stem cells of the permanent meristems of plants do not age, for the number of experiences that they receive is limited, and there is little if any evocation of a change in their course of development as a consequence of the continual maintenance of the homoeostatic environment in which they reside. However, the stimulus to flower, for example, is a new experience and new potentialities are evoked from the stem cells and their more immediate descendants; as a result the length of their future existence is shortened. The cells of the apex age, and when the developmental climax has been accomplished they enter a phase of senescence.

With a similar line of reasoning we may consider the nature of impermanent determinate meristems. These meristems, be they in animals (e.g. at the tip of the developing wing of a chick (Lewis, 1976)) or plants (e.g. in developing leaves) emit cells that do not differentiate identically at different moments of time. This is because any cell and its descendants within the meristem, including the stem-cell zone, are continually receiving new experiences. These may, in part, be created by the tissues that have already been differentiated. Hence this is a system of continual change. We can see, therefore, that the permanence or impermanence of meristems are functions of the experiences (environmental changes) received by the dividing cells. It is the homoeostatic, or unchanging, nature of the permanent meristem unit and the unvarying quality of the experiences that they receive that favours the self-renewing and self-maintaining tissue unit. In impermanent meristems stem cells may exist for a time, but the tissue unit is not self-perpetuating because it is part of a qualitatively changing environment. Such changes cause the stem cells and their derivatives to use up their potentialities, i.e. they age, and ultimately cease dividing.

As a further aid to understanding the senescence of tissue units and the role of stem cells, a useful analogy can be found in mechanical systems so constructed that the rate of 'ageing' and failure of the components influences the reliability of the system. Flehinger (1960), in an analysis of system reliability, demonstrated that efficiency was improved by maintenance checks being made at regular intervals in which all defective parts were replaced; less efficient was the system in which there was no scheduled maintenance but in which parts were replaced only in the event of their failure. Stem cells may be viewed as a versatile cell-type that can replace any defective components of the tissue system. As stem cells do divide (though in the different systems the differential between the rates of division in the stem cell and in the remainder of the meristem may vary considerably) we may ask whether they do so in response to defects in the tissue system (system 1), or whether their divisions are programmed to provide a regular maintenance of the system (system 2). From the discussion in the previous sections we may conclude that both attributes obtain: system 1 in disturbed growth conditions, system 2 in steady-state conditions. Given that stem cells are 'unaged', or possess a maximum of potential, then the more frequently they divide so the more efficient the system to which they contribute is likely to be.

CONCLUDING REMARKS AND SUMMARY

The life-histories of higher plants and animals can be divided into three phases: embryogenesis, during which the young organism acquires a definite form as it grows from a zygote and develops within, and is nourished by, the maternal tissue; blastogenesis, during which the organism establishes its independence and continues development to acquire its final mature form; and teleogenesis, during which the organism, having acquired a mature adult form, achieves a developmental state which it will maintain for the rest of its life (see Souèges (1954) for a similar viewpoint). Unlike the situation during the two earlier phases, during teleogenesis tissues are present that are self-renewing. And it is in this phase that stem cells, although possibly present earlier, assume the important role of adding cells to the organism thereby assuring it of a continuing source of cells for specific pathways of differentiation.

Plants are self-maintaining systems in which growth with cell division continues for a long period of time. The permanent meristems confer this 'open' nature and so there must be a reservoir of stem cells that are the primary source of this character. Stem cell populations can be delineated in shoots and roots by their position, which also coincides with certain distinguishing cytological features; stem cells are less obvious in cambium, though they must be there too. In shoots and roots, stem cells divide infrequently, can regenerate the meristem in the event of its being damanged, and, if not totipotent, they are at least pluripotent. But as far as there are any relevant data on the subject, the cells are pluripotent only to the extent of being able to give rise to the other cell-types that make up the tissue unit. Thus, in a tissue unit, while the stem cells are an integral part of it, they perform a function that is neither more nor less than that required to perpetuate the unit.

Whether or not the stem cells of plants have a mechanism to retain a strand of DNA that is free of errors introduced during DNA replication is not known. If they do, there could be advantages to the species if stem cells were able to contribute directly to the formation of gametes. This may be the case in the central zone of the shoot apex. Furthermore, certain observations by Rutishauser & Röthlisberger (1966) on the transmission of supernumerary chromosomes from one generation to the next in *Crepis capillaris* suggest that in some circumstances the stem cells may serve not only as a repository of genetic information but even as an accumulator of additional genetic information that is ordinarily unavailable to other cells of somatic tissue. But even if they do not have such a mechanism, the fact that the stem cells have a long intermitotic period, of which the phase of DNA synthesis occupies a small fraction, minimises the rate at which such errors might accumulate. On the other hand, the longevity of many plants and their stem cells in some measure counteracts this benefit. In order to assess the true importance of stem cells it is necessary to know whether they have a greater resistance, as well as a greater ability to repair damage to the cell (e.g. damage in the form of thermal shocks, ionising or cosmic radiations). There are reports (e.g. Coggle, 1968) that so-called G_0 cells (a phase often associated with stem cells) are more resistant to X-rays, yet less able to repair radiation damage than cells in, say, G_1 phase. The former character may be related to the decreased volume (target size) of the nuclei of the G_0 cells and perhaps also to a smaller volume of cytoplasm per genome (see von Wangenheim (1975) for a discussion of this aspect of radiation response in plant systems), while the latter character may relate to their low degree of cellular metabolism. However, ionising radiation is not a hazard normally present in the environment of plants. Thermal shocks are, though, and the resistance of the quiescent centre to them is similar to its radio-resistance (Clowes, 1967); possibly this common response is mediated by similar cellular and nuclear properties.

The ease with which root and shoot meristems may be obtained, cultured and experimented upon, enables us to gain many viewpoints of the circumstances that maintain organised growth. But the role that the stem cells themselves play in creating the physiological milieu that will maintain meristematic activity and organ growth has hardly been explored. The quiescent centre of roots has been postulated to be a site of origin of growth regulators in addition to playing the role of a centre from which root organogenesis is directed (Barlow, 1976; Feldman & Torrey, 1976). Should this dual role be true of stem cells in general, then one is investigating a remarkable population of cells that function at a variety of developmental levels by virtue of their physiological, kinetic and organogenetic properties. Furthermore, within the context of the unit in which they exist, stem cells and the tissue derived from them together form a concinnous whole, for the latter serves the organism's contemporary needs, while the stem cells maintain this potential for the future. It is the stem cells' spatial location in the tissue unit that is able to reconcile these two otherwise unrelated aspects of immediacy of function, and persistence in time.

I am grateful to Drs D. N. Butcher, J. Cairns, Elizabeth G. Cutter, Daphne J. Osborne and K.-H. von Wangenheim for discussions relating to various aspects of the material presented here.

After this contribution went to press two papers were published that prompted the following addition:

The apical cell that is present at the tip of the shoot, root and leaves of some species of ferns (some of whose properties were mentioned on pp. 96–7 and 100–1) has been an object of interest to botanists for over one hundred years and continues to attract their attention. The generally held view of the apical cell of the root is that it is the initial cell (stem cell) for all the tissues of the root. Additional light on the role of this cell has, however, been cast by the observations of Kuligowski-Andrès (1977). She had studied the embryogenesis of the waterfern, *Marsilea vestita*, and comes to the opinion that the formation of an apical cell (which occurs at about the 32-cell stage) is essential for the induction of the root of the young embryo. Mme Kuligowski-Andrès also suggests in her paper that the apical cell may be the site of biosynthesis of some inductive substance. The evidence for an inductive role is circumstantial, but is extremely interesting, for it suggests that in addition to its stem-cell role, the apical cell may have an equally important role as a tissue organiser. Perhaps the same two characteristics apply to the quiescent centre of the roots of higher plants (cf. Barlow, 1976).

Meanwhile, Mme Sossountzov, after further study of the adult shoot apex of *Marsilea drummondii*, has not been able to adduce any evidence that the apical cell functions either as a stem cell or as an organiser (Sossountzov, 1976). She favours the idea that it is a neutral cell type, being both non-meristematic and non-differentiated.

REFERENCES

Al-Dewachi, H. S., Wright, N. A., Appleton, D. R. & Watson, A. J. (1975). Cell population kinetics in the mouse jejunal crypt. *Virchows Archiv*, **18B**, 225–42.

Arzee, T., Schwartz, M. & Cohen, L. (1977). A negative image of the quiescent centre in regenerating root apices of *Zea mays*. *Planta*, **133**, 207–8.

Bannan, M. W. (1955). The vascular cambium and radial growth in *Thuja occidentalis* L. *Canadian Journal of Botany*, **33**, 113–38.

Barlow, P. W. (1973). Mitotic cycles in root meristems. In *The cell cycle in development and differentiation*, ed. M. Balls & F. S. Billett, pp. 133–65. London: Cambridge University Press.

Barlow, P. (1974). Regeneration of the cap of primary roots of *Zea mays*. *New Phytologist*, **73**, 937–54.

Barlow, P. W. (1976). Towards an understanding of the behaviour of root meristems. *Journal of theoretical Biology*, **57**, 433–51.

Barlow, P. W. & Macdonald, P. D. M. (1973). An analysis of the mitotic cell cycle in the root meristem of *Zea mays*. *Proceedings of the Royal Society of London*, **183B**, 385–98.

Bartels, F. (1961). Zur Entwicklung der Keimpflanze von *Epilobium hirsutum*. IV. Der Nachweis eines Scheitelzellenwachstums. *Flora*, **150**, 552–71.

Beijerinck, M. W. (1887). Beobachtungen und Betrachtungen über Wurzelknopsen und Nebenwurzeln. *Verhandelingen der Koninklije Akademie van Wetenschappen*, **25**, 146 pp.

Booker, C. E. & Dwivedi, R. S. (1973). Ultrastructure of meristematic cells in dormant and released buds of *Tradescantia paludosa*. *Experimental Cell Research*, **82**, 255–61.

Bracker, C. E., Grove, S. N., Heintz, C. E. & Morré, D. J. (1971). Continuity between endomembrane components in hyphae of *Pythium* spp. *Cytobiologie*, **4**, 1–8.

Buvat, R. (1955). Le méristème apical de la tige. *Année biologique, 3ᵉ série*, **31**, 595–656.

Buvat, R. & Liard, O. (1953). Interprétation nouvelle du fonctionnement de l'apex d'*Equisetum arvense* L. *Comptes rendus hebdomadaires des Séances de l'Académie des Sciences, Paris*, **237**, 88–90.

Cairnie, A. B. & Millen, B. H. (1975). Fission of crypts in the small intestine of the irradiated mouse. *Cell and Tissue Kinetics*, **8**, 189–96.

Cairns, J. (1975). Mutation selection and the natural history of cancer. *Nature, London*, **255**, 197–200.

Catesson, A.-M. (1953). Structure, evolution et fonctionnement du point végétatif d'une monocotylédon: *Luzula pedemontana* Boiss. et Reut. (Joncacées). *Annales des Sciences naturelles: Botanique, Paris, 11ᵉ série*, **14**, 253–91.

Catesson, A. M. (1974). Cambial cells. In *Dynamic aspects of plant ultrastructure*, ed. A. W. Robards, pp. 358–90. Maidenhead: McGraw-Hill Book Company (UK) Ltd.

Champagnat, M. (1971). Recherches sur la multiplication végétative de *Neottia nidus-avis* Rich. *Annales des Sciences naturelles: Botanique et Biologie végétale, Paris, 12ᵉ série*, **12**, 209–47.

Chiang, S.-H. & Gifford, E. M. (1971). Development of the root of *Ceratopteris thalictroides* with special reference to apical segmentation. *Journal of the Indian botanical Society*, Golden Jubilee Volume, **50A**, 96–106.

Clowes, F. A. L. (1954). The promeristem and the minimal constructional centre in grass root apices. *New Phytologist*, **53**, 108–16.

Clowes, F. A. L. (1956). Nucleic acids in root apical meristems of *Zea*. *New Phytologist*, **55**, 29–35.

Clowes, F. A. L. (1959). Reorganization of root apices after irradiation. *Annals of Botany*, **23**, 205–10.

Clowes, F. A. L. (1961). *Apical meristems*. Oxford: Blackwell Scientific Publications.

Clowes, F. A. L. (1963). X-irradiation of root meristems. *Annals of Botany*, **27**, 343–52.

Clowes, F. A. L. (1964). The quiescent center in meristems and its behavior after irradiation. *Brookhaven Symposia in Biology*, **16**, 46–57.

Clowes, F. A. L. (1965). The duration of the G_1 phase of the mitotic cycle and its relation to radiosensitivity. *New Phytologist*, **64**, 355–9.

Clowes, F. A. L. (1967). The functioning of meristems. *Science Progress, Oxford*, **55**, 529–42.

Clowes, F. A. L. (1970). The immediate response of the quiescent centre to X-rays. *New Phytologist*, **69**, 1–18.

Clowes, F. A. L. (1971). The proportion of cells that divide in root meristems of *Zea mays* L. *Annals of Botany*, **35**, 249–61.

Clowes, F. A. L. (1972). Regulation of mitosis in roots by their caps. *Nature, New Biology, London*, **235**, 143–4.

Clowes, F. A. L. (1976). Meristems. In *Perspectives in experimental biology*, ed. N. Sunderland, vol. 2, pp. 25–32. Oxford: Pergamon Press.

Coggle, J. E. (1968). Effect of cell cycle on recovery from radiation damage in the mouse liver. *Nature, London*, **217**, 180–2.

D'Amato, F. (1975). Recent findings on the organization of apical meristems with single apical cells. *Giornale botanico italiano*, **109**, 321–34.

Davis, E. L. & Steeves, T. A. (1977). Experimental studies on the shoot apex of *Helianthus annuus*: the effect of surgical bisection on quiescent cells in the apex. *Canadian Journal of Botany*, **55**, 606–14.

DuPraw, E. J. (1970). *DNA and chromosomes*. New York: Holt, Rinehart & Winston, Inc.

Feldman, L. J. (1976). The *de novo* origin of the quiescent center regenerating root apices in *Zea mays*. *Planta*, **128**, 207–12.

Feldman, L. J. & Torrey, J. G. (1976). The isolation and culture in vitro of the quiescent center of *Zea mays*. *American Journal of Botany*, **63**, 345–55.

Flehinger, B. J. (1960). System reliability as a function of system age; effects of intermittent component usage and periodic maintenance. *Operations Research*, **8**, 30–44.

Hayflick, L. (1973). The biology of human aging. *American Journal of medical Science*, **265**, 432–45.

Herreros, B. & Giannelli, F. (1967). Spatial distribution of old and new chromatid sub-units and frequency of chromatid exchanges in induced human lymphocyte endoreduplications. *Nature, London*, **216**, 286–8.

Iqbal, J. (1972). Effects of acute gamma radiation on the survival, growth and radiosensitivity of the apical meristem of *Capsicum annuum* L. at different stages of seedling development. *Radiation Botany*, **10**, 337–43.

Kadej, F. (1960). Regeneration der Scheitelzelle bei Farnwurzeln. *Acta Societatis botanicorum poloniae*, **29**, 363–8.

Kuligowski-Andrès, J. E. (1977). Etude de l'organogenèse radiculaire chez le *Marsilea vestita*. *Flora*, **166**, 333–56.

Langenauer, H. D. & Davis, E. L. (1973). *Helianthus annuus* responses to acute X-irradiation. I. Damage and recovery in the vegetative apex and effects on development. *Botanical Gazette*, **134**, 301–16.

Lewis, J. H. (1976). Rules for building the chick wing: discrete and continuous aspects of morphogenesis. In *Automata, languages, development*, ed. A. Lindenmayer & G. Rozenberg, pp. 97–108. Amsterdam: North-Holland Publishing Co.

Lyndon, R. F. (1976). The shoot apex. In *Cell division in higher plants*, ed. M. M. Yeoman, pp. 285–314. New York & London: Academic Press.

McGinnis, R. C. (1956). Cytological studies of chromosomes of rust fungi. III. The relationship of chromosome number to sexuality in *Puccinia*. *Journal of Heredity*, **47**, 255–9.

Molisch, H. (1929). *Die Lebensdauer der Pflanze*. Jena: G. Fischer Verlag.

Morré, D. J. & Mollenhauer, H. H. (1974). The endomembrane concept: a functional integration of endoplasmic reticulum and Golgi apparatus. In *Dynamic aspects of plant ultrastructure*, ed. A. W. Robards, pp. 84–137. Maidenhead: McGraw-Hill Book Company (UK) Ltd.

Newman, I. V. (1956). Pattern in the meristems of vascular plants. I. Cell partition in living apices and in the cambial zone in relation to the concepts of initial cells and apical cells. *Phytomorphology*, **6**, 1–19.

Newman, I. V. (1961). Pattern in the meristems of vascular plants. II. A review of shoot apical meristems of gymnosperms, with comments on apical biology and taxonomy, and a statement of some fundamental concepts. *Proceedings of the Linnean Society of New South Wales*, **86**, 9–59.

Newman, I. V. (1965). Pattern in the meristems of vascular plants. III. Pursuing the patterns in the apical meristem where no cell is a permanent cell. *Journal of the Linnean Society of London (Botany)*, **59**, 185–214.

Nougarède, A. (1977). Infrastructure des axillaires cotylédonaires du *Pisum sativum* L. (var. nain hâtif d'Annonay) durant le blocage en phase G_1 (état inhibé) et après la reprise d'activité. *Comptes rendus hebdomadaires des Séances de l'Académie des Sciences, Paris, Series D*, **284**, 25–8.

Nougarède, A. & Rondet, P. (1977). Les histones nucléaires et leurs variations dans les méristèmes axillaires du pois (*Pisum sativum* L., var. nain hâtif d'Annoney), soumis ou non à la dominance apicale. *Comptes rendus hebdomadaires des Séances de l'Académie des Sciences, Paris, Series D*, **284**, 623–6.

Paolillo, D. J. & Gifford, E. M. (1961). Plastochronic changes and the concept of apical initials in *Ephedra altissima*. *American Journal of Botany*, **48**, 8–16.

Peterson, R. L. (1970). Bud development at the root apex of *Ophioglossum petiolatum*. *Phytomorphology*, **20**, 183–90.

Potten, C. S. (1974). The epidermal proliferative unit: the possible role of the central basal cell. *Cell and Tissue Kinetics*, **7**, 77–88.

Potten, C. S. (1975). Kinetics and possible regulation of crypt cell populations under normal and stress conditions. *Bulletin du Cancer*, **62**, 419–30.

Priestley, J. H. & Swingle, C. F. (1929). *Vegetative propagation from the standpoint of plant anatomy. United States Department of Agriculture Technical Bulletin*, No. 151, 98 pp.

Rosenberger, R. F. & Kessel, M. (1968). Nonrandom sister chromatid segregation and nuclear migration in hyphae of *Aspergillus nidulans*. *Journal of Bacteriology*, **96**, 1208–13.

Rossini, M., Lin, J.-C. & Baserga, R. (1976). Effects of prolonged quiescence on nuclei and chromatin of WI-38 fibroblasts. *Journal of cellular Physiology*, **88**, 1–11.

Rostowzew, S. (1890). Beiträge zur Kenntniss der Gefasskryptogamen. I. Umbildung von Wurzeln in Sprosse. *Flora*, **48**, 155–68.

RUTISHAUSER, A. & RÖTHLISBERGER, E. (1966). Boosting mechanism of B-chromosomes in *Crepis capillaris*. *Chromosomes Today*, **1**, 28–30.

SANIO, K. (1873). Anatomie der gemeinen Kiefer (*Pinus silvestris* L.). *Jahrbücher für wissenschaftliche Botanik*, **9**, 50–126.

SEKIGUCHI, R., YAMAKAWA, K. & YAMAGUCHI, H. (1971). Radiation damage in shoot apical meristems of *Antirrhinum majus* and somatic mutations in regenerated buds. *Radiation Botany*, **11**, 157–69.

SMITH, J. A. & MARTIN, L. (1973). Do cells cycle? *Proceedings of the National Academy of Sciences, USA*, **70**, 1263–7.

SOMA, K. & BALL, E. (1964). Studies of the surface growth of the shoot apex of *Lupinus albus*. *Brookhaven Symposia in Biology*, **16**, 13–43.

SOSSOUNTZOV, L. (1972). Structure et fonctionnement du méristème apical des Ptéridophytes: présent et avenir. *Bulletin de la Société botanique de France*, **119**, 341–52.

SOSSOUNTZOV, L. (1976). Infrastructure comparée de l'apex de bourgeons en activité et de bourgeons au repos chez une fougère *Marsilea drummondii* A. Br. *La Cellule*, **71**, 275–307.

SOUÈGES, R. (1954). *La Cinématique de la Vie*. Paris: Flammarion.

STEWART, R. N. & DERMEN, H. (1970). Determination of number and mitotic activity of shoot apical initial cells by analysis of mericlinal chimeras. *American Journal of Botany*, **57**, 816–26.

SUSSEX, I. & ROSENTHAL, D. (1973). Differential ^{3}H-thymidine labeling of nuclei in the shoot apical meristem of *Nicotiana*. *Botanical Gazette*, **134**, 295–301.

SUSSEX, I. & STEEVES, T. A. (1967). Apical initials and the concept of promeristem. *Phytomorphology*, **17**, 387–91.

THOMPSON, J. & CLOWES, F. A. L. (1968). The quiescent centre and rates of mitosis in the root meristem of *Allium sativum*. *Annals of Botany*, **32**, 1–13.

TUCKER, S. C. (1968). Meristem, determinate. In *McGraw-Hill Yearbook of Science and Technology (1968)*, pp. 250–3. New York: McGraw-Hill.

WAGENAAR, E. B. (1969). End-to-end chromosome attachments in mitotic metaphase and their possible significance to meiotic chromosome pairing. *Chromosoma*, **26**, 410–26.

WANGENHEIM, K.-H. von (1975). A major component of the radiation effect: interference with endocellular control of cell proliferation and differentiation. *International Journal of Radiation Biology*, **27**, 7–30.

WARDLAW, C. W. (1949). Further experimental observations on the shoot apex of *Dryopteris aristata* Druce. *Philosophical Transactions of the Royal Society of London*, **233B**, 415–51.

WARDLAW, C. W. (1957). Experimental and analytical studies of Pteridophytes. XXXV. The effects of direct applications of various substances to the shoot apex of *Dryopteris austriaca* (*D. aristata*). *Annals of Botany*, **21**, 85–120.

WILSON, B. F. (1964). A model for cell production by the cambium of conifers. In *The formation of wood in forest trees*, ed. M. H. Zimmermann, pp. 19–36. New York & London: Academic Press.

YUN, K.-B. & NAYLOR, J. M. (1973). Regulation of cell reproduction in bud meristems of *Tradescantia paludosa*. *Canadian Journal of Botany*, **51**, 1137–45.

Divergence and convergence in lens cell differentiation: regulation of the formation and specific content of lens fibre cells

R M. CLAYTON

Institute of Animal Genetics, University of Edinburgh,
West Mains Road, Edinburgh EH9 3JN, UK

The lens of the vertebrate eye develops from a vesicle formed from an invagination of the competent head ectoderm after induction by the eye cup. From this stage onwards it is essentially a closed cellular system: the anterior part of the vesicle, facing outwards, becomes the future lens epithelium, and the posterior part, facing towards the developing retina, becomes the elongated primary fibres. All subsequent growth must depend on a population of dividing stem cells within the lens itself. Secondary fibres are recruited throughout life, although at a declining rate. These fibres differentiate from the equatorial margin of the anterior epithelium, and are successively wrapped around those formed earlier. The lens thus grows concentrically: and no cells are lost from it. The whole is enclosed in the acellular capsule which is synthesised by the epithelial cells. The central fibres, which are the earliest to be formed, are small: they are surrounded by fibres of increasing length, until, towards the equator of the lens, the most recently formed fibres, which are not yet mature, are again shorter in length. The development of functionally appropriate shapes and spatial relationships between the tissues of the eye has been reviewed by Coulombre (1965). The function of the lens is to provide a transparent structure which focuses light on the retina behind it, and which will continue to do so as the whole eye increases in size, from the neonate to the adult. This requires not only continuing adjustment of the dimensions of the cornea, lens and retina so that the optical functions of the whole are unimpaired throughout growth, but also a fall in the refractive index of the lens from centre to periphery, which is associated with a parallel fall in the protein concentration. The optical properties and high protein concentration imply an ordered assembly at molecular and ultrastructural levels. The earliest-formed, central fibres, which come to have the highest protein concentration, also have an array of specific lens proteins (the crystallins) which differs from that of the outer fibres, formed later in life. Certain of the crystallins are successively replaced by related members of the series during ontogeny, while other crystallins are synthesised throughout life, and it would seem likely that the high refractive index and high protein concentration of the nucleus (or centre of the lens) are made possible by the preferential synthesis of crystallin polypeptide chains capable of entering into closer molecular packing than those characteristic of the cortex (or periphery) (Clayton, 1970, 1974).

Thus in examining any problem concerning the stem cell population of the lens we must consider not only the signals which initiate the differentiation of the terminal cell, and the subsequent events during fibre maturation, but also the nature and regulation of the overall ontogenic changes in the composition of successive fibres.

A further, and perhaps unique, problem is offered by this system. A lens fibre cell, with its characteristic ultrastructure and specific lens protein content, may be derived from one of several possible alternative pathways from ocular tissues, which are separate anatomically and of different embryonic history and function (Figs. 1 and 2). Wolffian regeneration, in which an apparently normal lens is regenerated from the dorsal edge of the iris after lentectomy in newts was the first to be discovered (reviewed Yamada, 1967), but other extra-lenticular sources of lens fibre cells are also known.

Recently the differentiation of lens fibre cells has been obtained from cell cultures *in vitro*, from lens epithelial cells, and also from several of these possible alternative origins, such as the neural retina and the pigmented epithelium of the retina (reviewed by Okada, 1976; Eguchi, 1976). The crystallins themselves are readily detected and identified and crystallin mRNAs have been studied and isolated in several laboratories. Developmental processes, cell properties and crystallin composition may be investigated in mice and chicks with genetic modifications of lens structure, so that the provocative nature of the questions opened by developmental studies are matched by the powerful experimental techniques which are now available.

A balanced review of the literature is not possible in this short

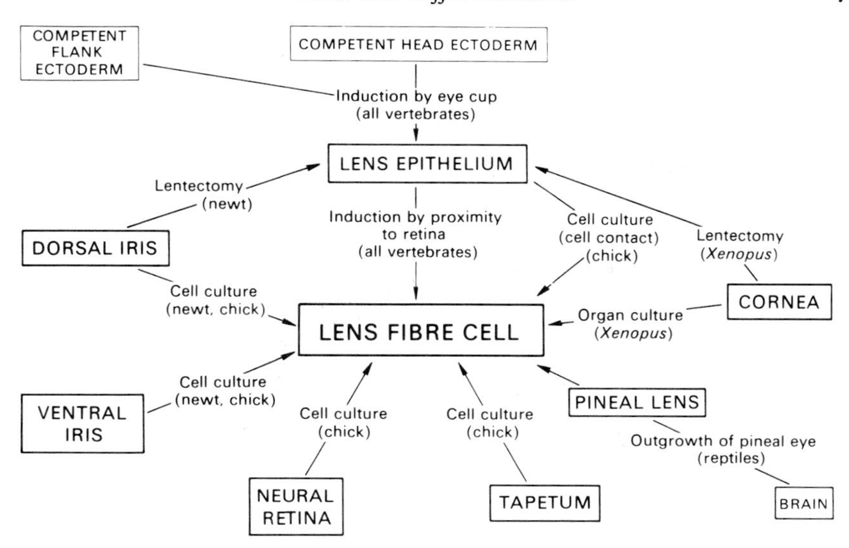

Fig. 1. The different routes which may lead to crystallin-synthesising fibre or fibre-like cells. (This diagram does not indicate other modifications which some of these cell types may undergo: for example, neural retina and pigmented epithelium are each capable of regenerating the other in embryos; pigmented cells develop in cell cultures of chick embryo neural retina; and cornea of some species may revert to skin or form other skin derivatives under various experimental conditions.) (Data from Coulombre, 1965; Eguchi, 1976; and this article.)

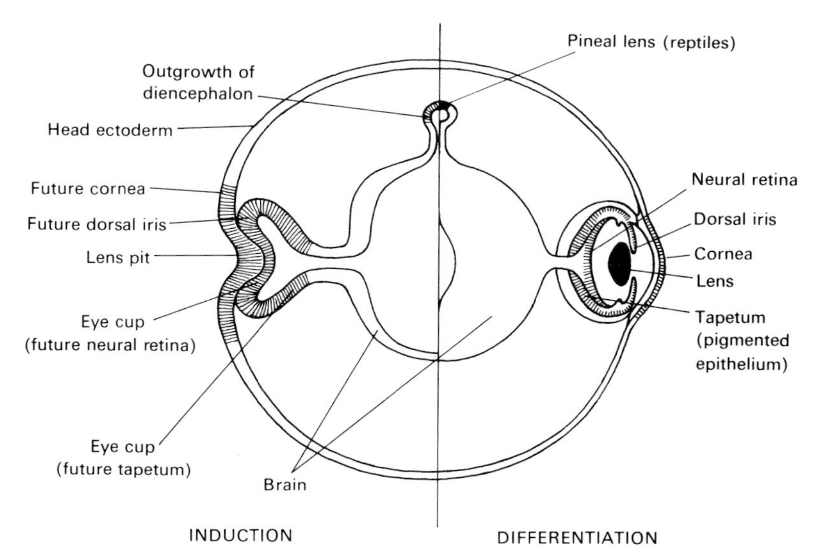

Fig. 2. Diagram showing the developmental relationships of crystallin-synthesising cells (black) or tissues capable of giving rise to such cells (hatched). *Left*, early induction; *right*, differentiated structures. (See text for discussion.)

article which merely outlines two problems: the regulation of fibre formation, and of its crystallin content.

Restrictions on space limit the number of references that can be cited. Reviews and recent references have therefore been given in most cases, although this has the distressing effect of omitting many important references to the first observations of various phenomena; these references are available in the works cited.

TRANSPARENCY AND ORGANISATION OF CRYSTALLINS

The transparency of the lens is associated with an ordered assembly of proteins at a high concentration (Bettelheim, 1975). The refractive index and the protein concentration fall from centre to periphery (Philipson, 1969; Bettelheim & Wang, 1974). The ultrastructural appearance of the constituents of chick fibre cells spread on an air/water interface (Perry, 1976) indicates that some at least of the different crystallins must associate in specific relationships *in situ*. Total chick crystallins can reassociate after urea dissociation and form mixed complexes of unusual composition (Clayton & Truman, 1967) as can total bovine crystallins (Bloemendal, Zweers, Benedetti & Walters, 1975) and α crystalline subunits (van Kemp, van Kleef & Hoenders, 1974*b*). Crystallins are required not only to form ordered assemblies but to permit some gradual ontogenic modification of these assemblies. It has been suggested (Clayton, 1970, 1974) that the extreme evolutionary conservatism of crystallin subunits is associated with the mutual constraints exerted by their association together in a range of organised complexes. These characteristics together with the formation *in vitro* of complexes not found *in vivo* point to the existence of precise but non-coordinated mechanisms regulating crystallin synthesis.

CRYSTALLIN ONTOGENY

Lens fibres are terminal cells: they develop after a final mitosis in the germinal zone but their crystallin composition depends on the age of the animal at the time (reviewed by Clayton, 1970, 1974). Gamma crystallins are characteristic of fibres of pre-natal and very young animals, and the members of the γ gene family evidently have slightly differing roles, since there is a gradual replacement in the bovine lens of γIV by γIII, γII and finally, as γ synthesis stops, by β^S (Slingsby & Croft, 1973). Beta crystallins increase in importance as development proceeds, there is a shift in α crystallin subunit proportions, and a loss of two embryonic α crystallin subunits, αA^X αB^X (van Kamp *et al.*, 1974*a*, *b*). The ratio of αA_2 to αB_2 chains also changes during fibre formation (Delcour & Papaconstantinou, 1974). In the chick the early, very high proportions of δ crystallin drop steadily and its synthesis eventually stops; later-developed fibres contain only α and β crystallins and there is an increase in the ratio of anodal to cathodal β crystallins (Rabaey, 1962; Truman, Brown & Campbell, 1972; Genis-Galvez, 1975). The time of appearance of the different crystallins in the embryo has been studied using immunofluorescence (Zwaan, 1968; Brahma & van Doorenmaalen, 1971; Waggoner, Lieska, Alcala & Maisel, 1976). Synthesis of δ crystallin increases more rapidly than the synthesis of other crystallins and non-crystallins in explants of 6-day embryo epithelium, and the proportion of δ crystallin RNA is still about 80% of the total in fibres of 15-day embryos (Piatigorsky *et al.*, 1973, 1976; Milstone, Zelenka & Piatigorsky, 1976). However, we find that 15-day embryo lens epithelium mRNA synthesises appreciable levels of α and β crystallins, as well as δ crystallin, and lower amounts of α and β crystallins from fibre mRNA, and by 1 day post-hatch the level of δ synthesis has declined (Thomson, Wilkinson, Jackson, de Pomerai, Truman, Clayton & Williamson, unpublished data).

FIBRE DIFFERENTIATION

Both the normal lens and the regenerating lens (compared in Fig. 3) require proximity to the retina for fibre elicitation *in vivo*, and fibres only form from the epithelial cells closest to retina. This was demonstrated by rotation of the embryo chick lens (Coulombre & Coulombre, 1963; Genis-Galvez, 1964), by implanting lenses into eyes from which retina has been extirpated (Yamamoto, 1976) and by displacement experiments: a lens in the anterior chamber forms no fibres but one close to the retina is wholly converted into fibres (Mikami, 1941). Proximity to retina is also a requirement for explanted whole lenses (Takeichi, 1970) and for the formation of lens from dorsal iris in a variety of experimental conditions. In all these experimental situations, both for normal and for regenerate lenses,

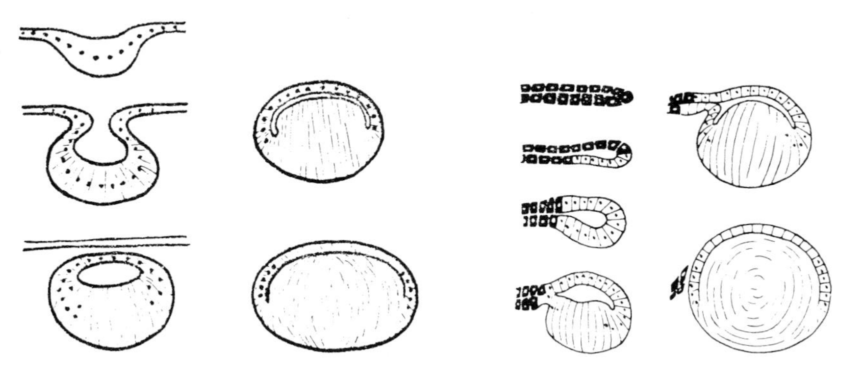

Fig. 3. Diagram showing the early stages in the formation of a normal lens (*left*) and one formed from dorsal iris (*right*). Nucleus of cells which undergo mitosis are indicated. (Not to scale.)

the fibres orient towards the inducing tissue. The need for an inductor applies to lens fibre formation *in vivo* or in organ culture but it is bypassed in cell culture conditions (Okada, Eguchi & Takeichi, 1971; Eguchi, Clayton & Perry, 1975). After the cultures become confluent, scattered lentoids appear, formed from groups of elongated cells with typical fibre ultrastructure.

In cell culture only cells surrounded by others in a cell sheet contain detectable crystallins (Clayton *et al.*, 1976*a*). Lentoids tend to appear first in the ridges formed where two colonies meet (Okada, 1976; D. J. Pritchard, personal communication).

There is also some genetic evidence relating cell surface composition or behaviour with fibre differentiation. Two inbred strains of chicks, Hy-1 and Hy-2, are each characterised by epithelial multilayering in the lens. These epithelial cells, except for those in the outermost layer, have no free boundary, and tend to form short fibre-like cells over the anterior face of the lens (Clayton, 1975). The cultured cells are mutually adhesive, forming clusters and multilayers, and differentiate precociously and also more frequently into lentoids (Eguchi *et al.*, 1975; Clayton *et al.*, 1976*a*). The ultrastructure and cell membrane polypeptide composition are modified in these genotypes (Clayton *et al.*, 1976*b*, *c*; P. G. Odeigah, R. M. Clayton & D. E. S. Truman, unpublished data). If the cell sheet in transdifferentiating cultured chick embryo neural retina is folded, thus increasing the area of cell–cell contact but not the total cell number, lentoids develop at 21 days instead of at 26 days, and Hy-1 cells are more responsive to this treatment than normal cells (Clayton, de Pomerai & Pritchard, 1977).

These data suggest that a contact-mediated cell surface effect can initiate fibre differentiation, and that the function of neural retina in fibre initiation is to supply a diffusible effector for such change in the epithelial cell surface configuration or composition. Such regulatory mechanisms have been found in other systems (O'Malley & Means, 1974).

However, cell contact *per se* is evidently insufficient in some situations. Epithelial nodules and masses can form in certain mutant or pathological lenses, often following fibre breakdown, for example the Cat^{fr} and the *Sey* mouse mutants (Zwaan & Williams, 1968; Clayton, 1970; Pritchard, Clayton & Cunningham, 1974). Here the increased cell contact in the epithelium is associated with excess capsule formation. However there may be an additional factor; improper orientation of the epithelial cells can prevent fibre differentiation in explants (McLean & Finnegan, 1973) and also in the mutant *aphakia* in which crystallins are not detectable (Zwaan, 1975; Kirkland & Zwaan, 1975).

DORSAL AND VENTRAL IRIS

Adult newt dorsal iris can regenerate a lens *in situ* following lentectomy (Wolffian regeneration). The subject is reviewed by Reyer (1962), Stone (1965), Scheib (1965) and Yamada (1967). Retinal tissue is normally essential for lens formation from dorsal iris *in vivo*, in organ culture, (Yamada, Reese & McDevitt, 1973) and in tissue culture (Eguchi, 1967). It may be replaced, in the eye or in culture, by other tissues such as pituitary (Zalokar, 1944; Connelly, Ortiz & Yamada, 1973; Powell & Segil, 1976) and is not required for dorsal iris transplanted into a limb blastema (Reyer, Woolfitt & Withersty, 1973). Ventral iris does not normally give rise to lens tissue in any of these situations, but the difference between dorsal and ventral iris is evidently not one of potential for cellular transdifferentiation, rather of factors associated with the stability and coherence of the tissue. Multiple lens formation can be elicited from ventral iris *in vivo* by *N*-methyl-*N*-nitro-*N*-nitrosoguanidine (Eguchi & Watanabe, 1973) and lentoid colonies can be obtained in cell culture in the absence of retina from dissociated cells of either dorsal or ventral iris (Eguchi, Abe & Watanabe, 1974).

There are no dividing cells in adult newt iris (Yamada, Roesel & Beauchamp, 1975). The cells are released from G_0 by lentectomy, explantation or dissociation. Both in-vitro and in-vivo depigmentation is an essential prerequisite for lens fibre transdifferentiation but many cells merely depigment reversibly and Yamada (1976) considers that cells with a mitotic cycle of 46 hours will eventually transform into lens while those with a 76-hour cycle will undergo a reversible dilution of pigment.

The in-vivo regenerate lens contains α, β and γ crystallins, γ crystallin being confined to the fibres (Takata, Albright & Yamada, 1966). It is possible that the subunit composition of the crystallins may not be identical with that in the original lens, since modifications of the immunoelectrophoretic pattern were recorded (Ogawa, 1967). Transdifferentiated iris cultures contain α, β and γ crystallins (Eguchi *et al.*, 1974).

PIGMENTED EPITHELIUM

Eguchi & Okada (1973) showed by establishing single-cell clones, that the progeny of a pigmented epithelial cell of the chick embryo retina could transform ('transdifferentiate') into crystallin-synthesising cells. Only a small proportion of clones depigment and transdifferentiate into lentoids.

Transdifferentiation from tapetal cells occurs after 60–90 days in chick, 35–40 in newt, and 25 days in quail, or in chick cells grown with phenylthiourea (Eguchi, 1976) which inhibits melanin synthesis (Eppig, 1970). In the case of chick only some cells in a transforming clone will form lentoid cells; in the case of newt all cells in a transforming clone do so together, after 7–10 divisions (Eguchi, 1976).

NEURAL RETINA

Chick embryo neural retina grown in cell culture can give rise to pigmented cells and to lens fibre cells containing crystallins (Okada, Ito, Watanabe & Eguchi, 1975; Itoh, Okada, Ide & Eguchi, 1975). Neuroblast-like and fibroblast-like cells may also develop. Cells which transdifferentiate into lens may be different from those which transdifferentiate into pigment cells. Okada (1976) reports that no single-cell clone in secondary culture gives rise to both pigment and lens cells.

Modifications of culture conditions affect the incidence of lentoid and pigment cell formation in neural retina culture. The rate of both transformations is increased by ascorbic acid (Itoh, 1976) and totally suppressed by conditioned Hams F12 medium (Okada, 1976). Pigment cell formation is favoured by initial plating in Hams F12 (Okada, 1976) and strongly favoured by Eagles minimal essential medium (Clayton, de Pomerai & Pritchard, 1977). Lentoid formation is favoured by conditioned Eagles medium (Okada, 1976), by the Hy-1 strain, and is accelerated by folding the cell sheet (Clayton, de Pomerai & Pritchard, 1977). We also find that cell density has an effect. At initial densities of 10^5 cells per dish, from 20 days onwards only pigment cells appeared; at 10^6 cells per dish numerous pigment cells and a few lentoids appeared from 20 days onwards. At an initial density of 10^7 cells numerous lentoids developed from 20 days and a few pigment cells after 36 days. The proportions and changes in the times of appearance of these two cell types, and the effect caused by multilayering, both suggest that mere selection between precursor cell types is too simple an explanation.

Table 1. *The crystallin composition of lentoids of different origins*

Crystallin class		% concentration of crystallin					
		N strain[a]			Hy-1 strain		
		Day-old LE freshly excised	Terminal cultures: LE[b] (21-day)	Terminal cultures: NE[c] (42-day)	Day-old LE freshly excised	Terminal cultures: LE (21-day)	Terminal cultures: NR (42-day)
	Reference extract						
α	12	68	26.7	6.3	54	22.4	26.6
δ	58	0.3	6.5	9.2	0.9	21.6	17.2
Cβ[d]	20	22.7	44.5	79.2	30	37.3	53.2
Aβ[e]	10	9 ± 3	22.3 ± 6.7	5.3 ± 2.0	15.0 ± 4.5	18.7 ± 5.6	30.0 ± 1.2

[a] N strain, a strain of inbred birds with normal lens morphology, used as controls.
[b] LE, lens epithelium.
[c] NE, neural retina epithelium.
[d] Cβ, cathodal β crystallins.
[e] Aβ, anodal β crystallins.

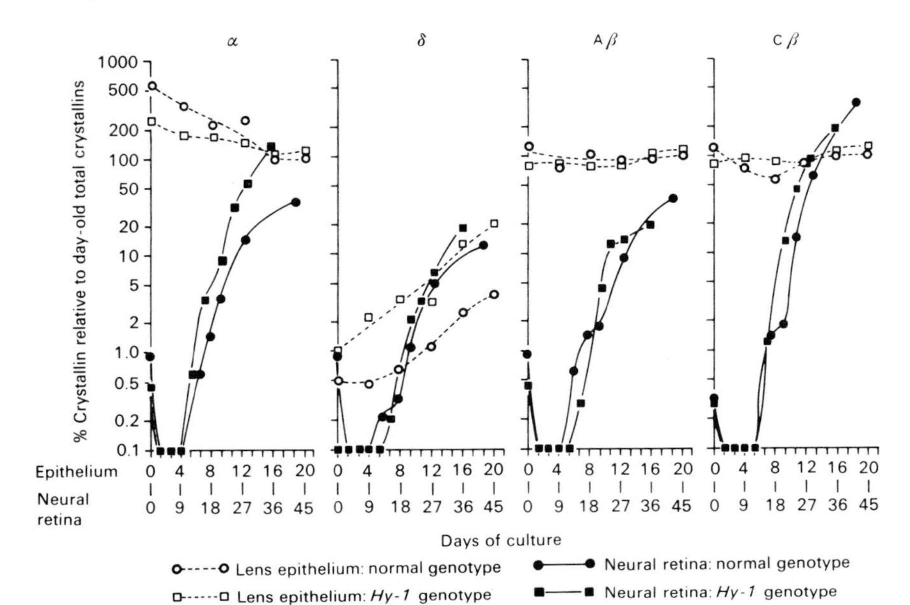

Fig. 4. Haemagglutination inhibition assays (Evans, Steel & Arthur, 1974) with antisera monospecific for α δ, and anodal (Aβ) and cathodal (Cβ) β crystallins. Cultures of epithelium of day-old chick lenses were assayed from explantation to day 20. Lentoids form from day 7 from *Hy-1* genotype cells and day 9 from normal cells. Neural retina cultures were assayed from explantation to day 45. Lentoids appear from about 26 days. (From data presented in de Pomerai, Pritchard & Clayton, 1977.)

Using antisera specific for α, δ, anodal and cathodal β crystallins respectively, and the extremely sensitive and quantifiable haemagglutination inhibition assay, crystallins were detected by the twelfth day of culture. They increased at different rates, and the final composition differed quantitatively from that in lens epithelial fibres (Fig. 4; Table 1). A genotype-specific effect was observed (de Pomerai, Pritchard & Clayton, 1977).

Total mRNA from freshly isolated 8-day embryo chick neural retina when translated in a reticulocyte lysate system primes for the synthesis of a large number of retinal proteins among which crystallins are not distinguishable. In 42-day terminally transdifferentiated neural retina cultures the predominant mRNA species prime for crystallins, and are closely comparable with the translation products of total crystallin mRNA with a few minor additional components (Fig. 5).

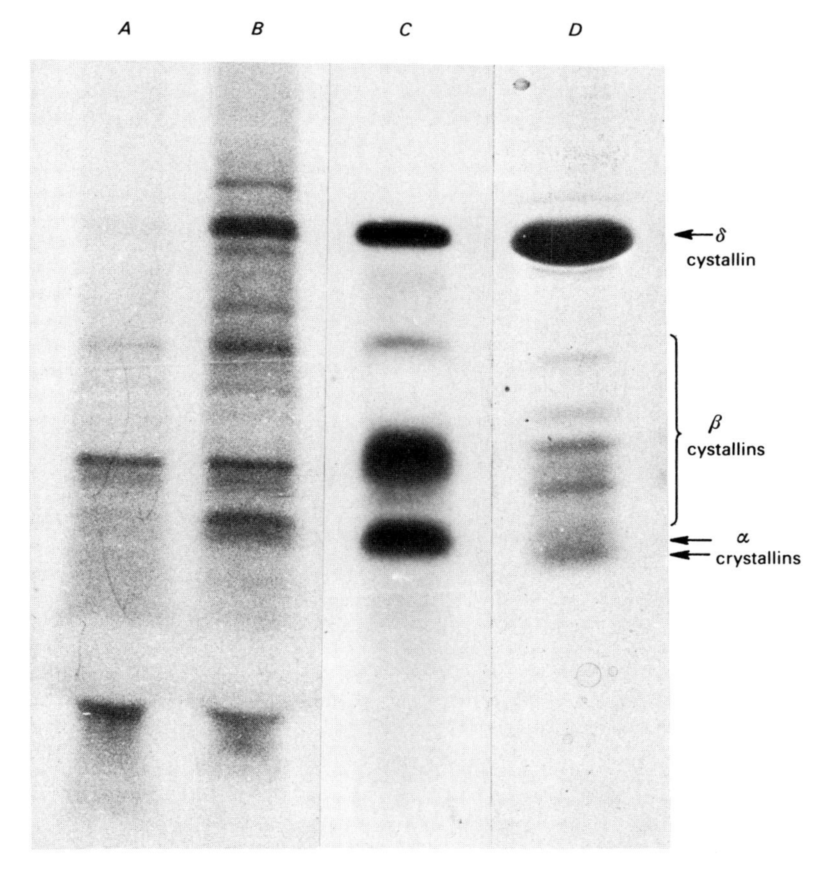

Fig. 5. *A*, *B* and *C* are fluorographs (Bonner & Laskey, 1974) prepared from polyacrylamide gel SDS electrophoretic separation (Laskey & Mills, 1975) of products synthesised in a mRNA-dependent cell-free lysate system (Pelham & Jackson, 1976). Sources of mRNA: *A*, non-polysomal mRNA, day-old chick lens; *B*, mRNA from chick embryo terminal neural retina culture, containing lentoids; *C*, total polysomal mRNA, day-old chick lens; *D*, crystallins from day-old chick lens, electrophoresed under identical conditions, stained with Coomassie blue. (I am grateful to Dr I. Thomson for these photographs.)

THE PINEAL LENS

The median pineal eye of reptiles is formed from a mid-dorsal evagination of the diencephalon, and the transparent pineal lens is formed from cells in the uppermost area (Eakin, 1973). The pineal lens of *Anolis carolinensis* contains antigens which are the same as, or strongly cross-reactive with, some at least of the crystallins of the lateral eye (McDevitt, 1972).

CORNEA

Regeneration of the lens in *Xenopus laevis* takes place from the cornea following lentectomy (Freeman, 1963; Campbell, 1965). *Xenopus laevis* cornea can give rise to lens cells under a range of conditions such as transplantation into a lentectomised eye (Reyer, 1962), into limb blastema (Waggoner, 1973) or in organ culture *in vitro* (Campbell & Jones, 1968). The last two do not require retinal tissue. Immunofluorescence with antibodies to crystallins shows that the lens regenerate contains crystallins (Campbell, 1965; Brahma & McDevitt, 1974).

Lens regenerates formed from cornea *in vivo* and lentoids formed from cultured cornea have been compared with the normal contralateral control lens (Campbell & Jones, 1968; Clayton, 1970; Campbell & Truman, 1977). The crystallin composition of the lentoids in culture was markedly deficient, and a variable number of β crystallins were absent from a proportion of regenerate lenses. Since β crystallin is the major component in the *Xenopus laevis* lens (Campbell, Clayton & Truman, 1968; Brahma & McDevitt, 1974) problems of detection do not provide an obvious explanation, and we must consider two possibilities: that some β crystallins are facultative and not obligate gene products in lens fibre cells, or that minor variations in morphology or size are associated with an unrepresentative assemblage of fibres, reflected in specific deficiencies in crystallin composition.

SPECIFICITY OF LENS AND RETINAL TISSUES

It has been proposed that tissue specificity is not necessarily characterised by the possession of unique constituents but that proteins present at very high levels in one cell type might be present in trace amount in other tissues, much tissue specificity therefore being combinatorial and quantitative in nature (Clayton, 1964, 1970; Clayton, Campbell & Truman, 1968). Clayton *et al.* (1968) analysed cross-reactions between lens and several other tissues and reviewed the literature on cross-reactions to that date, and Bours (1973) refers to more recent publications. Cross-reactivity between lens, iris and retina is most frequently reported: whether these cross-reactions are non-specific (Mikhailov & Barabanov, 1975) or crystallin-like (Bours, 1973) is not clear. On the other hand, several investigators (Takata *et al.*, 1966; Zwaan, 1968; McDevitt *et al.*, 1969) have used antisera with no such cross-reactivity, their data supporting the concept of complete organ specificity for the crystallins.

Several possible explanations of these divergent results were suggested in Clayton *et al.* (1968) and Clayton (1970). An absence of cross-reactivity may be due to: (1) a low level of cross-reactive material owing to an insufficient concentration of the test extract; (2) differences in specificity of the antisera and in titres with respect to relevant molecules or particular molecular determinants; (3) the sensitivity of the test employed. Positive cross-reactions may be due to: (1) the presence of cross-reactive but genetically distinct proteins; (2) the presence of a crystallin at very low levels in tissues other than the lens; (3) firm binding of crystallins to structural proteins not restricted to lens cells, an antiserum apparently specific to a crystallin being therefore also reactive with some non-crystallin. The lens is known to contain collagen (Grant, Kefalides & Prockop, 1972), capsular glycoprotein (Hughes, Laurent, Lonchampt & Courtois, 1975), microtubules (Lonchampt *et al.*, 1976) and probably spectrin (P. G. Odeigah, personal communication). Finally, α and δ crystallins appear to form part of the bovine and chick cell membrane complex (Broekhuyse & Kuhlmann, 1974; Dunia *et al.*, 1974; Maisel, Alcalá & Lieska, 1976).

Many proteins once considered to be cell-specific are known to be present in very low amounts in many other tissues. For example actin, once considered to be specific to muscle, has been found in platelets, fibroblasts, spermatozoa, nerve cells, intestinal microvilli and dividing cells where it may be organised into filaments or dispersed (reviewed Pollard & Weihing, 1974; Tilney, 1975). Hybridisation studies with globin cDNA (or complementary DNA, synthesised *in vitro*, complementary to isolated mRNA) show that globin mRNA is found in large amounts in the reticulocyte but very low levels may be found in a number of other cell types (Humphries, Windass & Williamson, 1976).

We are using cDNA to total lens mRNA and to the most abundant sequences (putative crystallin mRNA) to test for transcribed crystallin mRNA in retina before and during transdifferentiation. Total

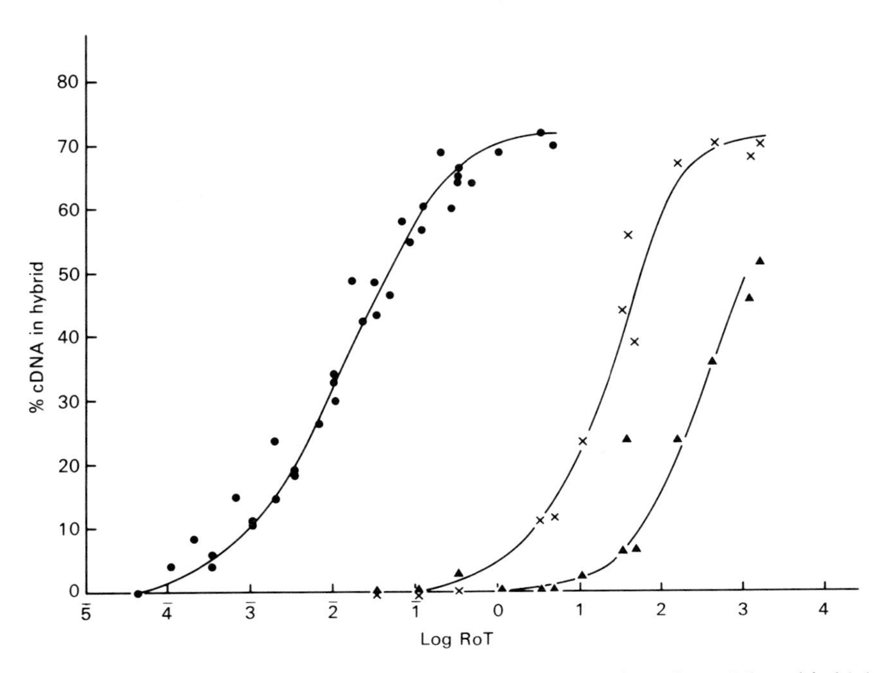

Fig. 6. Hybridisation studies on cDNA to the most abundant class of day-old chick lens mRNA with excess mRNA. ●, mRNA from day-old chick lens; ×, mRNA from pigmented epithelium of 8-day chick embryo retina; ▲, mRNA from 8-day chick embryo neural retina. There is no detectable hybridisation with chick embryo muscle mRNA or from headless embryos (*E. coli* mRNA, negative control). (I am grateful to Mr J. Jackson for this figure.)

lens cDNA hybridised to total lens mRNA shows three transitions, the first representing not less than four sequences at an average of 28×10^{15} copies per lens, the second about 160 sequences at an average of 37×10^{14} copies per lens and the third about 5600 sequences at an average of 11×10^{13} copies per lens.

cDNA for the most abundant mRNA species was isolated by the method of Hastie & Bishop (1976). Seventy-five per cent of this cDNA hybridised to total lens mRNA at a RoT of 4.9×10^{-1}. (RoT½ is a measure, derived from the rate of hybridisation of cDNA to the corresponding mRNA in specified conditions and mRNA excess, which can be used to estimate the number and frequency of mRNA sequences involved.) The reaction shows first-order kinetics similar to that of cDNA to δ crystallin mRNA with total lens mRNA, and a computer-derived estimate of the RoT½ corresponds to four sequences in the most abundant mRNA class of day-old lens. This is lower than the number of identified crystallins translated from total mRNA from day-old lens, and since only crystallins are synthesised in detectable amounts in a cell-free system (Fig. 5) we have assumed that this class of cDNA may be considered to be essentially crystallin-specific. However, crystallin amino acid sequences are very conservative (see Clayton, 1974). If the nucleotide sequences contain conservative regions cross-hybridisation will lead to an underestimation of the numbers.

The kinetics of hybridisation of lens mRNA to total cDNA of neural retina and also a pigment epithelium show that while neither tissue contains any mRNA sequences not also present in lens, no lens sequences are in a high abundance class in these tissues. However, cDNA to the most abundant (putative crystallin) mRNA sequences of lens has been hybridised to pigment epithelium and shows that similar sequences are present in this tissue at 1/2500 of the level found in lens (Fig. 6). We have not calculated these values on a per cell basis because of the cellular heterogeneity of both sources of mRNA.

POSSIBLE REGULATORY MECHANISMS

Age-dependent variables which might operate during development to regulate the crystallin profile include the frequency and distribution of mitosis in the epithelium (Modak, Morris & Yamada, 1968), the rate of fibre maturation (Hanna & Keatts, 1966), the total cell mass, the composition of the subjacent fibres and, in the earlier stages, energy metabolism.

The age of the individual from which epithelium is taken affects the cultured cells: terminal lentoids in cultures of embryo lens epithelium cells contain more δ crystallin than those from older lenses, but we have not yet investigated the stability of these commitments. Experimental modifications of the mitotic rate, of the energy and transport functions in the lens, and of cell contact and position, affect the level of synthesis of each crystallin subunit independently of the others, and each experimental condition tested produced its own distinctive pattern (Clayton *et al.*, 1976*b*). The *Hy-1* and *Hy-2* genotypes described above produce genotype-specific modifications of these patterns of response (Fig. 7). The wide range of crystallin compositions shown by terminal lentoids (Table 1) also leads to the conclusion that non-coordinate regulation occurs and

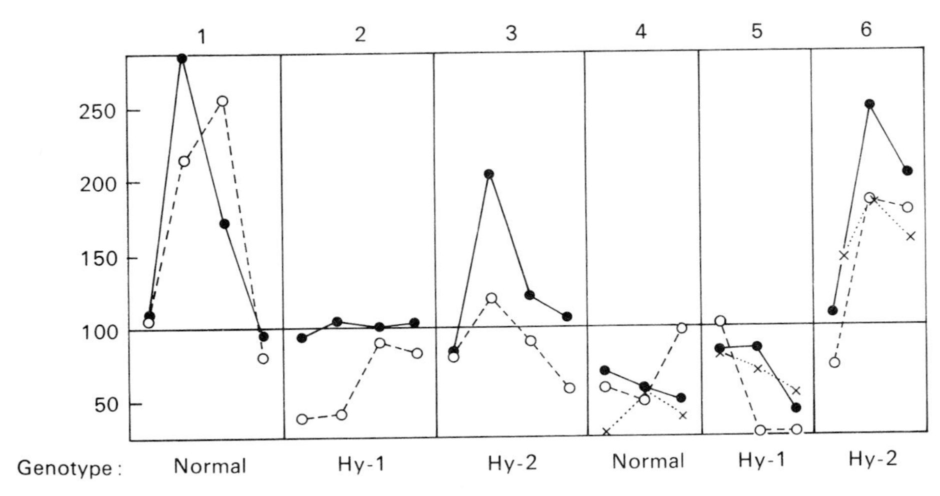

Fig. 7. The effect of various growth conditions and of genotypes on the rate of synthesis of different β crystallin subunits as a percentage of the rate of synthesis under controlled conditions in day-old chick lens epithelium. Key for 1, 2 and 3: ●——●, free epithelial sheet; ○----○, epithelium cultured on cortical fibres before removal. Key for 4, 5 and 6: ●——●, medium, no foetal calf serum; ○----○, control medium plus Daunomycin; ×····×, control medium plus insulin. (Data from Clayton *et al.*, 1976*b*.)

that the outcome depends on tissue of origin (lens or neural retina) and on membrane genotype.

The proportion of clones capable of transdifferentiation gives evidence for heterogeneity in the initial cell population, and some apparently regulatory conditions may act selectively at this point. Lentoid formation in transdifferentiating systems requires times ranging from 21 to 90 days according to the tissue of origin and the experimental conditions. The requirement for depigmentation must account for some initial delay, but if cell division were required solely to permit the derepression of crystallin genes we could expect that some cells should be able to transdifferentiate following the first mitotis after the loss of melanin. It may be necessary to distinguish between the requirement for cell divisions before transdifferentiation and the requirement for cell division in the post-transdifferentiated cell. Our experiments on multilayering suggests that cells which could form lentoids may not always do so, implying a translational control.

Zalik, Scott & Dimitrov (1976) have found a series of changes in the cell surface during transdifferentiation: the rates of synthesis and

Table 2. *Crystallin ontogeny in the developing chick lens: immunofluorescence with specific antisera*

Embryo age (days)	N strain[a]								Hy-1 strain							
	Epithelium				Fibre				Epithelium				Fibre			
	α	δ	Cβ[b]	Aβ[c]	α	δ	Cβ	Aβ	α	δ	Cβ	Aβ	α	δ	Cβ	Aβ
2½			+			+	+	+		⊕			⊕	+	+	+
3		+		+												
4					+						⊕	⊕				
5									⊕							
8	+															

From McDevitt & Clayton (1975 and unpublished data).
Circled figures indicate crystallins appearing at different times in Hy-1 lenses compared to normal lenses.
[a] N strain, a strain of inbred birds with normal lens morphology, used as controls.
[b] Cβ, cathodal β crystallins.
[c] Aβ, anodal β crystallins.

turnover of nuclear, ribosomal and tRNA and also mRNA stability in the lens (Truman, Clayton, Gillies & Mackenzie, 1976) and the apparent ontogenic sequence of crystallins (Table 2) are also associated with cell membrane modifications. Initiation of δ crystallin synthesis is also associated with changes at a cell–cell interface (Hendrix & Zwaan, 1975; Piatigorsky *et al.*, 1976). Our working hypothesis is that the membrane imposes some modulation on the signals received and transmits these to effector systems in the cell. In any case the wide range of non-coordinate responses indicates that there must be several levels at which regulatory mechanism operate.

Piatigorsky and co-workers have demonstrated the relationship between the rate of δ crystallin synthesis and the cell content of δ crystallin mRNA (reviewed Piatigorsky *et al.*, 1976); in agreement with this is our finding that the quantitative profiles of crystallins synthesised in a cell-free system by mRNAs from day-old fibre and epithelial cells respectively are similar to the profiles of crystallin synthesis in the intact tissues. The levels of specific mRNAs may be regulated by unequal rates of transcription but differential stability is likely to be a significant control mechanism. Crystallin mRNA stabilises during development and fibre maturation (Scott & Bell,

1965; Stewart & Papaconstantinou, 1967; Yoshida & Katoh, 1971) but the rate and degree of stabilisation are mRNA-specific and distinguish the different crystallins (Clayton, 1970; Clayton *et al.*, 1972, 1974; Delcour, Odaert & Bouchet, 1976). The stability of mRNA is also modified by the *Hy-1* and *Hy-2* genotypes (Truman *et al.*, 1976) and the times of appearance of the crystallins are each modified from normal in *Hy-1* (Table 2) (McDevitt & Clayton, 1975, and unpublished data).

We also find (Fig. 5) that the mRNA in the post-polysomal supernatant is non-representative, synthesising mainly β crystallins, while the polysomal mRNA from the same preparation synthesises α and δ crystallins in addition. This suggests a possible mechanism of translational regulation.

A hierarchy of control mechanisms is overwhelmingly likely; however difficult these prove to elucidate, the lens system must continue to have its attractions.

It is a pleasure to thank my friends and colleagues in Edinburgh for their collaboration in the experiments and for very many enjoyable discussions: Dr J. C. Campbell, Mr J. Cuthbert, Mr J. Jackson, Mr P. G. Odeigah, Dr D. I. de Pomerai, Dr D. J. Pritchard, Dr I. Thompson and Dr D. E. S. Truman. They are not responsible, however, for any obscurity of exposition or eccentricity of view expressed here.

I am glad to thank Miss L. Ritchie for her cheerful help with the preparation of the manuscript, and Mr E. D. Roberts for diagrams.

I am grateful to the Medical Research Council and the Cancer Research Campaign for continued support.

REFERENCES

Bettelheim, F. A. (1975). On the optical anisotropy of lens fibre cells. *Experimental Eye Research*, **21**, 231–4.

Bettelheim, F. A. & Wang, T. J. Y. (1974). Topographic distribution of refractive indices in bovine lenses. *Experimental Eye Research*, **18**, 351–6.

Bloemendal, H., Zweers, A., Benedetti, E. S. & Walters, H. (1975). Selective reassociation of the crystallins. *Experimental Eye Research*, **20**, 463–78.

Bonner, W. M. & Laskey, R. A. (1974). A film detection method for tritium-labelled proteins and nucleic acids in polyacrylamide gels. *European Journal of Biochemistry*, **67**, 247–56.

Bours, J. (1973). The presence of lens crystallins as well as albumin and other serum proteins in chick iris tissue extracts. *Experimental Eye Research*, **15**, 299–319.

BRAHMA, S. K. & VAN DOORENMAALEN, W. J. (1971). Immunofluorescence studies of chick lens FISC and α-crystallin antigens during lens morphogenesis and development. *Ophthalmological Research*, **2**, 344–57.
BRAHMA, S. A. & MCDEVITT, D. S. (1974). Ontogeny and localisation of the lens crystallins in *Xenopus laevis* lens regeneration. *Journal of Embryology and experimental Morphology*, **32**, 783–94.
BROEKHUYSE, R. M. & KUHLMANN, E. D. (1974). Lens membrane. I. Composition of urea treated plasma membranes from calf lens. *Experimental Eye Research*, **19**, 297–302.
CAMPBELL, J. C. (1965). An immunofluorescent study of lens regeneration in larval *Xenopus laevis*. *Journal of Embryology and experimental Morphology*, **13**, 171–9.
CAMPBELL, J. C., CLAYTON, R. M. & TRUMAN, D. E. S. (1968). Antigens of the lens of *Xenopus laevis*. *Experimental Eye Research*, **7**, 4–10.
CAMPBELL, J. C. & JONES, K. W. (1968). The in vitro development of lens from cornea of larval *Xenopus laevis*. *Developmental Biology*, **17**, 1–15.
CAMPBELL, J. C. & TRUMAN, D. E. S. (1977). Variations in differentiation in the regenerating lens of *Xenopus laevis*. *Experimental Eye Research*, **25**, 99–100.
CLAYTON, R. M. (1964). Differentiation. In *Penguin Science Survey, Series B*, ed. A. McLaren & A. Barnett, pp. 68–88. London: Penguin.
CLAYTON, R. M. (1970). Problems of differentiation in the vertebrate lens. *Current Topics in developmental Biology*, **5**, 115–80.
CLAYTON, R. M. (1974). Comparative aspects of lens proteins. In *The Eye*, vol. 5, ed. H. Davson & L. T. Graham, pp. 399–494. New York & London: Academic Press.
CLAYTON, R. M. (1975). Failure of growth regulation of the lens epithelium in strains of fast-growing chicks. *Genetical Research, Cambridge*, **25**, 79–82.
CLAYTON, R. M., CAMPBELL, J. C. & TRUMAN, D. E. S. (1968). A re-examination of the organ specificity of lens antigen. *Experimental Eye Research*, **7**, 11–29.
CLAYTON, R. M., EGUCHI, G., TRUMAN, D. E. S., PERRY, M. M., JACOB, J. & FLINT, O. P. (1976*a*). Abnormalities in the differentiation and cellular properties of hyperplastic lens epithelium from strains of chickens selected for high growth rate. *Journal of Embryology and experimental Morphology*, **35**, 1–23.
CLAYTON, R. M., ODEIGAH, P. G., DE POMERAI, D. I., PRITCHARD, D. J., THOMSON, I. & TRUMAN, D. E. S. (1976*b*). Experimental modifications of the quantitative pattern of crystallin synthesis in normal and hyperplastic lens epithelia. In *Biology of the lens epithelial cell in relation to development, ageing and cataract*, ed. Y. Courtois & F. Regnault, pp. 123–36. Paris: INSERM.
CLAYTON, R. M., DE POMERAI, D. I. & PRITCHARD, D. J. (1977). Experimental manipulations of alternative pathways of differentiation in cultures of embryonic chick neural retina. *Development, Growth and Differentiation*, **19**, in press.
CLAYTON, R. M. & TRUMAN, D. E. S. (1967). Molecular structure and antigenicity of lens proteins. *Nature, London*, **214**, 1202–4.
CLAYTON, R. M., TRUMAN, D. E. S. & CAMPBELL, J. C. (1972). A method for direct assay of messenger RNA turnover for different crystallins in the chick lens. *Cell Differentiation*, **1**, 25–35.
CLAYTON, R. M., TRUMAN, D. E. S. & HANNAH, A. I. (1974). RNA turnover and

translational regulation of specific crystallin synthesis. *Cell Differentiation*, **3**, 135–45.

CLAYTON, R. M., TRUMAN, D. E. S., HUNTER, J., ODEIGAH, P. G. & DE POMERAI, D. I. (1976*c*). Protein synthesis and its regulation in the lenses of normal chicks and in two strains of chicks with hyperplasia of the lens epithelium. *Documenta ophthalmologica, Proceedings Series*, **8**, 27–37.

CONNELLY, T. G., ORTIZ, J. R. & YAMADA, T. (1973). Influence of the pituitary on Wolffian lens regeneration. *Developmental Biology*, **31**, 301–15.

COULOMBRE, A. J. (1965). The eye. In *Organogenesis*, ed. R. L. DeHaan, pp. 219–51. New York: Holt, Rinehart & Winston.

COULOMBRE, J. L. & COULOMBRE, A. J. (1963). Lens development: fibre elongation and lens orientation. *Science*, **142**, 1489–90.

CROFT, L. (1973). Low molecular weight proteins in the human lens. In *The human lens in relation to cataract*, ed. K. Elliott & D. W. Fitzsimons, *CIBA Symposium 19*, pp. 207–24. Amsterdam: Associated Scientific Publishers.

DELCOUR, J., ODAERT, S. & BOUCHET, H. (1976). Stability of the protein synthesis machinery in the bovine lens. In *Biology of the lens epithelial in relation to development, ageing and cataract*, ed. Y. Courtois & F. Regnault, pp. 39–52. Paris: INSERM.

DELCOUR, J. & PAPACONSTANTINOU, J. (1974). A change in the stoichiometry of assembly of bovine lens α-crystallin subunits in relation to cellular differentiation. *Biochemical and biophysical Research Communications*, **57**, 134–41.

DUNIA, I., SEN GHOSH, C., BENEDETTI, E. L., ZWEERS, A. & BLOEMENDAL, H. (1974). Isolation and protein patterns of eye lens fibre junctions. *FEBS Letters*, **48**, 139–44.

EAKIN, R. M. (1973). *The third eye*. Berkeley: University of California Press.

EGUCHI, G. (1963). Electron microscopic studies on lens regeneration. I. Mechanism of depigmentation of the iris. *Embryologia*, **8**, 47–62.

EGUCHI, G. (1967). In vitro analysis of Wolffian lens regeneration: differentiation of the regenerating lens rudiment of the newt, *Triturus pyrrhogaster*. *Embryologia*, **9**, 246–66.

EGUCHI, G. (1976). Transdifferentation of vertebrate cells in cell culture. In *Embryogenesis in mammals, Ciba Symposium*, **40**, pp. 241–58. Amsterdam: Elsevier Excerpta Medica.

EGUCHI, G., ABE, S. & WATANABE, K. (1974). Differentiation of lens-like structures from newt epithelial cells *in vitro*. *Proceedings of the National Academy of Sciences, USA*, **71**, 5052–6.

EGUCHI, G., CLAYTON, R. M. & PERRY, M. M. (1975). Comparison of the growth and differentiation of epithelial cells from normal and hyperplastic lenses of the chick: studies of in vitro cell cultures. *Development, Growth and Differentiation*, **17**, 395–413.

EGUCHI, G. & OKADA, T. S. (1973). Differentiation of lens tissue from the progeny of chick retinal pigment cells cultured *in vitro*: a demonstration of a switch of cell type in clonal cell culture. *Proceedings of the National Academy of Sciences, USA*, **70**, 1495–9.

EGUCHI, G. & WATANABE, K. (1973). Elicitation of lens formation from the 'ventral iris' epithelium of the newt by a carcinogen, N-methyl-N-nitro-N-nitrosoguanidine. *Journal of Embryology and experimental Morphology*, **30**, 63–71.

EPPIG, J. J. (1970). Melanogenesis in amphibians. I. A study of the fine structure of the normal and PTU-treated pigmented epithelium in *Rana pipiens* tadpole eyes. *Zeitschrift für Zellforschung und mikroskopische Anatomie*, **103**, 238–46.

EVANS, J., STEEL, M. & ARTHUR, F. (1974). A haemagglutination inhibition technique for detection of immunoglobulins in supernatants of human lymphoblastoid cell lines. *Cell*, **3**, 153–8.

FREEMAN, G. (1963). Lens regeneration from cornea in *Xenopus laevis*. *Journal of experimental Zoology*, **154**, 39–65.

GENIS-GALVEZ, J. M. (1964). Quelques aspects de la differentiation et détermination du crystallin. *Bulletin de l'Association des Anatomistes*, **49**, 642–53.

GENIS-GALVEZ, I. MA. (1975). Lens differentiation: immunological changes in the equatorial and cortical cells. *Cell Differentiation*, **4**, 49–55.

GRANT, M. E., KEFALIDES, N. A. & PROCKOP, D. J. (1972). The biosynthesis of basement membrane collagen in embryonic chick lens. I. Delay between the synthesis of polypeptide chains and the secretion of collagen by matrix-free cells. *Journal of biological Chemistry*, **247**, 3539–44.

HANNA, A. & KEATTS, H. C. (1966). Chicken lens development. Epithelial cell production and migration. *Experimental Eye Research*, **5**, 111–15.

HASTIE, N. D. & BISHOP, J. O. (1976). The expression of three abundance classes of messenger RNA in mouse tissues. *Cell*, **9**, 761–74.

HENDRIX, R. W. & ZWAAN, J. (1975). The matrix of the optic vesicle presumptive lens interface during induction of the lens in the chicken embryo. *Journal of Embryology and experimental Morphology*, **33**, 1023–49.

HUGHES, C., LAURENT, M., LONCHAMPT, M. O. & COURTOIS, Y. (1975). Lens glycoproteins: biosynthesis in cultured lens epithelial cells of bovine lens. *European Journal of Biochemistry*, **52**, 143–55.

HUMPHRIES, S., WINDASS, J. & WILLIAMSON, R. (1976). Mouse globin gene expression in erythroid and non-erythroid tissues. *Cell*, **7**, 267–77.

ITOH, Y. (1976). Enhancement of differentiation of lens and pigment cells by ascorbic acid in cultures of neural retinal cells of chick embryos. *Developmental Biology*, **54**, 157–62.

ITOH, Y., OKADA, T. S., IDE, H. & EGUCHI, G. (1975). The differentiation of pigment cells in cultures of chick embryonic neural retinae. *Development, Growth and Differentiation*, **17**, 39–50.

VAN KAMP, G. J., STRUYKER-BOUDIER, H. A. J. & HOENDERS, H. J. (1974*a*). The soluble proteins of the prenatal bovine eye lens. *Comparative Biochemistry and Physiology*, **49B**, 445–56.

VAN KAMP, G. J., VAN KLEEF, F. S. M. & HOENDERS, H. J. (1974*b*). Reaggregation studies on the polypeptide chains of calf lens α-crystallin. *Biochemica et biophysica Acta*, **342**, 89–96.

KIRKLAND, B. M. & ZWAAN, J. (1975). Malorientation of mitotic figures in early lens rudiments of Aphakia mouse embryos. *Anatomical Research*, **182**, 345–54.

LONCHAMPT, M. O., LAURENT, M., COURTOIS, Y., TRENCHEV, P. & HUGHES, R. C. (1976). Microtubules and microfilaments of bovine lens epithelial cells. Electron miscroscopy and immunofluorescence staining with specific antibodies. *Experimental Eye Research*, **23**, 505–18.

MCLEAN, B. G. & FINNEGAN, C. V. (1973). Observations of chick-embryo lens histogenesis *in vivo* and *in vitro*. *Canadian Journal of Zoology*, **52**, 345–52.

McDevitt, D. S. (1972). Presence of lateral eye lens crystallins in the median eye of the American chameleon. *Science*, **175**, 763–4.

McDevitt, D. S. & Clayton, R. M. (1975). Ontogeny and localization of α, β and δ-crystallins in the Hy-1 chick lens. *Journal of Cell Biology*, **67**, 273A.

McDevitt, D. S., Meza, I. & Yamada, T. (1969). Immunofluorescence localization of the crystallins in amphibian lens development, with special reference to the γ-crystallins. *Developmental Biology*, **19**, 581–607.

Maisel, H., Alcalà, J. & Lieska, N. (1976). The protein structure of chick lens fiber cell membranes and intracellular matrix. *Documenta ophthalmologica, Proceedings Series*, **8**, 121–33.

Mikami, Y. (1941). Experimental analysis of the Wolffian lens regeneration in adult newt, *Triturus pyrrhogaster*. *Japanese Journal of Zoology*, **9**, 269–302.

Mikhailov, A. T. & Barabanov, B. M. (1975). Immunochemical analysis of water-soluble antigens of chick retina in the course of embryogenesis. *Journal of Embryology and experimental Morphology*, **34**, 531–57.

Milstone, L. M. & Piatigorsky, J. (1975). Rates of protein synthesis in explanted embryonic chick lens epithelium: differential stimulation of δ-crystallin synthesis. *Developmental Biology*, **43**, 91–100.

Milstone, L. M., Zelenka, P. & Piatigorsky, J. (1976). δ-Crystallin mRNA in chick lens cells: mRNA accumulates during differential stimulation of δ-crystallin synthesis in cultured cells. *Developmental Biology*, **48**, 197–204.

Modak, S. P., Morris, G. & Yamada, Y. (1968). DNA synthesis and mitotic activity during early development of chick lens. *Developmental Biology*, **17**, 544–61.

Ogawa, T. (1967). The similarity between antigens in the embryonic lens and in the lens regenerate of the newt. *Embryologia*, **9**, 295–305.

Okada, T. S. (1976). Transdifferentiation of cells of specialised eye tissue in cell culture. In *Tests of teratogenicity in vitro*, ed. J. D. Ebert & M. Marois, pp. 91–105. Amsterdam: North-Holland.

Okada, T. S., Eguchi, G. & Takeichi, M. (1971). The expression of differentiation by chicken lens epithelium in in vitro cell culture. *Development, Growth and Differentiation*, **13**, 323–36.

Okada, T. S., Eguchi, G. & Takeichi, M. (1973). The retention of differentiated properties by lens epithelial cells in clonal cell culture. *Developmental Biology*, **34**, 321–33.

Okada, T. S., Itoh, Y., Watanabe, K. & Eguchi, G. (1975). Differentiation of lens in cultures of neutral retinal cells of chick embryos. *Developmental Biology*, **45**, 318–29.

O'Malley, B. W. & Means, A. R. (1974). Female steroid hormones and target cell nuclei. *Science*, **183**, 610–20.

Pelham, H. R. B. & Jackson, R. J. (1976). An efficient mRNA dependent translation system from reticulocyte lysates. *European Journal of Biochemistry*, **67**, 247–56.

Perry, M. M. (1976). A method to demonstrate the fine structural components of lens fibre cells. *Experimental Eye Research*, **22**, 125–28.

Philipson, B. (1969). Distribution of protein within the normal rat lens. *Investigative Ophthalmology*, **8**, 258–70.

Piatigorsky, J., Beebe, B. C., Zelenka, P., Milstone, S. M. & Shinohara,

T. (1976). Regulation of δ-crystallin gene expression during development of the embryonic chick lens. In *Biology of the lens epithelial cell, in relation to development, ageing and cataract*, ed. Y. Courtois & F. Regnault, pp. 85–112. Paris: INSERM.

Piatigorsky, J., Rothschild, S. S. & Milstone, L. M. (1973). Differentiation of lens fibres in explanted embryonic chick lens epithelia. *Developmental Biology*, **34**, 334–45.

Pollard, T. D. & Weihing, T. T. (1974). Actin and myosin and cell movement. *Critical Reviews of Biochemistry*, **2**, 1–65.

de Pomerai, D. I., Pritchard, D. J. & Clayton, R. M. (1977). Biochemical and immunological studies of lentoid formation in cultures of embryonic chick neural retina and day-old chick lens epithelium. *Developmental Biology*, **60**, 416–27.

Powell, J. A. & Segil, N. (1976). Secondary lens formation caused by implantation of pituitary into the eye of the newt *Notophthalmus*. *Developmental Biology*, **52**, 128–40.

Pritchard, D. J., Clayton, R. M. & Cunningham, W. L. (1974). Abnormal lens capsule carbohydrate associated with the dominant gene small eye (*Sey*) in the mouse. *Experimental Eye Research*, **19**, 335–40.

Rabaey, M. (1962). Electrophoretic and immunoelectrophoretic studies in the soluble proteins in developing lens of birds. *Experimental Eye Research*, **1**, 310–16.

Reyer, R. W. (1962). Regeneration in the amphibian eye. *Symposium of the Society for the Study of Development and Growth*, **20**, 211–65.

Reyer, R. W., Woolfitt, R. A. & Withersty, L. T. (1973). Stimulation of lens regeneration from the newt dorsal iris when implanted into the blastema of the regenerating limb. *Developmental Biology*, **32**, 258–81.

Scheib, D. (1965). Recherches recentes sur la régénération du cristallin chez les vertébrés. Evolution du problème entre 1931 et 1963. *Ergebnisse der Anatomie und Entwicklungsgeschichte*, **38**, 47–114.

Scott, R. B. & Bell, E. (1965). Messenger RNA utilization during development of chick embryo lens. *Science*, **147**, 405–7.

Slingsby, C. & Croft, L. R. (1973). Developmental changes in the low molecular weight proteins of the bovine lens. *Experimental Eye Research*, **17**, 369–76.

Stewart, J. A. & Papaconstantinou, J. (1967). A stabilisation of RNA templates in lens cell differentiation. *Proceedings of the National Academy of Sciences, USA*, **58**, 95–102.

Stone, L. S. (1965). The regeneration of the crystalline lens. *Investigative Ophthalmology*, **4**, 420–32.

Takata, C., Albright, J. F. & Yamada, T. (1966). Gamma crystallin in Wolffian lens regeneration demonstrated by immunofluorescence. *Developmental Biology*, **14**, 382–400.

Takeichi, M. (1970). Growth of the chicken embryonic lens transplanted onto the chorioallantoic membrane. *Development, Growth and Differentiation*, **12**, 21–30.

Tilney, L. G. (1975). The role of actin in non-muscle cell motility. In *Molecules and cell movement*, ed. S. Inoue & R. E. Stephens, pp. 339–88. Raven Press.

Truman, D. E. S., Brown, A. G. & Campbell, J. C. (1972). The relationship between the ontogeny of antigens and of the polypeptide chains of the crystallins during chick lens development. *Experimental Eye Research*, **13**, 58–69.

Truman, D. E. S., Clayton, R. M., Gillies, A. & Mackenzie, H. (1976). RNA synthesis in the lenses of normal chicks and in two strains of chick with hyperplasia of the lens epithelium. *Documenta Ophthalmologica, Proceedings Series*, **8**, 17–26.

WAGGONER, P. R. (1973). Lens differentiation from the cornea following lens extirpation or cornea transplantation in *Xenopus laevis*. *Journal of experimental Zoology*, **186**, 97–109.

WAGGONER, P. R., LIESKA, N. ALCALA, J. & MAISEL, H. (1976). Ontogeny of chick lens β-crystallin polypeptides by immunofluorescence. *Ophthalmological Research*, **8**, 292–301.

YAMADA, T. (1967). Cellular and subcellular events in Wolffian lens regeneration. *Current Topics in developmental Biology*, **2**, 247–83.

YAMADA, T. (1976). Dedifferentiation associated with cell-type conversion in the newt lens regenerating system. A review. In *Progress in differentiation research*, ed. N. Muller-Berat, pp. 355–60. Amsterdam: North-Holland.

YAMADA, T., REESE, T. H. & MCDEVITT, D. S. (1973). Transformation of iris *in vitro* and its dependency on neural retina. *Differentiation*, **1**, 65–82.

YAMADA, T. & ROESEL, M. (1971). Control of mitotic activity in Wolffian lens regeneration. *Journal of experimental Zoology*, **177**, 119–28.

YAMADA, T., ROESEL, M. & BEAUCHAMP, J. J. (1975). Cell cycle parameters in dedifferentiating iris epithelial cells. *Journal of Embryology and experimental Morphology*, **34**, 497–510.

YAMAMOTO, Y. (1976). Growth of lens and ocular environment. Role of neural retina. The growth of mouse lens as revealed by an implantation experiment. *Development, Growth and Differentiation*, **18**, 273–8.

YOSHIDA, K. & KATOH, A. (1971). Crystallin synthesis by chick lens. Changes in synthetic activities of epithelial and fiber cells during embryonic development. *Experimental Eye Research*, **11**, 184–94.

ZALIK, S. E., SCOTT, V. & DIMITROV, E. (1976). Changes at the cell surface during *in vivo* and *in vitro* dedifferentiation in cellular metaphasia. *Progress in differentiation research*, ed. N. Muller-Berat, pp. 361–7. Amsterdam: North-Holland.

ZALOKAR, M. (1944). Contribution a l'étude de la régeneration du cristallin chez le Triton. *Revue suisse de zoologie*, **51**, 443–521.

ZWAAN, J. (1968). Lens-specific antigens and cytodifferentiation in the developing lens. *Journal of Cellular Physiology*, **72**, suppl. 1, 47–72.

ZWAAN, J. (1975). Immunofluorescent studies on Aphikia, a mutation of a gene involved in the control of lens differentiation in the mouse embryo. *Developmental Biology*, **44**, 306–12.

ZWAAN, J. & WILLIAMS, R. M. (1968). Morphogenesis of the eye lens in a mouse strain with hereditary cataracts. *Journal of experimental Zoology*, **169**, 407–22.

Stem cell functions and the clonal haemopathies of man

E. A. McCULLOCH

The Ontario Cancer Institute and The Institute of Medical Science,
University of Toronto, Toronto, Ontario, Canada M5S 1A1

The concept is emerging that metazoan cellular communities function co-operatively by reason of interactions mediated by proteins with recognition functions. Certain of these systems also depend on renewal of their cellular elements and this function is determined principally by the regulators intrinsic to proliferation. In mammals, the haemopoietic system provides suitable material for probing both communication mechanisms and cellular proliferation. Haemopoietic populations exist together almost as free-living cells, not constrained by organ-defining structures, such as basement membranes, required where specific anatomical relationships underline function. Freed of such obligatory associations, haemopoietic populations may be defined in functional terms; they can be separated into subpopulations, and the effects of both such separation and subsequent recombination determined. Developmental techniques are available for the study of proliferation both in culture and *in vivo*, particularly after cell transfer. The purpose of the present paper is to consider stem cells in haemopoietic tissue with particular reference to their role as units of regulation. Such consideration includes a discussion of events occurring during clonal expansion from individual stem cells and regulatory mechanisms developing within and between such clones. Consideration will include both the roles of specific mutations and the clonal haemopathies of man in the definition of regulatory mechanisms.

STEM CELLS AND CELL RENEWAL

Cell renewal is necessary in those systems where function is subserved by populations with shorter lifespans than the animals they serve. The replacement of such functional cells is dependent upon proliferation occurring in a minority population, capable of long survival because its members have the capacity to renew themselves as well as the capacity for producing differentiated descendants. Self-renewal provides stem cell populations with the independence that differentiated cells relinquished during the acquisition of structures or molecules adapted for specific functions. Independence through self-renewal is the essential property that defines stem cells (Siminovitch, McCulloch & Till, 1963). Further aspects of stem cell definition relate to specific cell renewal systems. For normal haemopoiesis, two other properties are required: these are capacity for differentiation (Siminovitch *et al.*, 1963) and sensitivity to regulatory mechanisms (Becker, McCulloch, Siminovitch & Till, 1965).

Present understanding permits a description of stem cell properties in quantitative rather than qualitative terms. Even self-renewal appears to be indefinite only following malignant transformation; for normal cells, capacity for proliferation, upon which independence depends, is limited. Diploid cells in culture are capable of only approximately 50 doublings and haemopoietic tissue can be retransplanted successfully for only a limited number of transfer generations (Siminovitch, Till & McCulloch, 1964), although no evidence exists indicating that similar mechanisms underlie the limitation of proliferation observed in culture and *in vivo*. Differentiation capacity can also be expressed quantitatively, as the number of lineages derived from a single stem cell. A number of at least 2 defines pluripotent stem cells. The stem cells of specific renewal systems in adult life are distinguished from fertilised ova because the number of differentiation pathways is limited. This limitation is the basis for considering individual stem cells, each with the potential of developing into a self-maintaining clone, as units of regulation within specific cell-renewal systems.

Sensitivity to regulatory mechanisms is not a unique stem cell function; rather, mechanisms, different at least in detail, operate along a differentiation lineage. At each stage these mechanisms are multiple; consequently, stem cell responsiveness to regulation, like other properties, may be considered quantitatively, as the number of potential responses varies. A major difference between normal and neoplastic stem cells resides in their sensitivity to regulation; complete loss of such responsiveness would define truly autonomous neoplasms, while varying but incomplete loss yields dependent neoplasms.

STEM CELLS AND THE REGULATION OF HAEMOPOIESIS

Cellular organisation in myelopoiesis

Adult haemopoiesis originates in a heterogeneous class of pluripotent stem cells. In rodents, members of this class are detected by their capacity to form macroscopic spleen colonies, containing granulopoietic, erythropoietic and megakaryocytic cells, the three lineages of myelopoiesis (Till & McCulloch, 1961; Wu, Till, Siminovitch & McCulloch, 1967). Each of the lineages is headed by one or more classes of progenitors, distinguished from their pluripotent ancestors on the basis of physical, physiological and genetic criteria (for a review see McCulloch, Mak, Price & Till, 1974). These committed progenitors provide a cellular basis for independent regulation of the different differentiation processes occurring within a single clone. Concomitant with loss of capacity for independent self-renewal the committed progenitors have new or increased sensitivity to regulatory mechanisms with specificity for that lineage.

Within a myelopoietic lineage, considerable fine structure can be detected. Identified classes of cells in the erythropoietic arm of myelopoiesis are listed in Table 1. Prior to the emergence of recognisable differentiated erythroblasts, three erythropoietic committed populations can be identified. The progenitor class of transient erythropoietic colonies illustrates the pitfalls of attempting to order progenitors in strict linearity. This class is identified by its capacity to form small, but macroscopic colonies of erythroblasts that persist for only a few days in the spleens of sublethally irradiated animals subjected to erythropoietic stimulation either by bleeding or erythropoietin administration (Gregory, McCulloch & Till, 1975*a*). This particular erythropoietic proliferation is detected only under these extreme conditions of stimulation and genetic evidence (see below) is available to indicate a functional role for transient erythropoietic colony formation specifically under conditions of severe erythropoietic demand (Gregory, McCulloch & Till, 1975*b*).

Table 1. *The origins of erythropoiesis*

Name	Assay	Significance	Reference
CFU-S	Mixed colonies at 8–12 days in irradiated or W/W^v mice	Pluripotent stem cell	Till & McCulloch (1961)
TE-CFU	Endogenous erythropoietic colonies at day 5–6 in irradiated, stimulated mice	Cell required for rapid erythropoietic response	Gregory, McCulloch & Till (1975*a*, *b*)
BFU-E	'Burst' formation in cultures at high erythropoietin levels	Early committed progenitor of erythropoiesis	Heath, Axelrad, McLeod & Shreeve (1976)
CFU-E	Colony formation in cultures at low erythropoietin levels	Late, committed progenitor of erythropoiesis	Stephenson, Axelrad, McLeod & Shreeve (1971)
Pronormoblast	Identified morphologically	Earliest recognisable erythropoietic precursor	—

It is possible, therefore, that the progenitors of transient erythropoietic colonies represent a process in erythropoiesis parallel to, rather than linear with, that occurring in steady state differentiation. The two remaining classes of early erythropoietic progenitors are detected in culture. They differ in proliferative potential; however, the larger colonies derived from the less differentiated progenitors appear to be made up of several units, identical with the smaller colonies that develop from more differentiated progenitors. Thus, the actual processes of growth underlying the functions of each are common rather than uniquely different.

Granulopoiesis is headed by committed progenitors detected in culture (Pluznik & Sachs, 1965; Bradley & Metcalf, 1966). This population is also known to be heterogeneous, in size and in responsiveness to sources of granulopoietic stimulation in culture (Messner, Till & McCulloch, 1973). The functional significance of this heterogeneity remains unexplained. It may represent a quantitative variation in stem cell function, particularly proliferative potential.

Genetic analysis of haemopoietic regulation

The most persuasive insights into haemopoietic regulation have derived from studies of genetically determined anaemias in mice (Russell & Bernstein, 1966). Anaemic animals of genotype W/W^{v} have defective stem cells, lacking sufficient proliferative potential to form macroscopic spleen colonies (McCulloch, Siminovitch & Till, 1964). The genetically determined lesion does not abolish stem cell function; microscopic collections of transplanted cells are observed in irradiated recipients of W/W^{v} marrow (Lewis & Trobaugh, 1964) and occasionally complete repopulation has been documented using cells of this source (Sutherland, Till & McCulloch, 1970). It is probable that this genetic locus codes for regulatory messages rather than for structural proteins. Mice of genotype Sl/Sl^{d} contain normal stem cells; their haemopoietic spaces are defective in their capacity to support rapid repopulation following cell transfer (McCulloch *et al.*, 1965; McCulloch, Gregory & Till, 1973). The Sl/Sl^{d} locus clearly codes for a regulatory function. These genetic anaemias, each with a basis in stem cell function, served as examples of the two major classes of regulatory mechanisms affecting haemopoiesis. The first, exemplified by the function of the *W* locus, concerns mechanisms operative intracellularly; the second, exemplified by the

Sl locus, consists of intercellular regulation, relating stem cell function to other cell populations within haemopoietic organs.

A third genetically determined anaemia is that observed in the embryos of mice of genotype *f/f* (Gruneberg, 1942). The anaemia disappears spontaneously but is recapitulated as reduced erythropoiesis during the exponential growth phase of marrow grafts. The cellular basis of the anaemia has been associated with erythropoietic progenitors responsible for transient erythropoietic colony formation (Gregory *et al.*, 1975*b*). Since this process is a part of the physiology of rapidly regenerating populations, the *f* locus may also be considered to encode regulatory messages. Table 2 summarises the nature of the three anaemias.

Intracellular regulation and clonal expansion

Function of the *W* locus provides a specific example of intracellular regulation. But the genome must contain information required for many processes occurring during development of clones from individual stem cells. These processes include a number of alternative avenues; for example, stem cells may either differentiate or renew themselves. If the choice is towards differentiation, a pathway must be selected. Both pluripotent stem cells (Becker *et al.*, 1965) and their committed descendants (Iscove, Till & McCulloch, 1970) can alter their physiological states, either proliferating rapidly or slowly (a genuine resting state termed G_0 has been postulated (Quastler, 1963) and recently has been documented formally for cells in culture by Chang & Baserga (1977)). Evidence is lacking that the choices between these various alternative pathways are determined rigidly. Rather, analysis of individual haemopoietic clones provides data compatible with relaxed intracellular regulation. The observation is that individual spleen colonies, each resulting from clonal expansion, vary greatly in their content of the different cellular classes that developed during the processes of stem cell renewal, differentiation and proliferation (Siminovitch *et al.*, 1963; Gregory, McCulloch & Till, 1973). Colony-to-colony variation of this degree might be anticipated if alternate cellular pathways were selected randomly rather than directions determined by a specific regulatory message (Till, McCulloch & Siminovitch, 1964; Korn, Henkelman, Ottensmeyer & Till, 1973). The concept of random events leading to heterogeneity during clonal expansion has been used as the basis of a proposed technique for arranging cell classes within differentia-

Table 2. *Three genes regulating myelopoiesis*

Gene symbol	Effect on CFU-S	Effect on growth	Interpretation	Reference
W	W/W^v marrow cells do not form macroscopic colonies	Normal cells form colonies in unirradiated W/W^v mice	*W* is an intrinsic regulator of CFU-S	McCulloch *et al.* (1964)
Sl	None	Normal cells grow poorly in irradiated Sl/Sl^d hosts	*Sl* is an extrinsic regulator of CFU-S	McCulloch & Till (1977); Bernstein (1970)
f	Defective erythropoiesis and reduced transient erythropoietic colony formation		Defective erythropoietic progenitors required for rapid response	Gregory *et al.* (1974); Gregory *et al.* (1975*a*, *b*)

Mice of genotype Sl/Sl^d and W/W^v are superficially similar: both are black-eyed whites, sterile, unusually radiosensitive and have a severe macrocytic anaemia. Mice of genotype *f/f* (flexed) have kinked tails. (Russell & Bernstein, 1966.)

tion lineages (Gregory, McCulloch & Till, 1973). The method is based on measuring statistical correlations between different populations within clones. If little or no correlation is found, it is considered that the cell classes belong either to different pathways or are widely separated along the same pathway; correlation is considered to be lost because numerous randomising events intervene between the two populations during clonal expansion. Conversely, close correlation occurs when few events separate populations and is interpreted to mean either close association on the same differentiation pathway or with a progenitor common to two lineages. Measurements of correlations yield lineage maps compatible with those based on other considerations such as proliferative potential deduced from clonal size, or differentiation potential deduced from clonal composition. The erythropoietic lineage tabulated earlier (Table 1), was ordered on the basis of these latter considerations and supported by data derived from measurements of correlations.

Intercellular regulation

Studies of genetically anaemic mice of genotype Sl/Sl^{d} have yielded the most conclusive evidence for regulatory interactions between stem cells and other elements in haemopoietic tissue. Cells bearing the information encoded by *Sl* function without proliferation, since their regulatory capacity survives large doses of ionising radiation. Their function is, however, dependent on organ integrity; the abnormal Sl/Sl^{d} phenotype cannot be transferred using suspended cells but organ fragments can be successfully engrafted (Bernstein, 1970).

The *Sl* gene product stands alone as a fully documented external regulator of pluripotent stem cell function. It has been postulated, however, that two other genetically determined surface components have stem cell controlling functions; these are gene products of the H-2 complex and products of a different gene, not yet assigned to a linkage group. A regulatory role for these gene products has been postulated on the basis of analysis of the reduced growth observed when marrow cells from certain parental genotypes are transplanted into heavily irradiated F_1 recipients (McCulloch *et al.*, 1973). Reservations concerning the significance of the findings in respect to normal stem cell regulation are based on failure to observe similar effects in the absence of genetic incompatability.

Studies of regulation at the level of stem cells have been restricted largely to genetic analysis because the cells of interest have neither been purified nor successfully grown in culture. The great power of culture methods for analysing cell–cell interactions has been demonstrated by studies of granulopoiesis. Under appropriate culture conditions, human granulopoietic progenitors will form colonies containing differentiated cells if stimulated either by managerial cells or bioactive molecules derived from their surfaces. Quantitation is possible for each of the three components of the system: the proliferative progenitors, nonproliferative managerial cells and molecular species (collectively termed colony stimulating activity or CSA) that mediate the interaction between the two cell populations (for a review see McCulloch, 1975). From such studies, two findings stand out particularly; first, CSA contains components cross-reactive immunologically with β_2 microglobulin (Price, McCulloch & Till, 1976). This ubiquitous, nonpolymorphic protein is found in association with histocompatibility gene products (Berggard & Bearn, 1968; Grey *et al.*, 1973; Tanigaki & Pressman, 1974). Its association with CSA relates this regulator of granulopoiesis to general mechanisms of recognition on the one hand, and on the other, provides support for a regulatory role associated with H-2 gene products.

Secondly, capacity to stimulate granulopoiesis or to release CSA into culture media was found to be associated with specific subpopulations of haemopoietic cells (Messner *et al.*, 1974). This apparent specificity, however, was much less obvious when the lectin phytohaemagglutinin (PHA) was added to cultures and CSA release into media measured (Prival, Paran, Gallo & Wu, 1974). These data were interpreted as indicating the facilitation of CSA release by PHA which might be accomplished if the location of CSA-proteins within the membrane was altered by the lectin. Such considerations were used as the basis of a detailed model of granulopoietic regulation in culture. It was proposed that bioactive CSA molecules were present in association with the membranes of many cell types, but that their anatomical position determined their availability for interaction with CSA-receptors on granulopoietic progenitors. External substances (for example bacterial products) could interact with cell membranes, changing the availability of granulopoietic regulatory molecules for interaction. Thus, signals from the environment, either near or distant, could change the number of mature granulocytes available for specific function (McCulloch, 1975).

THE CLONAL HAEMOPATHIES OF MAN

It has been emphasised that perturbations of stem cell function resulting from specific mutations have provided powerful insights into normal function. Traditionally, perturbations introduced by disease, the 'experiments of nature' have been equally powerful investigative tools. In man, it has been feasible to trace certain diseases as originating in pluripotent stem cells even though a direct assay for human stem cells is not available. Two genetic techniques have been used to identify cells belonging to single clones. First, chromosomal abnormalities, particularly those generated at random, are convincing clonal markers. The second technique involves the use of the isoenzyme of glucose-6-phosphate dehydrogenase (G-6-PD) (Fialkow, Gartler & Yoshida, 1967). The G-6-PD gene is X-linked and the somatic cell populations of heterozygous females are mosaics; each individual cell contains only a single isoenzyme, but the two are distributed in approximately equal proportions within the somatic population. This distribution results from the random inactivation of X-chromosomes during embryonic development (Lyonisation). If a disease originates in a single stem cell after X-inactivation, all of the descendants of the originally affected cell will contain only the isoenzyme coded by the gene carried by the active X-chromosome. When such an abnormal clone dominates haemopoietic tissues, the difference between the presence of a single isoenzyme in haemopoietic tissue and both isoenzymes in other tissues serves to demonstrate the clonal nature of the abnormality. Three human diseases have been identified as clonal by this means: these are chronic myeloblastic leukaemia (CML) (Fialkow *et al.* 1967), polycythaemia vera (P-vera) (Adamson *et al.*, 1976), and idiopathic myelofibrosis (IMF) (Jacobson & Fialkow, 1976). In affected females with these diseases only a single G-6-PD isoenzyme is found in the three lineages of myelopoiesis, indicating both the origin of the disease in a single pluripotent stem cell and also the subsequent expansion of the clone to occupy the whole haemopoietic system.

Chromosomal abnormalities have been used to establish the stem cell origin of the myeloblastic leukaemias. The Philadelphia chromosome, characteristic of CML, can be identified in granulopoietic, erythropoietic and megakaryocytic cells (Whang-Peng *et al.*, 1963). This evidence, while consistent with the clonal origin of CML, is not conclusive; since the Philadelphia chromosome is not a unique randomly generated marker, but rather reappears repeatedly in each new CML patient, the possibility exists that the same marker might be generated several times in different lineages within a single patient. Fortunately the studies with G-6-PD isoenzymes, described earlier, make this an unlikely possibility.

The stem cell origin of acute myeloblastic leukaemia (AML) may be demonstrated in those patients where unique chromosomal markers exist. Such markers have been found in the cells of origin of granulopoietic colonies from marrow or blood of patients with AML (Duttera, Whang-Peng, Bull & Carbone, 1972; Moore, Williams & Metcalf, 1972; Aye, Till & McCulloch, 1974). Erythropoietic differentiation in AML has been identified using a combination of radioactive iron radioautography and chromosomal analysis (Blackstock & Garson, 1974). Since chromosomal aberrations in AML are not highly repetitive, the data may be interpreted with some confidence as indicating the origin of AML from pluripotent stem cells. However, direct evidence is not available for leukaemic transformation occurring in a single cell in each patient. Where chromosomal markers are found they usually predominate; but, chemotherapeutically induced remissions are associated with the emergence of karyotypically normal cells. Since chromosomal abnormalities are not found consistently in AML, karyotypic normality cannot be advanced confidently as evidence that the cells are not leukaemic. Alternatively, they may be members of an ancestral clone from which the chromosomally-marked clone evolved. Such evolution occurs regularly in myeloblastic leukaemia; for example, blast transformation of CML is frequently associated with new chromosomal rearrangements (Hossfeld, 1975). The precedent of CML makes it attractive to consider that the genetic transformation leading to AML occurs in one or a small number of pluripotent stem cells, and, as in CML, P-vera and IMF, the resulting clone becomes predominant. These studies have permitted a new grouping of diseases based upon their origin from pluripotent stem cells. The term clonal haemopathy (McCulloch & Till, 1977*a*) has been suggested for the group. Table 3 contains a summary of the clonal haemopathies, along with evidence for their assignation to that grouping.

The concept of relaxed intracellular regulation of clonal expansion can be used as a basis for an interpretation of cell culture data

Table 3. *The clonal haemopathies*

Name	Evidence for clonal origin	Reference
Chronic myeloblastic leukaemia	(1) Philadelphia chromosome in granulopoietic, erythropoietic and megakaryocytic cells	Whang-Peng *et al.* (1963)
	(2) Single G-6-PD isoenzyme in erythrocytes and granulocytes of heterozygous females	Fialkow *et al.* (1967)
Acute myeloblastic leukaemia	Clonal chromosome marker in myelopoietic and erythropoietic cells	Blackstock & Garson (1974)
Polycythaemia vera	Single G-6-PD isoenzyme in red cells and granulocytes of heterozygous females	Adamson *et al.* (1976)
Idiopathic myelofibrosis	Single G-6-PD isoenzyme in red cells and granulocytes of heterozygous females	Jacobson & Fialkow (1976)

obtained from studies of the marrow of patients with AML. The technique for obtaining granulopoietic colonies in culture has been applied widely in the study of this disease. Chromosomal evidence is available that at least some granulopoietic colonies belong to leukaemic clones (Duttera *et al.*, 1972; Moore *et al.*, 1972; Aye *et al.*, 1974). But, consistent growth patterns have not been observed, even in patients with similar clinical presentations. Rather, great patient-to-patient variation has been reported; marrow cells from many patients yield few if any colonies and all gradations are seen between this and colony formation greatly in excess of normal (Cowan *et al.*, 1972). These findings might be interpreted as indicating great heterogeneity within AML, previously not disclosed by the methods available for clinical evaluation. Alternatively, the variation might be likened to that observed when haemopoietic clones in mice are examined. From this point of view, random events during clonal expansion rather than heterogeneity of the leukaemic phenotype may account for the observations in AML.

Studies of clonal haemopathies provide evidence for interactions between haemopoietic stem cells occurring *in vivo*. For example, in P-vera circulating erythrocytes, granulocytes and platelets all belong to the same clone as determined by G-6-PD isoenzyme studies. Erythropoietic colonies growing in cultures in the presence of low concentrations of erythropoietin also belong to the dominant P-vera clone; however, when larger concentrations of erythropoietin are added to cultures, progenitors belonging to at least one different clone express their proliferative potential (Adamson *et al.*, 1976). Thus, *in vivo*, P-vera clones not only expand preferentially but also have the capacity to repress growth potential in other stem cells. In contrast with this repressive interaction, the abnormal clone of IMF stimulate proliferation in the stromal fibroblasts of marrow. Indeed, marrow fibrosis often dominates the clinical picture, yet isoenzyme studies show that haemopoietic cells belong to a single clone while the proliferating fibroblasts are polyclonal. These findings in the clonal haemopathies, added to the genetic evidence derived from mice of genotype Sl/Sl^{d}, provide strong evidence for cell interactions as important determinants of haemopoietic stem cell function *in vivo* as well as in culture.

PERSPECTIVES

Certain perspectives for the future emerge from present research trends. From these are derived perceptions of new technologies and new organisations now being developed as present approaches reach the limit of their resolving power.

First, the available information about stem cells derives almost entirely from developmental techniques: in these, stem cells are not assayed directly; rather, their numbers and properties are deduced from clonal collections of their descendants held together as colonies under appropriate experimental conditions. The indirection inherent in this methodology places inescapable limits on its application. For example, unknown factors such as plating efficiency in culture reduce the reliability of estimates of stem cell frequency based upon colony formation; and analysis of the cellular composition of colonies provides only a minimum estimate of progenitor capacity for proliferation and differentiation. An approach to avoiding this indirection may be emerging from studies of surface markers detected by immunological means. For example, antisera raised in rabbits against brain cells may have sufficient specificity to distinguish between pluripotent stem cells and their most immediate granulopoietic committed descendants (Van de Engh & Golub, 1974). Ia-like surface proteins have also been suggested as markers of the undifferentiated state (Schlossman, Chess, Humphreys & Strominger, 1976). Antibodies with stem cell specificity may be becoming available through such avenues, and these combined with rapidly advancing cell sorting capabilities may yield direct assays. Studies of distributions of such markers already provide evidence that surface proteins suggested as markers of primitive cells may be preserved to varying extents during differentiation. Thus, Ia-derived determinants are found on both primitive cells and mature macrophages (Hammerling, 1976).

Studies of clonal haemopathies may be expected to continue to illuminate stem cell function *in vivo*. For example, the relationship between the origins of myelopoiesis and lymphopoiesis remains controversial. Recent observations of surface proteins (Beard *et al.*, 1976; Janossy *et al.*, 1976) and the enzyme terminal transferase (Sarin & Gallo, 1974) marker usually associated with lymphopoietic cells, emerging during blast transformation in CML, have led to a hypothesis (Janossy, Roberts & Greaves, 1976) concerning the differential potentials of haemopoietic stem cells. Tests of these hypotheses will centre on direct study of this important human material.

Finally, it is hoped that the overview of stem cell research contained herein will illustrate an example for investigators. Present knowledge of stem cell function is derived from the work of geneticists, biochemists, cell biologists, instrument designers and clinicians. The developing logic of the biology, rather than any organisation initiative, has dictated these multidisciplinary endeavours. Their continuing necessity makes stem cell studies a natural basis upon which stimulating associations of talents may be centred, leading to new insights into the organised cellular communities of Metazoa.

REFERENCES

ADAMSON, J. W., FIALKOW, P. J., MURPHY, S., PRCHAL, J. F. & STEINMANN, L. (1976). Polycythemia vera: stem cell and probable clonal origin of the disease. *New England Journal of Medicine*, **295**, 913–16.

AYE, M. T., TILL, J. E. & MCCULLOCH, E. A. (1974). Cytological studies of colonies in culture derived from the peripheral blood cells of two patients with acute leukemia. *Experimental Hematology*, **2**, 362–71.

BEARD, M. E. J., DURRANT, J., CATOVSKY, D., WILTSHAW, E., AMESS, J. L., BREARLEY, R. L., KIRK, B., WRIGLEY, P., JANOSSY, G., GREAVES, M. F. & GALTON, D. A. G. (1976). Blast crisis of chronic myeloid leukaemia (CML). I. Presentation of simulating acute lymphoid leukaemia (ALL). *British Journal of Haematology*, **34**, 167–78.

BECKER, A. J., MCCULLOCH, E. A., SIMINOVITCH, L. & TILL, J. E. (1965). The effect of differing demands for blood cell production on DNA synthesis by hemopoietic colony forming cells of mice. *Blood*, **26**, 296–308.

BERGGARD, I. & BEARN, A. G. (1968). Isolation and properties of a low molecular weight B_2-globulin occurring in human biological fluids. *Journal of biological Chemistry*, **243**, 4095–103.

BERNSTEIN, S. E. (1970). Tissue transplantation as an analytic and therapeutic tool in hereditary anemias. *American Journal of Surgery*, **119**, 448–51.

BLACKSTOCK, A. M. & GARSON, O. M. (1974). Direct evidence for involvement of erythroid cells in acute myeloblastic leukemia. *Lancet*, **ii**, 1178–9.

BRADLEY, T. R. & METCALF, D. (1966). The growth of mouse bone marrow cells *in vitro*. *Australian Journal of experimental Biology and medical Science*, **44**, 287–300.

CHANG, H. L. & BASERGA, R. (1977). Demonstration of a G_0 state in mammalian cells by the use of temperature sensitive mutant cell lines. *Journal of cellular Physiology*, (in press).

COWAN, D. H., CLARYSSE, A., ABU-ZAHRA, H., SENN, J. S. & MCCULLOCH, E. A. (1972). The effect of remission-induction in acute myeloblastic leukemia on colony formation in culture. *Series Haematology*, **5**, 179–88.

Duttera, M. J., Whang-Peng, J., Bull, J. M. C. & Carbone, P. P. (1972). Cytogenetically abnormal cells *in vitro* in acute leukemia. *Lancet*, **i**, 715–18.

Fialkow, P. J., Gartler, S. M. & Yoshida, A. (1967). Clonal origin of chronic myelogenous leukemia in man. *Proceedings of the National Academy of Sciences, USA*, **58**, 1468–71.

Gregory, C. J., McCulloch, E. A. & Till, J. E. (1973). Erythropoietic progenitors capable of colony formation in culture: state of differentiation. *Journal of cellular Physiology*, **81**, 411–20.

Gregory, C. J., McCulloch, E. A. & Till, J. E. (1975*a*). Transient erythropoietic spleen colonies: effects of erythropoietin in normal and genetically anemic W/W^v mice. *Journal of cellular Physiology*, **86**, 1–8.

Gregory, C. J., McCulloch, E. A. & Till, J. E. (1975*b*). The cellular basis for the defect in haemopoiesis in flexed tailed mice. III. Restriction of the defect to erythropoietic progenitors capable of transient colony formation *in vivo*. *British Journal of Haematology*, **30**, 401–10.

Gregory, C. J., Tepperman, A. D., McCulloch, E. A. & Till, J. E. (1974). Erythropoietic progenitors capable of colony formation in culture: response of normal and genetically anemic *W/W* mice to manipulations of the erythron. *Journal of cellular Physiology*, **84**, 1–12.

Grey, H. M., Kuba, T. R., Colon, S. M., Poulik, M. D., Cresswell, P., Springer, T., Turner, M. & Strominger, J. I. (1973). The small subunit of HL-A antigens is B_2-microglobulin. *Journal of experimental Medicine*, **138**, 1608–12.

Gruneberg, H. (1942). The anemia of flexed-tailed mice (*Mus musculus* L.). II. Siderocytes. *Journal of Genetics*, **44**, 246–51.

Hammerling, G. J. (1976). Tissue distribution of Ia antigens and their expression on lymphocyte subpopulations. *Transplantation Review*, **30**, 64–82.

Heath, D. S., Axelrad, A. A., McLeod, D. L. & Shreeve, M. (1976). Separation of the erythropoietin-responsive progenitors BFU-E and CFU-E in mouse bone marrow by unit gravity sedimentation. *Blood*, **47**, 777–92.

Hossfeld, D. K. (1975). Chronic myelocytic leukemia: cytogeneic findings and their relations to pathogenesis and clinic. *Series Haematology*, **8**, 53–72.

Iscove, N. N., Till, J. E. & McCulloch, E. A. (1970). The proliferative states of mouse granulopoietic progenitor cells. *Proceedings of the Society for experimental Biology and Medicine*, **134**, 33–6.

Jacobson, R. J. & Fialkow, P. J. (1976). Idiopathic myelofibrosis: stem cell abnormality and probable neoplastic origin. *Clinical Research*, **24**, 439A (abstr.).

Janossy, G., Greaves, M. F., Revesz, T., Lister, T. A., Roberts, M., Durrant, J., Kirk, B., Catovsky, D. & Beard, M. E. J. (1976). Blast crisis of chronic myeloid leukaemia (CML). II. Cell surface marker analysis of lymphoid and myeloid cases. *British Journal of Haematology*, **34**, 179–92.

Janossy, G., Roberts, M. & Greaves, M. F. (1976). Target cell in chronic myeloid leukaemia and its relationship to acute lymphoid leukaemia. *Lancet*, **ii**, 1058–60.

Korn, A. P., Henkelman, R. M., Ottensmeyer, F. P. & Till, J. E. (1973). Investigations of a stochastic model of haemopoiesis. *Experimental Haematology*, **35b**, 1–14.

LEWIS, J. P. & TROBAUGH, F. E., Jr (1964). Hematopoietic stem cells. *Nature, London*, **204**, 589–90.

MCCULLOCH, E. A. (1975). Granulopoiesis in cultures of human hemopoietic cells. *Clinics in Haematology*, **4**, 509–33.

MCCULLOCH, E. A., GREGORY, C. J. & TILL, J. E. (1973). Cellular communication early in haemopoietic differentiation. *Ciba Foundation Symposium*, **13**, 183–204.

MCCULLOCH, E. A., MAK, T. W., PRICE, G. B. & TILL, J. E. (1974). Organization and communication in populations of normal and leukemic hemopoietic cells. *Biochimica et biophysica Acta*, **355**, 260–99.

MCCULLOCH, E. A., SIMINOVITCH, L. & TILL, J. E. (1964). Spleen colony formation in anemic mice of genotype W/W^v. *Science*, **144**, 844–6.

MCCULLOCH, E. A., SIMINOVITCH, L., TILL, J. E., RUSSELL, E. S. & BERNSTEIN, S. E. (1965). The cellular basis of the genetically determined hemopoietic defect in anemic mice of genotype Sl/Sl^d. *Blood*, **26**, 399–410.

MCCULLOCH, E. A. & TILL, J. E. (1977*a*). Stem cells in normal early haemopoiesis and certain clonal haemopathies. *Recent Advances in Haematology*, **2**, 85–110.

MCCULLOCH, E. A. & TILL, J. E. (1977*b*). Interacting cell populations in cultures of leukocytes from normal and leukemic peripheral blood. *Blood*, **49**, 269–80.

MESSNER, H. A., TILL, J. E. & MCCULLOCH, E. A. (1973). Interacting cell populations affecting granulopoietic colony formation by normal and leukemic human marrow cells. *Blood*, **42**, 701–10.

MOORE, M. A. S., WILLIAMS, N. & METCALF, D. (1972). Characterization of *in vitro* colony forming cells in acute and chronic myeloid leukemia. In *The nature of leukemia*, ed. P. Vincent, pp. 135–249. *Proceedings of the International Cancer Conference, Sydney, Australia.*

PLUZNIK, D. H. & SACHS, L. (1965). The cloning of normal 'mast' cells in tissue culture. *Journal of cellular and comparative Physiology*, **66**, 319–24.

PRICE, G. B., MCCULLOCH, E. A. & TILL, J. E. (1976). Cross reactivity of human B_2-microglobulin with human granulocyte colony stimulating activity. *Journal of Immunology*, **117**, 416–18.

PRIVAL, J. T., PARAN, M., GALLO, R. C. & WU, A. M. (1974). Colony stimulating factors produced from human peripheral blood cells in culture. *Journal of the National Cancer Institute*, **53**, 1583–8.

QUASTLER, H. (1963). The analysis of cell population kinetics. In *Cell proliferation*, vol. 18, ed. L. F. Lamerton & R. J. Fry, pp. 18–34. Oxford: Blackwell.

RUSSELL, E. S. & BERNSTEIN, S. E. (1966). Blood and blood formation. In *The biology of the laboratory mouse*, ed. E. L. Green, pp. 351–72. New York: McGraw-Hill.

SARIN, P. S. & GALLO, R. C. (1974). Terminal deoxynucleotidyltransferase in chronic myelogenous leukemia. *Journal of biological Chemistry*, **249**, 8051.

SCHLOSSMAN, S. F., CHESS, L., HUMPHREYS, R. E. & STROMINGER, J. L. (1976). Distribution of Ia-like molecules on the surface of normal and leukemic human cells. *Proceedings of the National Academy of Sciences, USA*, **73**, 1288–92.

SIMINOVITCH, L., MCCULLOCH, E. A. & TILL, J. E. (1963). The distribution of colony forming cells among spleen colonies. *Journal of cellular and comparative Physiology*, **62**, 327–36.

SIMINOVITCH, L., TILL, J. E. & MCCULLOCH, E. A. (1964). Decline in colony forming ability of marrow cells subjected to serial transplantation into irradiated mice. *Journal of cellular and comparative Physiology*, **64**, 23–31.

STEPHENSON, J. R., AXELRAD, A. A., McLEOD, D. C. & SHREEVE, M. M. (1971). Induction of colonies of hemoglobin-synthesizing cells by erythropoietin *in vitro*. *Proceedings of the National Academy of Sciences, USA*, **68**, 1542–6.

SUTHERLAND, D. J. A., TILL, J. E. & McCULLOCH, E. A. (1970). A kinetic study of the genetic control of hemopoietic progenitor cells assayed in culture and *in vivo*. *Journal of cellular Physiology*, **75**, 267–74.

TANIGAKI, N. & PRESSMAN, D. (1974). The basic structure and the antigenic characteristics of HL-A antigens. *Transplantation Review*, **21**, 15–34.

TILL, J. E. & McCULLOCH, E. A. (1961). A direct measurement of the radiation sensitivity of normal mouse bone marrow cells. *Radiation Research*, **14**, 213–22.

TILL, J. E., McCULLOCH, E. A. & SIMINOVITCH, L. (1964). A stochastic model of stem cell proliferation, based on the growth of spleen colony forming cells. *Proceedings of the National Academy of Sciences, USA*, **51**, 29–36.

VAN DEN ENGH, G. J. & GOLUB, E. (1974). Antigenic differences between hemopoietic stem cells and myeloid progenitors. *Journal of experimental Medicine*, **139**, 1621–27.

WHANG-PENG, J., FREI, E., TJIO, J. H., CARBONE, P. P. & BRECHER, G. (1963). The distribution of the Philadelphia chromosome in patients with chronic myelogenous leukemia. *Blood*, **22**, 664–73.

WU, A. M., TILL, J. E., SIMINOVITCH, L. & McCULLOCH, E. A. (1967). A cytological study of the capacity for differentiation of normal hemopoietic colony forming cells. *Journal of cellular Physiology*, **69**, 177–84.

Lymphocytes: their diversity, production and homeostasis

H. S. MICKLEM
Immunobiology Unit, Department of Zoology,
University of Edinburgh, Edinburgh EH9 3JT, UK

It is not so many years since it was possible to speak of 'the small lymphocyte'. It seemed, wrote Trowell (1958) 'a poor sort of cell, characterized by mainly negative attributes. . .'. That those days of innocence have so quickly passed away is due to the development and exploitation of a large range of experimental analytical techniques. Some of the most important steps were the demonstration of lymphocyte recirculation; the establishment of the thymus as a site of lymphocyte differentiation and of the genesis of immunocompetence; the realisation that the thymus is, nevertheless, not all-important; the gradual identification of cell surface and other markers and physical characteristics, which can differentiate subclasses of lymphocytes and thus add biochemical and functional dimensions to their unremarkable morphology; and the continuing discovery (it is too soon to say elucidation) of the complex cellular interactions which contribute to immune responses.

These advances have brought their problems. Lymphocyte subclasses, and claims of interactions between them, have been proliferating at an alarming rate, and the process still continues. A recent paper in this field (Cohen & Fairchild, 1976) formally thanked William of Occam for the use of his razor, but adherents of the good doctor have in general been having a hard time. Nevertheless, whatever the present confusions and uncertainties, lymphocytes are already among the most fully characterised of cells, display an unrivalled wealth of molecular and functional variety, and offer a happy hunting-ground for students of differentiation.

The purpose of this paper is briefly, by way of introduction, to summarise some aspects of the diversity of lymphocytes, and to discuss our still fragmentary understanding of the ontogeny and maturation of these cells and of the factors which govern their numbers. It is not a comprehensive review, but it attempts to bring together some threads which have generally remained unconnected, and to introduce the most recent literature.

THE DIVERSITY OF LYMPHOCYTES

The T–B dichotomy

The division into 'T' and 'B' lymphocytes (Roitt *et al.*, 1969) is now generally accepted as a significant one, at least in mammals and birds. 'T' denotes derivation from, or at least influence by, the thymus gland. 'B' denotes derivation (or influence) from the bursa of Fabricius in birds, or from some analogue of the bursa in mammals. The terms are sometimes diluted in meaning to include cells in the thymus and bursa themselves, but this is undesirable. The functional dichotomy is seen particularly clearly in birds, where the consequences of removing the bursa can be explored (Glick, Chang & Jaap, 1956; Cooper, Peterson, South & Good, 1966). In mammals only the thymus, not the bursa analogue (see below), can be effectively removed.

Work on mammals has generally supported the T–B distinction, B-cells being the potential antibody factories and T-cells performing a variety of regulatory and other functions. There is little evidence that any interconversion can take place between mature B- and T-lymphocytes. It is assumed that their progenitors diverge at some stage subsequent to the pluripotent stem cell, but it is not clear whether a cell exists which is committed to lymphoid development but in which the T versus B decision has not yet been reached. It is interesting that phylogenetically, the distinction between T-cells and B-cells seems less clear-cut. For example, thymocytes of bony (Emmrich, Richter & Ambrosius, 1975) and cartilaginous (Ellis & Parkhouse, 1975) fish have clearly detectable immunoglobulin (Ig) on their surface, a feature regarded as diagnostic for B-lymphocytes in birds and mammals. Amphibian tadpoles carry large concentrations of Ig on the thymocyte surface which become reduced as development proceeds (Du Pasquier, Weiss & Loor, 1972). According to Warr, Decker & Marchalonis (1976), bony fishes nevertheless possess lymphocytes which show homologies both functional and molecular with the T-lymphocytes of mammals. If this is true, the divergence between T- and B-lymphocytes may have begun as long ago as the Silurian period, among the placoderm fishes which were probably the latest common ancestors of modern bony fish and of mammals.

While it is generally agreed that B-lymphocytes 'recognise' the antigens to which they are equipped to react by means of specific immunoglobulin (antibody) receptors on their surface, it is less certain how T-lymphocytes do so. The difficulty has been that many workers have failed to identify Ig molecules on the surface of T-cells, while others have found them but shown them to be passively adsorbed. If immunoglobulin receptors were entirely absent, this would deprive T-cells of the one established molecule which is capable of specific recognition of a wide variety of antigenic determinants. The possession by T-cells of idiotypic specificities, which can be apparently identical to those on B-cells and which function as antigen receptors, implies the presence of at least the variable (N-terminal) portion of an Ig heavy chain (Binz, Kimura & Wigzell, 1975; Binz & Wigzell, 1975; Eichmann & Rajewsky, 1975; Black *et al.*, 1976).

Subpopulations of lymphocytes have been delineated on the basis of, amongst other criteria, function, lifespan, tissue-distribution, radiosensitivity, sensitivity to cortisone, size, density, electrophoretic mobility, adhesiveness, surface antigens and receptors for antigens, antigen–antibody complexes and complement. This sounds, and is, a daunting list. For our present purposes the most generally useful criteria are probably function, and surface antigens and receptors, and I shall have largely to ignore the rest.

Antigens and receptors which have been used to differentiate between T- and B-cells and their subsets are listed in Table 1.

Table 1. *Distinguishing surface characteristics of T- and B-lymphocytes*

	B	T
Surface Ig readily detectable	+	–
Thy1 (theta) alloantigen	–	+
Ly1 and/or 2 alloantigens	–	+
Receptors for C3[a]	+(most)	–

[a] Third component of complement.

Functions and surface characteristics of T-lymphocytes

Early studies established two functions for T-lymphocytes: as co-operators ('helpers') with B-cells in the production of antibody (Miller & Mitchell, 1969; Claman & Chaperon, 1969; Taylor, 1969), and as killer cells in cell-mediated immunity (Cerottini & Brunner, 1974). More recently it has emerged that T-lymphocytes as a whole have a regulatory role: this may manifest itself, according to experimental circumstances, as co-operation or as suppression (Gershon, 1975; Miller, 1975). It is becoming increasingly clear that these are functions of distinct T-cell subpopulations. This has emerged principally from studies of surface alloantigens Ly 1, 2 and 3, which are present on cells of the T-lineage in mice (Cantor & Boyse, 1975*a*, *b*; Feldmann, Beverley, Dunkley & Kontiainen, 1975; Cantor, Shen & Boyse, 1976; Vadas *et al.*, 1976). Most thymocytes and some peripheral T-cells express all these three Ly antigens. Other peripheral T-cells and some of the relatively cortisone-insensitive minority of thymocytes express either Ly1 or a combination of Ly2 and 3. $Ly1^{+},23^{-}$ cells function as helpers (T_H), while $Ly1^{-},23^{+}$ cells function as suppressors (T_S). Huber, Cantor, Shen & Boyse (1976) found no evidence of interconversion between these classes and suggested that they represent separate lines of differentiation from $Ly123^{+}$ precursors. In cell-mediated immunity, where cytotoxic (T_C) cells are assisted by amplifier (T_A) cells, T_C show the same Ly phenotype as T_S, and T_A the same as T_H (Stout & Herzenberg, 1975*a*, *b*; Stout, Waksal & Herzenberg, 1976). These authors have made further distinctions based on the presence or absence of receptors for the Fc fragment of IgG (see Table 2), although it is not yet clear whether these describe separate subsets or are merely different maturational stages.

It is apparent from the preceding paragraph that the Ly antigens of T-cells have great potential importance as analytical and preparative tools. Unfortunately, they are but feebly immunogenic and antisera are difficult to prepare, so that full realisation of their potential may be long delayed.

The Thy1 (otherwise known as theta) alloantigen of mice has been widely used as diagnostic for thymocytes and T-lymphocytes, and as a means of preparing pure B-cells. Subsets with high and low concentrations of Thy1 have been described (Roelants *et al.*, 1975; Schlesinger, Israel & Gery, 1976; Cantor & Weissman, 1976).

Table 2. *Markers of T-lymphocyte subpopulations*

Marker	Functional correlates	Other correlates
Antigens		
Thy1	All	Ly^+ (1 and/or 2)
High density	T_S?	Non-recirculating, short-lived, mainly splenic
Low density	T_H	Recirculating, long-lived
TL	Precursors (?all)	Largely restricted to thymocytes
Ly		
$1^+, 2^-$	T_H, T_A	
$1^-, 2^+$	T_S, T_C	
$1^+, 2^+$	Precursors?	All thymocytes and about 50% of splenic T-cells
Ia		
Surface Ia-4	T_S	
Others	T_H	
Secreted only	T_H	$Ly1^+, 2^-$
Receptors		
Fc^-	T_H	$Ly1^+, 2^-$
	T_{CP}	$1^-, 2^+$
Fc^+	T_A	$1^+, 2^-$
	T_C	$1^-, 2^+$
Radioresistance		
High	T_H	
Low	T_S	

Abbreviations for T-cell subsets:
H, helper; S, suppressor } in antibody formation
A, amplifier; C, cytotoxic; CP, precursor of cytotoxic } in cell-mediated immunity

Ia antigens, coded by the I-region of the major histocompatibility complex, are of special interest because the genes involved are either identical or closely linked to the *Ir* genes which control the responses to certain antigens. They may, indeed, be vitally involved in the mechanics of cellular interaction (for reviews, see Katz & Benacerraf, 1976; Möller, 1976). They are found on the surface of at least some T-cells and secreted by others, which have the Ly phenotype 1^+23^- (McKenzie & Parish, 1976), and different specificities have been reported on T_H and T_S classes (Okumura *et al.*, 1976*a*). There is some relationship, falling short, however, of actual identity, between Fc-receptors and Ia-antigenic structures on the surface of T-cells (Dickler, Arbeit, Henkart & Sachs, 1976). Ia antigens are also found on B-cells, macrophages, epidermal cells, spermatozoa and teratoma cells (Hämmerling, 1976). It is still too early to assess the significance of this tissue-distribution, or to delineate the precise functions of Ia-carrying molecules associated with T-cells.

Surface and other characteristics of T-cells are summarised in Table 2.

Functions and surface characteristics of B-lymphocytes

In comparison with T-cells, B-cells at present seem relatively simple to classify – although this may well turn out to be an illusion based on ignorance. It is not even certain that there are any genuinely distinct subpopulations of B-cells. The differences that can be observed in surface antigens and receptors and in function may, as discussed later, be marks of different maturational stages, and are probably at least in part consequences of antigenic stimulation.

Playfair & Purves (1971) distinguished between bone marrow and splenic B-cells on the basis of their differing responsiveness to help from T-cells. Most B-cells carry not only surface Ig-receptors for antigen (Raff, 1970) and Ia antigens (Hämmerling, 1976), but also receptors for the third component of complement (C3) and for the Fc fragment of immunoglobulin (Parish, 1975). Fc-receptors and Ia antigens are closely associated (Schirrmacher, Halloran, Arnaiz-Villena & Festenstein, 1976). However, a minority of B-cells lacks some or all of these markers. The absence of surface Ig probably indicates immaturity and inability to respond to antigenic stimulus, while the absence of C3-receptors and Ia antigens seems to be correlated with an inability to respond to help from T-cells, and to give rise to IgG-forming plasmacytes (Arnaiz-Villena, Playfair & Roitt, 1975; Lewis, Ranken, Nitecki & Goodman, 1976; Press, Klinman & McDevitt, 1976). Those cells which lack C3-receptors and/or Ia antigens are able to respond to 'thymus-independent' antigens and to give rise to IgM-secreting cells.

The dominant immunoglobulin heavy chain expressed on the surface of B-lymphocytes is the μ-chain (as found also in serum IgM). The small minority of cells carrying γ-chains on the surface appear to be members of antigen-primed clones carrying memory for specific IgG antibody production (Okumura *et al.*, 1976*b*; Mason,

1976), although some IgG memory cells appear to carry surface IgM, not IgG, receptors for antigen (Abney, Keeler, Parkhouse & Willcox, 1976). These and other studies strongly suggest that clones of antigen-stimulated lymphocytes can switch from IgM to IgG production and possibly back again, so that there is no reason to think of IgM- and IgG-bearing cells as fundamentally distinct subpopulations.

THE GENERATION OF T-LYMPHOCYTES

Thymic cells

Although recent work has done much to amplify our picture of mature T-lymphocytes and their subpopulations in both structural and functional terms, there are many gaps in our understanding of how they are produced. Following the chain backwards, there is little doubt that the vast majority comes initially from the thymus, although there have been suggestions that in the absence of a thymus some other tissue(s) may, to a very limited extent, substitute for it. Most of the T-lymphocytes in the body are long-lived, recirculating cells; of the many thymocytes produced by mitosis in the thymic cortex only a very small proportion are privileged ultimately to join this pool; what happens to the rest is still unclear (Laissue *et al.*, 1976). Within the thymus itself there is a cortex and a medulla, with most of the cell proliferation occurring in the cortex.

It appears that some of the cells produced in the cortex migrate out of the thymus via the medulla (Weissman, 1973). However, other evidence indicates independent development of cortical and medullary thymocytes (Shortman & Jackson, 1974). Droege & Zucker (1975) classified mouse and chicken thymocytes into four types on the basis of preparative separation techniques (cell electrophoresis, density gradient centrifugation and velocity sedimentation) and size distribution analysis: they distinguished two distinct sets of cortical lymphocytes, medullary small lymphocytes, and larger cells. They adduced evidence that one of the cortical populations belonged to the lineage of suppressor T-cells, and summarised a body of data showing that other functions were associated with medullary small lymphocytes: reactivity to the plant lectin phytohaemagglutinin (PHA), graft-versus-host activity, helper activity, and presence of presursors for the killer cells in cell-mediated cytotoxicity reactions. They also raised the possibility that the medullary thymocyte population might be heterogeneous. This possibility is strongly supported by the results of Elliott (1973, 1977*a*, *b*) who found that most of the thymic cells which could be coaxed into dividing *in vitro* by PHA or concanavalin-A belonged to a long-lived sedentary population with a very slow rate of turnover. Assuming that some medullary thymocytes do in fact contribute to the peripheral T-lymphocyte pool, Elliott's results entail the existence of at least two medullary subpopulations.

Extensive studies of thymocyte subpopulations were also carried out by Shortman, von Boehmer, Lipp & Hopper (1975) and Fathman, Small, Herzenberg & Weissman (1975), using a variety of cell separation and analytical techniques. Both groups distinguished at least three subclasses of small thymocyte: one consisted of cells with a high density of Thy1 antigen, the second (also high-Thy1) was particularly fragile and short-lived *in vitro*, and the third had a low density of surface Thy1 antigen and was relatively resistant to the lytic effects of cortisone. There is agreement that the first and third classes are the products of largely separate and parallel maturation streams, although technical problems make it difficult to decide with certainty whether or not cells feed in to the low-Thy1 population from high-Thy1 precursors. More recently, Droege (1976) has compared the profiles (based on size and electrophoretic mobility) of murine thymus and peripheral T-lymphocytes. He concludes that both cortical populations have counterparts, possibly descendants, in the peripheral pool. Both are present in lymph nodes and spleen, and include cells which persist for short and long times after thymectomy. It will no doubt soon emerge whether there are correlations with cell-surface phenotype (Ly antigens, Fc-receptors, etc.) and with particular functions.

Progenitors of thymocytes

The extent to which the progenitors of thymocytes are recruited from outside the organ is a question still not fully resolved. The experiments of Auerbach (1960) indicated that thymocytes developed from epithelial cells under the influence of mesenchyme. On the other hand, it was found in the embryos of birds and mice (Moore & Owen, 1967), in adult birds (Weber, 1975) and in the mouse thymus regenerating after irradiation (Micklem, Ford, Evans & Gray, 1966; Ford *et al.*, 1966; Micklem *et al.*, 1975*b*) that the thymus was populated by extrinsic progenitors. These were shown to come from the yolk sac of birds, the yolk sac and foetal liver of embryonic mice and the bone marrow of adult mice and birds. Similarly, Leuchars, Morgan, Davies & Wallis (1967) showed that the mitotic cells of thymus grafts were replaced essentially completely by immigrants from the host, at least so far as mitotic cells were concerned. Micklem *et al.* (1975*b*) calculated that in the shielded thymus of part-body irradiated mice the mitotic cell population was replaced from bone marrow progenitors at the rate of some 8–10% per day. Even in completely non-irradiated mice, some progeny of intravenously injected bone marrow cells appeared in the thymus (Micklem, Clarke, Evans & Ford, 1968). Evidence such as this has led to a general acceptance that fairly extensive replacement of thymocyte progenitors occurs from the bone marrow throughout life. An experiment of Nature has demonstrated chimaerism in the thymus of synchorial, non-identical twin marmosets (*Oedipomidas oedipus*) (Ford, 1966); this shows that in ontogeny, at least some of the mitotic thymocyte population comes from extrinsic progenitors even in ostensibly normal animals.

However, there is other evidence which does not fit comfortably into this picture. Ford (1966) studied the incidence of partner-derived mitotic cells in parabiotic mice. Although the proportions were variable between individuals, the mean proportion of partner-derived mitotic cells in the thymus fell far short of the value of 50% expected on the assumption that thymocyte progenitors immigrate continuously into the organ. Moreover, nearly all of the exchange took place during the first 8 weeks of parabiosis, a period during which the mice had to make abnormal physiological adjustments. Calculations based on Ford's figures suggested that, at most, less than 2% of the mitotic cells in the parabiotic thymus were being replaced from externally derived progenitors each day – a figure considerably below that calculated for part-body irradiated mice (Micklem *et al.*, 1975*b*). The difference could be explained on the hypothesis that thymic progenitor cells spend a very short time in the bloodstream and are unlikely to negotiate the capillary network between the two partners. There is no evidence for this, and it remains possible that the real meaning of these results is that experiments with irradiated thymuses and with thymus grafts exaggerate the extent to which the thymus is replenished from outside during adult life.

The above evidence is derived from cytological examination of cells in the metaphase of mitosis. Although it is likely to provide a fair picture of what is happening in the thymic cortex where most cells are in rapid cycle, it may not properly represent cells in the medulla where mitosis is less frequent. Indeed, the results of Elliott (1973, 1977*a*, *b*), referred to above, uncovered a thymocyte population which divides, if at all, too slowly to be detected in chromosome spreads.

Turpen, Volpe & Cohen (1973) grafted thymic rudiments between diploid and triploid embryonic frogs and found no evidence of immigration of host progenitor cells into the graft. In birds, a different conclusion was reached by Le Douarin & Jotereau (1975), using interspecific chimaeras formed between chick and quail embryos. Quail cell nuclei were distinguished from those of the chick by the presence of a large mass of heterochromatin associated with the nucleolus, making the latter strongly Feulgen-positive. These experiments clearly showed that, both in the bursa and in the thymus, progenitor cells migrated in from the blood at a critical stage in development, thus confirming the conclusions of Moore & Owen (1967). They showed, moreover, that progenitors which had lodged in the bursa could be induced experimentally to migrate into the thymus. The authors suggested that at the stage of initial localisation the progenitors were uncommitted as between the T and the B pathway of development, and that commitment occurred in the environment of the thymus or bursa respectively. It is indeed possible that some avian thymocytes normally originate from bursal progenitors, as Droege & Zucker (1975) suggest for one subpopulation.

These data can perhaps be summed up in a simplified way by saying that at least in birds and mammals the initial development of thymocytes depends on immigrant progenitors, and that immigration can continue to occur in adult life. The extent to which it actually does so may depend very much on physiological circumstances, and may normally be quite small. This is another way of saying that the progenitors of thymocytes in the thymus may have considerable powers of self-maintenance.

Fig. 1 represents an attempt at synthesis of the information available on thymocyte subpopulations, their origin and destiny.

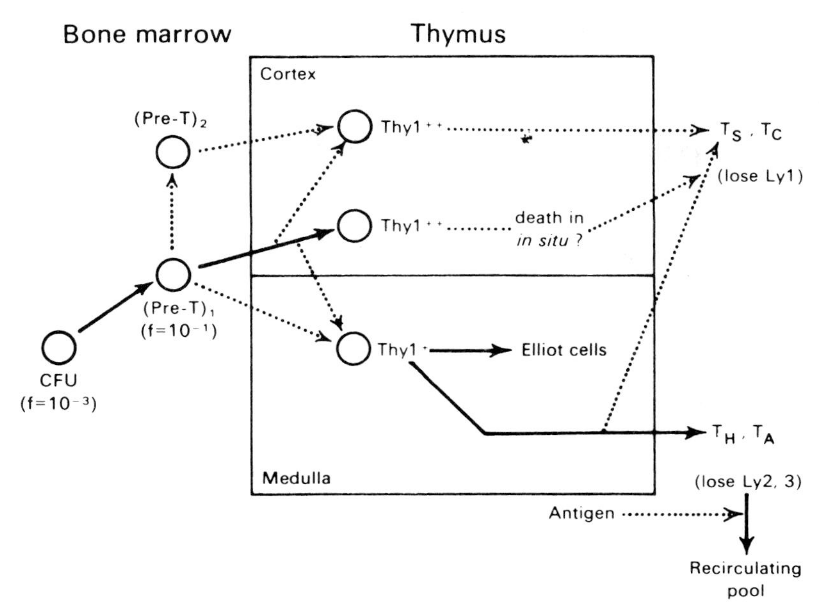

Fig. 1. Hypothetical scheme for the T-lymphocyte lineage, based on data discussed in the text. Arrows indicate maturational pathways, dotted lines being used for those pathways which are most debatable.

Identity of thymocyte progenitors

It is generally accepted that T-lymphocytes, like other differentiated cells in the blood, arise ultimately from pluripotent haematopoietic stem cells. The stages of differentiation between pluripotent stem cells and thymocytes have not, however, been clarified. The frequency of pluripotent stem cells in the thymus itself is extremely low (Micklem *et al.*, 1966; Micklem, Anderson, Ure & Jones, 1976). This is true also of the thymus of a lethally irradiated animal in the course of repopulation from injected bone marrow (Micklem *et al.*, 1966). There is evidence for the existence in bone marrow of a thymus-repopulating cell which is distinct both from pluripotent stem cells and from mature T-lymphocytes on the basis of average cell density (El-Arini & Osoba, 1973). Experiments in birds (Le Douarin & Jotereau, 1975) suggest that progenitor cells may lodge in the thymus while still in possession of the potential to develop into B-lymphocytes, the final commitment to T-cell differentiation presumably occurring within the thymus.

The thymus which is regenerating after irradiation appears to be repopulated by a small number of dividing cell clones (Micklem *et al.*, 1972, 1975*a*; Wallis, Leuchars, Chwalinski & Davies, 1975; Kadish & Basch, 1976). Wallis *et al.* (1975) also showed that the time of repopulation of irradiated thymus was related to the number of bone marrow cells injected, implying that the number of available thymocyte progenitors might be a limiting factor in the regeneration of the thymus. These data, however, only indicate that few progenitors enter the thymus *and found clones*; the possibility is not excluded that many progenitors enter the organ, but that the majority fail to become activated.

Other data suggest indirectly that the number of potential thymocyte progenitors in the bone marrow may, under appropriate experimental conditions, be quite large. For example, congenitally athymic nude mice were reported to have up to 18% of weakly Thy1-positive cells in their bone marrow and spleen (Roelants *et al.*, 1976). The TL antigen (a surface marker of cortical thymocytes) could be identified on some of these. Similar cells were rare in normal mice, but appeared after surgical removal of the thymus. The authors reported the existence of weakly Thy1-positive cells also in the thymus of 13-day, but not 15–16-day embryo mice. They suggested that these cells represented 'pre-thymocytes' and that their numbers in the bone marrow were under homeostatic feedback control by the thymus, a functional thymus suppressing the development of the cells, or at least the expression of the Thy1 and TL markers on their surface.

The acquirement of thymocyte surface characteristics (including Thy1 and TL antigens) by bone marrow cells treated with 'thymopoietin', a polypeptide isolated from calf thymus (Goldstein, 1974) has been interpreted as evidence that these cells are precursors of thymocytes (Komuro & Boyse, 1973*a*, *b*). Up to 10% of bone marrow cells were inducible – a figure whose resemblance to the figure for weakly Thy1-positive cells identified by Roelants, Mayor, Hägg & Loor (1976) may or may not be a matter of chance. In particular, Komuro, Goldstein & Boyse (1975) found that thymopoietin-induced bone marrow cells, subsequently treated with anti-Thy1 serum and complement, were depleted of cells capable of repopulating irradiated thymus. Other agents not specifically associated with the thymus can also induce T-cell surface characters, by interacting with β-adrenergic receptors (Scheid, Goldstein, Hämmerling & Boyse, 1975). These appear to have the capacity to activate cells in general, through the intermediary cyclic AMP, while thymopoietin is claimed to activate only thymocyte precursors. The authors consider that the precursors are being induced to embark on a differentiative programme to which they are already committed. The possibility that they are in reality post-thymic cells which are merely regaining lost T-markers is unlikely in view of their presence in foetal liver and in the bone marrow of athymic nude mice (Scheid *et al.*, 1975).

If, as seems possible, the thymopoietin-inducible cells are the same as those identified by Roelants *et al.* (1976), the control of the differentiative process appears to be complicated. Absence or removal of the thymus increases the number of weakly Thy1-positive, TL-positive cells in the bone marrow, while provision of thymopoietin in high concentrations increases the concentration of these antigens, perhaps mimicking what happens normally within the thymus. However, it remains to be shown conclusively that the bone marrow cells in which Thy1 and TL are present or inducible are the normal progenitors of thymocytes, or that they are committed to a T pathway of differentiation. Their relationship to pluripotent stem cells and their potential, if any, for self-renewal are also obscure.

It has to be emphasised that experiments in which cells are exposed *in vitro* to high concentrations of purified substances derived from the thymus, or indeed from any other source, are likely to be highly unphysiological. The isolation from normal serum of a nonapeptide, the presence of which is dependent on the thymus (Bach, Dardenne, Pleau & Bach, 1975; Bach, Dardenne, Pleau & Rosa, 1977), is of considerable interest in this context. In physiological concentrations, this substance influences the later stages of T-cell differentiation as monitored by various assays, but, on present evidence, probably not the differentiation of thymocyte progenitors.

Recently Silverstone, Cantor, Goldstein & Baltimore (1976) have reported that another apparent differentiation marker of cortical thymocytes can be induced in bone marrow by exposure to thymopoietin, namely terminal deoxynucleotidyl transferase activity. This interesting enzyme appears to be absent from mature T-cells and from the whole B-cell lineage. Its property of assembling DNA on a primer without any template invites speculation on its role in the generation of antibody diversity (Baltimore *et al.*, 1976).

The recent emergence of subpopulations of thymocytes and T-lymphocytes, discussed earlier, greatly complicates the identification of thymocyte progenitor cells. It may even turn out finally that everyone is right – that different subpopulations arise, respectively, from thymic epithelium, from self-maintaining progenitors derived originally from the yolk sac, and from continuously immigrating progenitors from the bone marrow.

However, as mentioned above, the available evidence suggests that the bulk of cells present in the thymus at any one time is the product of a small number of clones. This makes it improbable that significant diversification of receptors for antigen occurs in prethymocytes; diversification seems likely to take place within the thymus itself. Thus, even if the molecular basis of diversity is similar in T-cells and B-cells – i.e. the expression of immunoglobulin *V*-genes – the generation of diversity within the two populations probably proceeds independently.

GENERATION OF B-LYMPHOCYTES

In considering the production and maturation of B-lymphocytes, several distinguishable differentiative processes have to be taken into account. These include commitment to a lymphocytic (specifically, B-lymphocytic) pathway of differentiation, expansion of the population, the initiation of immunoglobulin synthesis, the placing of immunoglobulin molecules on the cell surface to function as antigen receptors, the acquirement of other receptors such as those for C3 and for the Fc portion of immunoglobulin, and finally the events which follow exposure to antigen and which culminate in the production of antibody-secreting plasmacytes. During the course of this sequence the lymphocytes diversify to produce immunoglobulins with a wide range of antigen-binding specificities, and also to produce antibodies of various heavy-chain classes (IgM, IgG, etc.).

The nature of the generator of diversity (GOD) for antibody molecules has long been a matter of controversy. The problem itself has diversified far beyond the original confrontation between germliners and somatic mutationists, and there is still little sign of a consensus. 'Critical' experiments have a way of not turning out to be critical after all, and proponents of each viewpoint can point to reasons why the others must be wrong. The field is well reviewed in a recent collection of papers edited by Cunningham (1976).

The avian bursa of Fabricius

The generation of B-lymphocytes is most conveniently studied in birds which, unlike mammals, possess a distinct organ, the bursa of Fabricius, in which most of the antigen-independent stages of B-cell maturation take place (Lydyard, Grossi & Cooper, 1976). The prevailing view is that, like the thymus, the embryonic bursa is populated by stem cells of yolk sac origin which enter the organ in the bloodstream (Moore & Owen, 1966; Owen, 1972). Le Douarin, Houssaint, Jotereau & Belo (1975) showed by means of the embryonic chicken–quail grafting system referred to earlier, that seeding began on the 7th to 8th day of embryonic life and was complete by about the 11th to 14th day. Individual follicles of grafts made between these times were often populated by donor or host cells exclusively, although many follicles containing a mixture of donor and host cells were also seen. Lydyard *et al.* (1976) interpreted these data in binomial terms as evidence for the monoclonality or oligoclonality of individual follicles. However, this conclusion is invalid if, as Le Douarin *et al.* (1975) state, initial seeding of the follicles is asynchronous. Cells capable of binding specific antigens arose singly or in small foci within scattered follicles (Lydyard *et al.*, 1976); these were detectable from the 16th day onwards. Cells carrying surface IgM were present from the 12th day and increased with a doubling time of about 10 hours. Small numbers of Ig-bearing cells appeared in the spleen on the 18th day and increased rapidly in number after hatching, a process which appeared to be independent of antigenic stimulation. Bursectomy at the time of hatching halted the appearance of IgM-bearing and antigen-binding cells in the spleen, and led to a long-standing, and probably permanent, deficit of these cells.

H-chain classes of B-lymphocyte surface Ig

These studies illuminate the bursa's organisation and function as the unique environment for the differentiation of B-lymphocytes in birds. They still leave open to argument the questions (*a*) by what stage the full repertoire of antibody specificities is developed, (*b*) at what stage the expression of non-μ (i.e. non-IgM) heavy chains occurs, and (*c*) whether either or both of these developments depends on antigenic stimulation. Lawton, Kincade & Cooper (1975) have data to show that the expression of μ, γ(IgG) and α(IgA) chains occurs in that sequence in the avian bursa, independently of any antigen. They also adduce some evidence for a sequential, antigen-independent development in the mouse. However, many other studies indicate that the receptors for antigens on the surface of 'virgin' B-cells are a form of IgM, with or without IgD. The germ-free, colostrum-deprived piglet provides an interesting model. Its B-cells possess surface IgM: γ-chains can be detected only after immunisation, and in conventionally reared piglets (Jarošková *et al.*, 1973). IgG receptors have been found to be associated with memory cells capable of developing into IgG-secreting cells in the rat (Mason, 1976) and mouse (Okumura *et al.*, 1976*b*).

Bursa analogues in mammals

The mammalian species in which the generation of B-lymphocytes has been studied in most detail is the mouse. In common with other mammals, the mouse has no obvious analogue of the avian bursa. No unique site for the differentiation of B-lymphocytes exists; rather, the necessary microenvironment (the properties of which have not been established) can apparently be provided by any haematopoietic tissue. In the foetal mouse the liver is the main site (Owen, Cooper & Raff, 1974), and in the adult the bone marrow (Osmond, 1975; Melchers, von Boehmer & Phillips, 1975; Miller & Phillips, 1975; Yoffey, 1975; Rosse, 1976). In both foetus and adult the spleen can function as an additional site. The fact that B-lymphocyte production is not separated anatomically from the production of erythrocytes, granulocytes and macrophages, complicates its study in mammals.

Development of B-lymphocytes from haemopoietic stem cells

The early work of Barnes, Ford, Gray & Loutit (1959) showed that a large proportion of the dividing cells in the bone marrow, spleen, thymus and lymph nodes of an individual mouse can belong (under special experimental conditions) to a single clone. Since then, numerous studies have suggested that B-lymphocytes are derived from pluripotent stem cells, or (to be more precise) from spleen-colony-forming cells or 'units' (CFU-S) (Till & McCulloch, 1961; Wu, Till, Siminovitch & McCulloch, 1968). These experiments have generally relied on karyotypic analysis to show that a unique pattern of chromosome rearrangements, induced originally by irradiation of an ancestral CFU, is present in lymphocytes as well as in the descendant CFUs. The difficulty has always been to identify the lymphocytes with certainty as functional B-cells. The field has been well reviewed by Miller & Phillips (1975). As they point out, one implication of these studies, in which the lymphoid system is largely repopulated by members of one, or very few, clones and can respond to various antigens (Trentin *et al.*, 1967; Yung, Wyn-Evans & Diener, 1973), is that the clone is capable of generating antibody diversity. The heterogeneity of antibodies which these animals can produce has not, however, been adequately examined.

After the pluripotent stem cell stage, the future B-lymphocyte enters a period of obscurity until it becomes identifiable as a cell synthesising immunoglobulin – more precisely, 7–8 S subunits of IgM (Melchers *et al.*, 1975). In the developing mouse embryo, cells containing intracellular IgM can be found in the liver as early as the 12th day. At that stage surface Ig is not detectable, but it appears within about 4 days (Nossal & Pike, 1973; Owen, Raff & Cooper, 1975). Some cells show both intracytoplasmic and surface Ig, suggesting a maturation sequence (Raff, Megson, Owen & Cooper, 1976). In-vitro studies reviewed by Owen *et al.* (1977) indicate that the murine foetal liver can provide an adequate milieu for the differentiation of B-lymphocytes from intracellular-Ig-containing cells. It seems very possible that these latter cells correspond functionally to the cells found by Melchers *et al.* (1975) to be rapidly synthesising and turning over 7–8 S IgM in adult bone marrow. These cells have some IgM accessible to surface radioiodination, but might be negative to immunofluorescence tests if the molecules were being rapidly degraded or shed (Melchers *et al.*, 1975).

Since a mouse's B-lymphocytes do not have lifespans approaching that of their owner (Strober, 1975; Elson, Jablonska & Taylor, 1976), their numbers must be constantly replenished. There is no reason to doubt that the process is essentially similar in the foetus and the adult. The genesis of B-lymphocytes in the bone marrow of mice has been extensively studied by Lafleur and his colleagues (1972*a*, *b*; 1973), Ryser & Vassalli (1974), Osmond and his colleagues (1974*a*, *b*; Stocker, Osmond & Nossal, 1974) and Rosse (1976). B-lymphocytes, as identified by surface Ig and other markers, were found to arise from non-proliferating precursors resembling small lymphocytes, but without detectable Ig. These were in turn derived from proliferating precursors, and 1½–2 days elapsed before the appearance of detectable Ig on the surface. There is, however, some doubt as to the stage at which Ig is first formed – a reflection, probably, of different techniques. Cells whose Ig is turning over rapidly, as identified in mouse bone marrow by Melchers *et al.* (1975), would not necessarily be detected as surface-Ig-positive by immunofluorescent or isotopic labelling. Lafleur *et al.* (1973) identified a 'pre-B' compartment consisting of large Ig-positive cells in bone marrow, their frequency being about 1 % of the total nucleated cells – some 10 times higher than that of CFU (pluripotent stem cells). They are distinguished from the latter also by their somewhat different sedimentation profile and their sensitivity of anti-μ serum. It is very likely that they are derived from CFU, and that they belong to the morphological category of transitional cells (Yoffey, 1975; Rosse, 1976), but their intermediate history – particularly the stage(s) at which they become committed to lymphocytic and specifically B-lymphocytic maturation, and become restricted in antibody specificity – remains completely unknown.

Self-maintenance of B-cell progenitors?

In birds, a self-maintaining population of committed B-progenitors exists in the bursa after hatching. It is not clear whether any such population exists in mammalian bone marrow or whether, as is more generally believed, only pluripotent stem cells have the capacity for self-renewal. There is evidence that in the bone marrow both of chromosomally distinguishable parabionts (Ford, 1966) and of part-body irradiated mice injected with chromosome-marked bone marrow (Micklem *et al.*, 1975*b*), there is a subpopulation of non-local origin. This has at least one characteristic of lymphoid cells – a tendency to home to the lymph nodes of recipients after intravenous injection. The data can be explained on the grounds either that the progenitors of this subpopulation (unlike other marrow cells) migrate to a considerable extent between bone marrow sites, or that the subpopulation has a capacity for self-renewal. More work is needed both to characterise this subpopulation and to establish its kinetics, but present information is consistent with the idea that a lympho-committed, indefinitely self-renewing cell line exists in mammalian bone marrow.

Maturation

Ig-positive cells are exported in large numbers (about 10^8 per day in young adult mice) from the bone marrow to the spleen and lymph nodes (Brahim & Osmond, 1970, 1976; Osmond, 1975). It may be that full maturation requires exposure to a splenic microenvironment. Playfair & Purves (1971) found that bone marrow cells, which they termed B_1, responded less to T-cell help than did splenic B_2 cells. C3-receptors were found to be absent from Ig^+ cells in foetal liver and spleen (Owen *et al.*, 1975) and almost absent in adult bone marrow, but present in adult spleen (Ryser & Vassalli, 1974); and the presence of these has since turned out to correlate with the ability to accept help from T-cells (Arnaiz-Villena *et al.*, 1975; Lewis *et al.*, 1976). Stocker *et al.* (1974) found that the adoptive immune response of bone marrow cells to a 'thymus-independent' antigen, initially poor, improved when a longer interval was left between injection of the cells and challenge of the irradiated recipients with antigen; the response of splenic cells fell off under the same conditions. Such data make it very likely that most bone marrow B-cells are functionally relatively immature, full responsiveness to T-help normally being acquired after emigration. However, it is still possible that T-responsive and T-unresponsive B-cells are products of separate differentiation pathways rather than stages of a single one.

HOMEOSTASIS OF LYMPHOCYTE NUMBERS

Antigen-independent regulation

Little is known about the factors which regulate the overall numbers of lymphocytes in the body. No broad regulatory mechanism analogous to that operating on erythropoiesis has been identified with certainty, and it may well be that no such mechanism exists. Wallis, Leuchars, Collavo & Davies (1976) studied the effects of injecting large numbers of mature lymphocytes repeatedly into mice. There appeared to be no inherent limit to the extent to which these cells could be incorporated into the recirculating lymphocyte pool of the recipients. The mice became 'hyperlymphoid', with more recirculating lymphocytes than their untreated controls. This state persisted for many months, during which time the proportion of donor PHA-responsive cells, identified by chromosome markers, declined slowly. The rate of decline seems to be consistent with a normal rate of turnover and replacement, but further study of this is needed. In the absence of the thymus, no decline in the proportion of donor cells was observed (V. J. Wallis, personal communication). These results, and earlier data obtained in rats (Vaughan & McGregor, 1972) suggest that an animal does not reduce its complement of long-lived lymphocytes if abnormally high numbers have been acquired by experimental intervention. Since recruitment of new T-cells appears to continue, there seems to be no effective negative feedback, such as limits erythrocyte production in polycythaemic animals. However, there are other mechanisms which limit the recruitment of newly formed T-lymphocytes into the recirculating pool, since mice carrying multiple thymus grafts, in which cell proliferation was active, did not become hyperlymphoid (Matsuyama, Wiadrowski & Metcalf, 1966). In both the bone marrow (Osmond, 1975) and the thymus (Cantor & Weissman, 1976), lymphocytopoiesis appears to be based on overproduction, followed by elimination of most of the cells formed. The homeostatic mechanisms implicit in this are evidently effective, but their nature remains unknown.

Antigen-dependent regulation

A little more is known about mechanisms which regulate lymphocyte numbers after immunisation and affect small subpopulations of lymphocytes carrying antigen-specific receptors. For example, recruitment of virgin B-lymphocytes into the recirculating pool of long-lived cells seems to depend at least partly on contact with their specific antigen (Strober, 1975). Antigen-specific clones of lymphocytes proliferate during an immune response; the extent to which they do so, and hence the size of the immune response, are under quite tight regulation, which can be mediated by such factors as the availability of antigen and the concentration of specific antibody (Bystryn, Graf & Uhr, 1970; Tew, Self, Harold & Stavitsky, 1973), and the activity of regulatory T-lymphocytes (Gershon, 1975; Miller, 1975; Waksman & Namba, 1976).

Allotype suppression and Ig class regulation

The phenomenon of chronic allotype suppression (Herzenberg, Jacobson, Herzenberg & Riblet, 1971) suggests the existence of another homeostatic mechanism. Briefly, if certain mice, heterozygous at a locus determining alloantigens on immunoglobulin-G molecules, are exposed neonatally to antibodies directed against one of the two alloantigens, production of molecules bearing that determinant is specifically suppressed. This is due to specific recognition by T_S cells. Recently Herzenberg *et al.* (1976) have also identified, in one strain of mice, allotype-specific T_H cells, and reported that the T_S cells act not directly on the B-cells which express the appropriate allotype, but by removing the specific T_H cells. These intriguing results raise several questions. (1) Since allotype-specific T_H and T_S cells were discovered by the use of artificial experimental procedures which are unlikely to be reproduced in nature, what is their biological significance? (2) How do the various cells involved recognise each other? Since allotypes are determinants situated on the constant regions of immunoglobulin heavy chains, their discrimination involves perception of this part of the immunoglobulin molecule. It may be, as recently discussed by Raff (1977), that allotype recognition is just an artificial by-product of immunoglobulin class recognition. There is little direct evidence for regulation by T-cells of the synthesis of individual immunoglobulin classes. However, T-cell help is needed for the production of IgG and IgA antibodies (Taylor & Wortis, 1968; Torrigiani, 1972; Pritchard, Riddaway & Micklem, 1973), and the existence of T_H cells specific for IgE-producing B-cells has been suggested (Kishimoto & Ishizaka, 1973). It seems a fair guess that the synthesis of antibodies in individual Ig classes will turn out to be under the homeostatic control of T-lymphocytes. Meanwhile, it will be important to see whether allotype-specific help, as well as suppression, from T-cells is demonstrable in other strains of mouse.

The Ly phenotypic identity (discussed above) between T_H and T_A cells, and between T_S and T_C cells, suggests another possible form of homeostatic interaction. Since T_A cells amplify the activity of T_C cells, it is very possible that T_H cells similarly influence T_S cells. As already noted, there is evidence that T_S suppress T_H cells. A T_H response to antigen could therefore be regulated by negative feedback via T_S cells.

Idiotype networks

Some years ago Jerne (1974) put forward a 'network' theory of the immune system, according to which the system consisted of a lattice of interacting lymphocytes recognising, on the one hand a vast variety of foreign antigenic determinants (epitopes), and on the other hand the antigenic configuration (idiotype) of the antigen receptors themselves. Since T-cells possess a range of idiotypic specificities which overlaps with, and may be identical to, that of B-cells (Binz & Wigzell, 1975; Eichmann & Rajewsky, 1975; Black *et al.*, 1976), the network must involve both of these lymphocyte classes. Macrophages, and the dendritic reticular cells of lymphoid tissue, must also be considered as components, since they are able to take up antigen–antibody complexes onto their surface and mediate some interactions between lymphocytes (Feldmann & Nossal, 1972).

Some direct evidence for mutual surveillance of the kind suggested by Jerne (1974) has since emerged. The experiments of Cosenza (1976) and Julius, Augustin & Cosenza (1977), for example, have made use of the fact that the response to the antigen phosphorylcholene involves an unusually small number of lymphocyte clones; the bulk of the antibody produced is homogeneous and possesses the same idiotype. These workers have been able to demonstrate the emergence of anti-idiotype antibodies during a response to phosphorylcholene – that is to say, antibodies directed at antigens associated with the specific anti-phosphorylcholene combining site. They have also obtained evidence for the existence of T-lymphocytes capable of recognising idiotypes on B-lymphocytes. The present status of the network theory and of the evidence related to it has recently been discussed by Raff (1977). It seems likely (though it may be difficult to prove) that the development of anti-idiotype reactivity during immune responses is a general phenomenon, existing even when the response involves multiple responding clones. In principle, the anti-idiotype responding cells may in turn evoke responses to their own idiotypic determinants, and so on. Such a process might continue indefinitely, but would presumably be limited in practice by the decreasing size, and hence decreasing immunogenicity, of each successive response.

If Jerne's ideas continue to stand up to experimental test, the response of the immune system to an antigen may be, in Raff's words, 'best understood in terms of a reverberating perturbation of the network, rather than as a response of individual antigen-sensitive lymphocytes'.

REFERENCES

ABNEY, E. R., KEELER, K. D., PARKHOUSE, R. M. E. & WILLCOX, H. N. A. (1976). Immunoglobulin M receptors on memory cells of immunoglobulin G antibody forming cell clones. *European Journal of Immunology*, **6**, 443–50.

ARNAIZ-VILLENA, A., PLAYFAIR, J. H. L. & ROITT, I. M. (1975). C3 receptor. A marker of a thymus-dependent B-cell subpopulation. *Clinical and experimental Immunology*, **20**, 375–8.

AUERBACH, R. (1960). Morphogenetic interactions in the development of the thymus gland. *Developmental Biology*, **2**, 271–84.

BACH, J. F., DARDENNE, M., PLEAU, J.-M. & BACH, M.-A. (1975). Isolation, biochemical characteristics, and biological activity of a circulating thymic hormone in the mouse and in the human. *Annals of the New York Academy of Sciences*, **249**, 186–210.

BACH, J.-F., DARDENNE, M., PLEAU, J.-M. & ROSA, J. (1977). Biochemical characterisation of a serum thymic factor. *Nature, London*, **266**, 55–6.

BALTIMORE, D., SILVERSTONE, A. E., KUNG, P. C., HARRISON, T. A. & MCCAFFREY, R. (1976). What cells contain terminal deoxynucleotidyl transferase? In *The generation of antibody diversity: a new look*, ed. A. Cunningham, pp. 21–30. New York & London: Academic Press.

BARNES, D. W. H., FORD, C. E., GRAY, S. M. & LOUTIT, J. F. (1959). Spontaneous and induced changes in cell populations in heavily irradiated mice. *Progress in nuclear Energy*, Series VI, **2**, 1–10.

BINZ, H., KIMURA, A. & WIGZELL, H. (1975). Idiotype-positive T lymphocytes. *Scandinavian Journal of Immunology*, **4**, 413–20.

BINZ, H. & WIGZELL, H. (1975). Shared idiotypic determinants on B and T lymphocytes reactive against the same antigenic determinants. *Scandinavian Journal of Immunology*, **4**, 591–600

BLACK, S. J., HÄMMERLING, G. J., BEREK, C., RAJEWSKY, K. & EICHMANN, K. (1976). Idiotypic analysis of lymphocytes *in vitro*. I. Specificity and heterogeneity of B and T lymphocytes reactive with anti-idiotypic antibody. *Journal of experimental Medicine*, **143**, 846–60.

BRAHIM, F. & OSMOND, D. G. (1970). Migration of bone marrow lymphocytes demonstrated by selective bone marrow labelling with thymidine-H3. *Anatomical Record*, **168**, 139–60.

BRAHIM, F. & OSMOND, D. G. (1976). Migration of newly formed small lymphocytes from bone marrow to lymph nodes during primary immune responses. *Clinical and experimental Immunology*, **24**, 515–26.

BYSTRYN J. C., GRAF, M. W. & UHR, J. W. (1970). Regulation of antibody formation by serum antibody. II. Removal of specific antibody by means of exchange transfusion. *Journal of experimental Medicine*, **132**, 1279–87.

CANTOR, H. & BOYSE, E. A. (1975*a*). Functional subclasses of T lymphocytes bearing different Ly antigens. I. The generation of functionally distinct T-cell subclasses is a differentiative process independent of antigen. *Journal of experimental Medicine*, **141**, 1376–89.

CANTOR, H. & BOYSE, E. A. (1975*b*). Functional subclasses of T lymphocytes

bearing different Ly antigens. II. Cooperation between subclasses of Ly+ cells in the generation of killer activity. *Journal of experimental Medicine*, **141**, 1390–9.
CANTOR, H., SHEN, F. W. & BOYSE, E. A. (1976). Separation of helper T cells from suppressor T cells expressing different Ly components. II. Activation by antigen: after immunization, antigen-specific suppressor and helper activities are mediated by distinct T-cell subclasses. *Journal of experimental Medicine*, **143**, 1391–1401.
CANTOR, H. & WEISSMAN, I. (1976). Development and function of subpopulations of thymocytes and T lymphocytes. *Progress in Allergy*, **20**, 1–64.
CEROTTINI, J. C. & BRUNNER, K. T. (1974). Cell-mediated cytotoxicity, allograft rejection and tumor immunity. *Advances in Immunology*, **18**, 67–132.
CLAMAN, H. N. & CHAPERON, E. A. (1969). Immunologic complementation between thymus and marrow cells – a model for the two cell theory of immunocompetence. *Transplantation Reviews*, **1**, 92–113.
COHEN, J. J. & FAIRCHILD, S. S. (1976). T-lymphocyte precursors. I. Synergy between precursor and mature T lymphocytes in the response to Concanavalin A. *Journal of experimental Medicine*, **144**, 456–66.
COOPER, M. D., PETERSON, R. D. A., SOUTH, M. A. & GOOD, R. A. (1966). The function of the thymus system and the bursa system in the chicken. *Journal of experimental Medicine*, **123**, 75–102.
COSENZA, H. (1976). Detection of anti-idiotype reactive cells in the response to phosphorylcholine. *European Journal of Immunology*, **6**, 114–16.
CUNNINGHAM, A. L. (ed.) (1976). *The generation of antibody diversity: a new look.* New York & London: Academic Press.
DICKLER, H. B., ARBEIT, R. D., HENKART, P. A. & SACHS, D. H. (1976). Association between Ia antigens and the Fc receptors of certain T lymphocytes. *Journal of experimental Medicine*, **144**, 282–7.
DROEGE, W. (1976). 'Early T cells' and 'late T cells'; suggestive evidence for two T cell lineages with separate developmental pathways. *European Journal of Immunology*, **6**, 763–8.
DROEGE, W. & ZUCKER, R. (1975). Lymphocyte subpopulations in the thymus. *Transplantation Reviews*, **25**, 3–25.
DU PASQUIER, L., WEISS, N. & LOOR, F. (1972). Direct evidence for immunoglobulins on the surface of thymus lymphocytes of amphibian larvae. *European Journal of Immunology*, **2**, 366–70.
EICHMANN, K. & RAJEWSKY, K. (1975). Induction of T and B cell immunity by anti-idiotypic antibody. *European Journal of Immunology*, **5**, 661–6.
EL-ARINI, M. O. & OSOBA, D. (1973). Differentiation of thymus-derived cells from precursors in mouse bone marrow. *Journal of experimental Medicine*, **137**, 821–37.
ELLIOTT, E. V. (1973). A persistent lymphoid cell population in the thymus. *Nature New Biology*, **242**, 150–2.
ELLIOTT, E. V. (1977*a*). The persistent PHA-responsive population in the mouse thymus. I. Characterization of the population. *Immunology*, **32**, 383–94.
ELLIOTT, E. V. (1977*b*). The persistent PHA-responsive population in the mouse thymus. II. Recirculatory characteristics and immunological properties. *Immunology*, **32**, 395–404.
ELLIS, A. E. & PARKHOUSE, R. M. E. (1975). Surface immunoglobulins on the lymphocytes of the skate *Raja naevus*. *European Journal of Immunology*, **5**, 726–8.

Elson, C. J., Jablonska, K. F. & Taylor, R. B. (1976). Functional half-title of virgin and primed B lymphocytes. *European Journal of Immunology*, **6**, 634–8.

Emmrich, F., Richter, R. F. & Ambrosius, H. (1975). Immunoglobulin determinants on the surface of lymphoid cells of carps. *European Journal of Immunology*, **5**, 76–8.

Fathman, C. G., Small, M., Herzenberg, L. A. & Weissman, I. L. (1975). Thymus cell maturation. II. Differentiation of three 'mature' subclasses *in vivo*. *Cellular Immunology*, **15**, 109–28.

Feldmann, M., Beverley, P. C. L., Dunkley, M. & Kontiainen, S. (1975). Different Ly antigen phenotypes of in vitro induced helper and suppressor cells. *Nature, London*, **258**, 614–16.

Feldmann, M. & Nossal, G. J. V. (1972). Tolerance, enhancement and the regulation of interactions between T cells, B cells and macrophages. *Transplantation Reviews*, **13**, 3–34.

Ford, C. E. (1966). Traffic of lymphoid cells in the body. In *Thymus: experimental and clinical studies*, ed. G. E. W. Wolstenholme & R. Porter, pp. 131–52. *Ciba Foundation Symposium*. London: Churchill.

Ford, C. E., Micklem, H. S., Evans, E. P., Gray, J. G. & Ogden, D. A. (1966). The inflow of bone marrow cells to the thymus. *Annals of the New York Academy of Sciences*, **129**, 283–97.

Gershon, R. K. (1975). A disquisition on suppressor T cells. *Transplantation Reviews*, **26**, 170–85.

Glick, B., Chang, T. S. & Jaap, R. G. (1956). The bursa of Fabricius and antibody production. *Poultry Science*, **35**, 224–45.

Goldstein, G. (1974). Isolation of bovine thymin: a polypeptide hormone of the thymus. *Nature, London*, **247**, 11–14.

Hämmerling, G. J. (1976). Tissue distribution of Ia antigens and their expression on lymphocyte subpopulations. *Transplantation Reviews*, **30**, 64–82.

Herzenberg, L. A., Jacobson, E. B., Herzenberg, L. A. & Riblet, R. J. (1971). Chronic allotype suppression in mice: an active regulatory process. *Annals of the New York Academy of Sciences*, **190**, 212–18.

Herzenberg, L. A., Okumura, K., Cantor, H., Sato, V. L., Shen, F.-W., Boyse, E. A. & Herzenberg, L. A. (1976). T-cell regulation of antibody responses: demonstration of allotype-specific helper T cells and their specific removal by suppressor T cells. *Journal of experimental Medicine*, **144**, 330–44.

Huber, B., Cantor, H., Shen, F.-W. & Boyse, E. A. (1976). Independent differentiative pathways of Ly1 and Ly23 subclasses of T cells. Experimental production of mice deprived of selected T-cell subclasses. *Journal of experimental Medicine*, **144**, 1128–33.

Jarošková, L., Trebichavský, I., Řiha, I., Kovářů, F. & Holub, M. (1973). Immunoglobulin determinants on lymphocytes in germ-free piglets. *European Journal of Immunology*, **3**, 818–24.

Jerne, N. K. (1974). Towards a network theory of the immune response. *Annales d'Immunologie, Paris*, **125C**, 373–89.

Julius, M. H., Augustin, A. A. & Cosenza, H. (1977). Recognition of a naturally occurring idiotype by autologous T cells. *Nature, London*, **265**, 251–3.

Kadish, J. L. & Basch, R. S. (1976). Hematopoietic thymocyte precursors. I.

Assay and kinetics of the appearance of progeny. *Journal of experimental Medicine*, **143**, 1082–99.

Katz, D. H. & Benacerraf, B. (eds.). (1976). *The role of products of the histocompatibility gene complex in immune responses.* New York & London: Academic Press.

Kishimoto, T. & Ishizaka, K. (1973). Regulation of antibody response *in vitro*. VI. Carrier-specific helper cells for IgG and IgE antibody responses. *Journal of Immunology*, **111**, 720–32.

Komuro, K. & Boyse, E. A. (1973*a*). Induction of T lymphocytes from precursor cells *in vitro* by a product of the thymus. *Journal of experimental Medicine*, **183**, 479–82.

Komuro, K. & Boyse, E. A. (1973*b*). *In vitro* demonstration of thymic hormone in the mouse by conversion of precursor cells into lymphocytes. *Lancet*, **i**, 740–3.

Komuro, K., Goldstein, G. & Boyse, E. A. (1975). Thymus-repopulating capacity of cells that can be induced to differentiate to T cells *in vitro*. *Journal of Immunology*, **115**, 195–8.

Lafleur, L., Miller, R. G. & Phillips, R. A. (1972*a*). A quantitative assay for the progenitors of bone marrow-associated lymphocytes. *Journal of experimental Medicine*, **135**, 1363–74.

Lafleur, L., Miller, R. G. & Phillips, R. A. (1973). Restriction of specificity in the precursors of bone marrow-associated lymphocytes. *Journal of experimental Medicine*, **137**, 954–66.

Lafleur, L., Underdown, B. J., Miller, R. G. & Phillips, R. A. (1972*b*). Differentiation of lymphocytes: characterization of early precursors of B-lymphocytes. *Series Haematologica*, **5**, 50–63.

Laissue, J. A., Chanana, A. D., Cottier, H., Cronkite, E. P. & Joel, D. D. (1976). The fate of thymic radioactivity after local labeling with ^{125}iododeoxyuridine. *Blood*, **47**, 21–30.

Lawton, A. R., Kincade, P. W. & Cooper, M. D. (1975). Sequential expression of germ line genes in development of immunoglobulin class diversity. *Federation Proceedings*, **34**, 33–9.

Le Douarin, N. M., Houssaint, E., Jotereau, J. & Belo, M. (1975). Origin of hemopoietic stem cells in embryonic bursa of Fabricius and bone marrow studied through interspecific chimeras. *Proceedings of the National Academy of Sciences, USA*, **72**, 2701–5.

Le Douarin, N. M. & Jotereau, F. V. (1975). Tracing of cells of the avian thymus through embryonic life in interspecific chimeras. *Journal of experimental Medicine*, **142**, 17–40.

Leuchars, E., Morgan, A., Davies, A. J. S. & Wallis, V. J. (1967). Thymus grafts in thymectomised and normal mice. *Nature, London*, **214**, 801–2.

Lewis, G. K., Ranken, R., Nitecki, D. E. & Goodman, J. W. (1976). Murine B-cell subpopulations responsive to T-dependent and T-independent antigens. *Journal of experimental Medicine*. **144**, 382–97.

Lydyard, P. M., Grossi, C. E. & Cooper, M. D. (1976). Ontogeny of B cells in the chicken. I. Sequential development of clonal diversity in the bursa. *Journal of experimental Medicine*, **144**, 79–97.

McKenzie, I. F. C. & Parish, C. R. (1976). Secretion of Ia antigens by a subpopulation of T cells which are Ly-1$^+$, Ly-2$^-$, and Ia$^-$. *Journal of experimental Medicine*, **144**, 847–51.

Mason, D. W. (1976). The class of surface immunoglobulin on cells carrying IgG memory in rat thoracic duct lymph: the size of the subpopulation mediating IgG memory. *Journal of experimental Medicine*, **143**, 1122–30.

Matsuyama, M., Wiadrowski, M. N. & Metcalf, D. (1966). Autoradiographic analysis of lymphopoiesis and lymphocyte migration in mice bearing multiple thymus grafts. *Journal of experimental Medicine*, **123**, 559–76.

Melchers, F., Boehmer, H. von & Phillips, R. A. (1975). B-lymphocyte subpopulations in the mouse. *Transplantation Reviews*, **25**, 26–58.

Micklem, H. S., Anderson, N., Ure, J. & Jones, H. P. (1976). Long-term immunoglobulin G production by transplanted thymus cells. *European Journal of Immunology*, **6**, 425–9.

Micklem, H. S., Clarke, C. M., Evans, E. P. & Ford, C. E. (1968). Fate of chromosome marked mouse bone marrow cells transfused into normal syngeneic recipients. *Transplantation*, **6**, 229–302.

Micklem, H. S., Ford, C. E., Evans, E. P. & Gray, J. G. (1966). Interrelationships of myeloid and lymphoid cells: studies with chromosome-marked cells transfused into lethally irradiated mice. *Proceedings of the Royal Society of London*, **165B**, 78–102.

Micklem, H. S., Ford, C. E., Evans, E. P. & Ogden, D. A. (1975*a*). Compartments and cell flows within the mouse haemopoietic system. I. Restricted interchange between haemopoietic sites. *Cell and Tissue Kinetics*, **8**, 219–32.

Micklem, H. S., Ford, C. E., Evans, E. P., Ogden, D. A. & Papworth, D. S. (1972). Competitive *in vivo* proliferation of foetal and adult haematopoietic cells in lethally irradiated mice. *Journal of cellular Physiology*, **79**, 293–8.

Micklem, H. S., Ogden, D. A., Evans, E. P., Ford, C. E. & Gray, J. G. (1975*b*). Compartments and cell flows within the mouse haemopoietic system. II. Estimated rates of interchange. *Cell and Tissue Kinetics*, **8**, 233–48.

Miller, J. F. A. P. (1975). T-cell regulation of immune responsiveness. *Annals of the New York Academy of Sciences*, **249**, 9–26.

Miller, J. F. A. P. & Mitchell, G. F. (1969). Thymus and antigen-reactive cells. *Transplantation Reviews*, **1**, 3–42.

Miller, R. G. & Phillips, R. A. (1975). Development of B lymphocytes. *Federation Proceedings*, **34**, 145–50.

Möller, G. (ed.) (1976). *Biochemistry and biology of Ia antigens. Transplantation Reviews*, **30**. Copenhagen: Munksgaard.

Moore, M. A. S. & Owen, J. J. T. (1966). Experimental studies on the development of the bursa of Fabricius. *Developmental Biology*, **14**, 40–51.

Moore, M. A. S. & Owen, J. J. T. (1967). Experimental studies on the development of the thymus. *Journal of experimental Medicine*, **126**, 715–26.

Nossal, G. J. V. & Pike, B. (1973). Studies on the differentiation of B lymphocytes in the mouse. *Immunology*, **25**, 33–45.

Okumura, K., Herzenberg, L. A., Murphy, D. B., McDevitt, H. O. & Herzenberg, L. A. (1976*a*). Selective expression of H-2 (I-region) loci controlling determinants on helper and suppressor T lymphocytes. *Journal of experimental medicine*, **144**, 685–98.

OKUMURA, K., JULIUS, M. H., TSU, T., HERZENBERG, L. A. & HERZENBERG, L. A. (1976*b*). Demonstration that IgG memory is carried by IgG-bearing cells. *European Journal of Immunology*, **6**, 467–72.

OSMOND, D. G. (1975). Formation and maturation of bone marrow lymphocytes. *Journal of the reticuloendothelial Society*, **17**, 99–114.

OSMOND, D. G. & NOSSAL, G. J. V. (1974*a*). Differentiation of lymphocytes in mouse bone marrow. I. Quantitative radioautographic studies of antiglobulin binding by lymphocytes in bone marrow and lymphoid tissues. *Cellular Immunology*, **13**, 117–31.

OSMOND, D. G. & NOSSAL, G. J. V. (1974*b*). Differentiation of lymphocytes in mouse bone marrow. II. Kinetics of maturation and renewal of antiglobulin binding cells studied by double labeling. *Cellular Immunology*, **13**, 132–45.

OWEN, J. J. T. (1972). The origins and development of lymphocyte populations. In *Ontogeny of acquired immunity*, ed. R. Porter & J. Knight, pp. 35–54. *Ciba Foundation Symposium*. Amsterdam: Associated Scientific Publishers.

OWEN, J. J. T., COOPER, M. D. & RAFF, M. C. (1974). *In vitro* generation of B lymphocytes in mouse foetal liver, a mammalian 'bursa equivalent'. *Nature, London*, **249**, 361–3.

OWEN, J. J. T., JORDAN, R. K., ROBINSON, J. H., SINGH, U. & WILLCOX, H. N. A. (1977). *In vitro* studies on the generation of lymphocyte diversity. *Cold Spring Harbor Symposia on quantitative Biology*, **41**, in press.

OWEN, J. J. T., RAFF, M. & COOPER, M. D. (1975). Generation of B lymphocytes in the mouse embryo. *European Journal of Immunology*, **5**, 468–73.

PARISH, C. R. (1975). Separation and functional analysis of subpopulations of lymphocytes bearing complement and Fc receptors. *Transplantation Reviews*, **25**, 98–120.

PLAYFAIR, J. H. L. & PURVES, E. C. (1971). Separate thymus dependent and thymus independent antibody forming cell precursors. *Nature New Biology*, **231**, 149–51.

PRESS, J. L., KLINMAN, N. R. & MCDEVITT, H. O. (1976). Expression of Ia antigens on hapten-specific B cells. I. Delineation of B cell subpopulations. *Journal of experimental Medicine*, **144**, 414–27.

PRITCHARD, H., RIDDAWAY, J. & MICKLEM, H. S. (1973). Immune responses in congenitally thymus-less mice. II. Quantitative studies of serum immunoglobulins, and antibody response to sheep erythrocytes and the effect of thymus allografting. *Clinical and experimental Immunology*, **13**, 125–38.

RAFF, M. C. (1970). Two distinct populations of peripheral lymphocytes in mice distinguishable by immunofluorescence. *Immunology*, **19**, 637–50.

RAFF, M. (1977). Immunological networks. *Nature, London*, **265**, 205–7.

RAFF, M., MEGSON, M., OWEN, J. J. T. & COOPER, M. D. (1976). Early production of intracellular IgM by B lymphocyte precursors in mouse. *Nature, London*, **259**, 224–6.

ROELANTS, G. E., LOOR, F., VON BOEHMER, H., SPRENT, J., HÄGG, L.-B., MAYOR, K. S. & RYDEN, A. (1975). Five types of lymphocytes characterized by double immunofluorescence and electrophoretic mobility. Organ distribution in normal and nude mice. *European Journal of Immunology*, **5**, 127–31.

ROELANTS, G. E., MAYOR, K. S., HÄGG, L.-B. & LOOR, F. (1976). Immature T lineage lymphocytes in athymic mice. Presence of TL, lifespan and homeostatic regulation. *European Journal of Immunology*, **6**, 75–81.

ROITT, I. M., GREAVES, M. F., TORRIGIANI, G., BROSTOFF, J. & PLAYFAIR, J. H. L. (1969). The cellular basis of immunological responses. *Lancet*, **ii**, 367–71.

ROSSE, C. (1976). Small lymphocyte and transitional cell populations of the bone marrow; their role in the mediation of immune and hemopoietic progenitor cell functions. *International Review of Cytology*, **45**, 155–290.

RYSER, J. E. & VASSALLI, P. (1974). Mouse bone marrow lymphocytes and their differentiation. *Journal of Immunology*, **113**, 719–28.

SCHEID, M. P., GOLDSTEIN, G., HÄMMERLING, U. & BOYSE, E. A. (1975). Lymphocyte differentiation from precursor cells *in vitro*. *Annals of the New York Academy of Sciences*, **249**, 531–8.

SCHIRRMACHER, V., HALLORAN, P., ARNAIZ-VILLENA, A. & FESTENSTEIN, H. (1976). Interaction of Fc and C3 receptors of lymphoid cells with antibodies against products of the major histocompatibility complex. *Transplantation Reviews*, **30**, 140–73.

SCHLESINGER, M., ISRAEL, E. & GERY, I. (1976). Antigenic properties of subsets of splenic T lymphocytes responding to lectins. *Immunology*, **30**, 865–72.

SHORTMAN, K., BOEHMER, H. VON, LIPP, J. & HOPPER, K. (1975). Subpopulations of T-lymphocytes. *Transplantation Reviews*, **25**, 163–210.

SHORTMAN, K. & JACKSON, H. (1974). The differentiation of T lymphocytes. I. Proliferation kinetics and interrelationships of subpopulations of mouse thymus cells. *Cellular Immunology*, **12**, 230–46.

SILVERSTONE, A. E., CANTOR, H., GOLDSTEIN, G. & BALTIMORE, D. (1976). Terminal deoxynucleotidyl transferase is found in prothymocytes. *Journal of experimental Medicine*, **144**, 543–8.

STOCKER, J. W., OSMOND, D. G. & NOSSAL, G. J. V. (1974). Differentiation of lymphocytes in the mouse bone marrow. III. The adoptive response of bone marrow cells to a thymus cell independent antigen. *Immunology*, **27**, 795–806.

STOUT, R. D. & HERZENBERG, L. A. (1975*a*). The Fc receptor on thymus-derived lymphocytes. I. Detection of a subpopulation of murine T-lymphocytes bearing the Fc receptor. *Journal of experimental Medicine*, **142**, 611–21.

STOUT, R. D. & HERZENBERG, L. A. (1975*b*). The Fc receptor on thymus-derived lymphocytes. II. Mitogen responsiveness of T lymphocytes bearing the Fc receptor. *Journal of experimental Medicine*, **142**, 1041–51.

STOUT, R. D., WAKSAL, S. D. & HERZENBERG, L. A. (1976). The Fc receptor on thymus-derived lymphocytes. III. Mixed lymphocyte reactivity and cell-mediated lympholytic activity of Fc^- and Fc^+ T lymphocytes. *Journal of experimental Medicine*, **144**, 54–68.

STROBER, S. (1975). Immune function, cell surface characteristics and maturation of B cell subpopulations. *Transplantation Reviews*, **24**, 84–112.

TAYLOR, R. B. (1969). Cellular cooperation in the antibody response of mice to two serum albumins: specific function of thymus cells. *Transplantation Reviews*, **1**, 114–49.

TAYLOR, R. B & WORTIS, H. H. (1968). Thymus dependence of antibody response: variation with dose of antigen and class of antibody. *Nature, London*, **220**, 927–8.

Tew, J. G., Self, C. H., Harold, W. W. & Stavitsky, A. B. (1973). The spontaneous induction of anamnestic antibody synthesis in lymph node cell cultures many months after primary immunization. *Journal of Immunology*, **111**, 416–23.

Till, J. E. & McCulloch, E. A. (1961). A direct measurement of the radiation sensitivity of normal mouse bone marrow cells. *Radiation Research*, **14**, 213–22.

Torrigiani, G. (1972). Quantitative estimation of antibody in the immunoglobulin classes of the mouse. II. Thymic dependence of the different classes. *Journal of Immunology*, **108**, 161–4.

Trentin, J. J., Wolfe, N., Cheng, V., Fahlberg, W., Weiss, D. & Bonhag, R. (1967). Antibody production by mice repopulated with limited numbers of clones of lymphoid cell precursors. *Journal of Immunology*, **98**, 1326–37.

Trowell, O. A. (1958). The lymphocyte. *International Review of Cytology*, **7**, 235–93.

Turpen, J. B., Volpe, E. P. & Cohen, N. (1973). Ontogeny and peripheralization of thymic lymphocytes. *Science*, **182**, 931–3.

Vadas, M. A., Miller, J. F. A. P., McKenzie, I. F. C., Chism, S. E., Shen, F. W., Boyse, E. A., Gamble, J. R. & Whitelaw, A. M. (1976). Ly and Ia antigen phenotypes of T cells involved in delayed-type hypersensitivity and in suppression. *Journal of experimental Medicine*, **144**, 10–19.

Vaughan, W. P. & McGregor, D. D. (1972). Lymphopoiesis in the rat. I. The effect of pool size on lymphocyte production. *Journal of cellular Physiology*, **80**, 1–12.

Waksman, B. H. & Namba, Y. (1976). On soluble mediators of immunologic regulation. *Cellular Immunology*, **21**, 161–76.

Wallis, V. J., Leuchars, E., Chwalinski, S. & Davies, A. J. S. (1975). On the sparse seeding of bone marrow and thymus in radiation chimaeras. *Transplantation*, **19**, 2–11.

Wallis, V. J., Leuchars, E., Collavo, D. & Davies, A. J. S. (1976). 'Hyperlymphoid' mice. *Advances in experimental Medicine and Biology*, **66**, 183–9.

Warr, G. W., Decker, J. M. & Marchalonis, J. J. (1976). Evolutionary and developmental aspects of T-cell recognition. *Immunological Communications*, **5**, 281–301.

Weber, W. T. (1975). Avian B lymphocyte subpopulations: origins and functional capacities. *Transplantation Reviews*, **24**, 113–58.

Weissman, I. L. (1973). Thymus cell maturation. Studies on the origin of cortisone-resistant thymic lymphocytes. *Journal of experimental Medicine*, **137**, 504–10.

Wu, A. M., Till, J. E., Siminovitch, L. & McCulloch, E. A. (1968). Cytological evidence for a relationship between normal hematopoietic colony-forming cells and cells of the lymphoid system. *Journal of experimental Medicine*, **127**, 455–63.

Yoffey, J. M. (1975). Histogenesis of B and T lymphocytes. *Israel Journal of medical Science*, **11**, 1230–41.

Yung, L. L. L., Wyn-Evans, T. C. & Diener, E. (1973). Ontogeny of the murine immune system: development of antigen recognition and immune responsiveness. *European Journal of Immunology*, **3**, 224–8.

Regulatory role of the macrophage in haemopoiesis

MALCOLM A. S. MOORE
Sloan-Kettering Institute for Cancer Research,
410 East 68th Street, New York, New York 10021, USA

The haemopoietic system exists as an unique hierarchy of pluripotent, committed and maturing cell populations continually differentiating into a minimum of nine different cell lineages and, indeed, considerably more if the functional heterogeneity of the T- and B-lymphocytes is considered. The system remains in exquisite homeostatic balance allowing for steady-state production of mature end cells yet retaining a rapid and specific responsiveness to altered demand for cell production in any one of the various cell lineages. The existence of in-vivo and in-vitro clonal assay systems for detection of early haemopoietic stem cells and progenitor cells has permitted a detailed dissection of stem cell organisation and regulation. The concepts evolving from such studies are probably equally applicable to understanding the nature of stem cells and their control in other renewing cell populations.

Pluripotential haemopoietic stem cells (CFU-S)

The pluripotential haemopoietic stem cell (CFU-S), defined by its capacity to form colonies in the spleens of lethally irradiated mice, is clearly capable of differentiating into all haemopoietic cell lineages including the T- and B-lymphocyte compartment (Abramson, Miller & Phillips, 1977). Furthermore, it is capable of extensive self-renewal in addition to differentiation (Siminovitch, McCulloch & Till, 1963; Metcalf & Moore, 1971). Restriction of the options available to the pluripotential haemopoietic stem cell resulting from differentiation into the various cell-line-specific committed stem cell compartments may be stochastic or may involve an interaction with a specific haemopoietic inductive microenvironment (Till, McCulloch & Siminovitch, 1964; Wolf & Trentin, 1968). Some insight into this question is provided by analysis of genetically anaemic W/W^{v} and Sl/Sl^{d} mice, since the lesion in *W* mice is due to an intrinsic defect in the pluripotential stem cell, whereas in *Sl* mice the defect is in the haemopoietic environment in which the stem cells reside (Russel & Bernstein, 1966; Metcalf & Moore, 1971). The haematological defect of *W* mice can, therefore, be cured by an inoculum of marrow cells from *Sl* mice and the macrocytic anaemia of *Sl* mice can be reversed following grafting of neonatal spleen or whole bone from *W* anaemic mice as a source of a non-defective haemopoietic microenvironment. Further insight into the haemopoietic microenvironment is provided by a liquid culture system in which the proliferation of murine pluripotential stem cells and production of differentiated cells of various lineages can be maintained *in vitro* for several months (Allen & Dexter, 1976; Dexter, Allen & Lajtha, 1977). This proliferation, differentiation and maturation of haemopoietic cells is dependent upon the formation of a bone-marrow-derived adherent population comprised of phagocytic mononuclear cells, 'epithelial' cells and giant fat-containing cells (Allen & Dexter, 1976). Within this adherent microenvironment extensive cellular interactions occur, leading to pluripotential stem cell renewal and differentiation.

Continuous haemopoiesis *in vitro* is markedly diminished when cultures are established with either W/W^{v} or Sl/Sl^{d} bone marrow, and in allogeneic combinations in which *Sl* marrow is used to establish the adherent microenvironment and the cultures subsequently inoculated with *W* marrow, no sustained haemopoiesis is observed (Dexter & Moore, 1977). In the reverse combination utilising *W* marrow to establish a non-defective adherent microenvironment with subsequent addition of *Sl* bone marrow as a source of normal stem cells, continuous haemopoiesis was observed. This in-vitro duplication and cure of the haemopoietic defects in genetically anaemic mice is of some significance since it allows direct analysis of the interaction of stem cells with a normal or defective microenvironment provided by a marrow adherent cell population.

Granulocyte–monocyte committed stem cells (*CFU-C*)

The in-vitro agar culture assay system developed by Bradley & Metcalf (1966) led to the recognition of a class of haemopoietic progenitor cells restricted to granulocyte–macrophage differentiation. Essential to the cloning technique is the presence of regulatory macromolecules which have been given the operational title 'colony stimulating factors' (CSF). A sigmoidal dose–response relationship exists between CSF concentration and the number of colonies developing in marrow culture and this provides a sensitive bioassay system capable of detecting as little as 10^{-11}–10^{-12} M CSF ml^{-1} (Stanley, Cifone, Heard & Defendi, 1976). The action of CSF *in vitro* is complex and it cannot be considered simply as an inducer of differentiation since it is necessary for every cell division and differentiation step in the granulocyte–macrophage lineage *in vitro* (Metcalf & Moore, 1971). In-vitro colony formation by activated peritoneal macrophages has been described and in this system the macrophage colonies develop after a lag period of 1–2 weeks in culture and are derived from cells with the adherence and phagocytic properties of macrophages (Lin & Stewart, 1973). It appears that this proliferation of mature macrophages is dependent upon a factor which is similar, if not identical, to CSF (Stanley *et al.*, 1976).

Considerable biophysical and functional heterogeneity of CFU-C has been reported. It seems likely that in the mouse three major populations of cells (with respect to buoyant density) can be stimulated to colony formation (Williams & Jackson, 1977). At day 6 of culture, all density subpopulations give rise to granulocyte, mixed and macrophage colonies and clearly no one progenitor cell population can specifically give rise to either granulocytes or macrophages. Nevertheless, studies comparing the cells stimulated by different kinds of CSF-containing preparations in the presence or absence of certain enhancing activities such as erythrocyte lysate or serum factors suggest that different activities influence in-vitro cloning of the various subpopulations (Metcalf & MacDonald, 1975; Williams & Van Den Engh, 1975; Williams & Jackson, 1977). This notion is supported by the observation that a titratable activity is present in mouse lung conditioned medium but not in pregnant mouse uterus extract and appears to be responsible for the differences in dose-responsiveness of bone marrow cells to these two CSF sources. This activity in lung conditioned medium stimulates the cloning of a discrete subpopulation of CFU-C of modal density 1.070 g cm^{-3} (Williams & Jackson, 1977). Metcalf & MacDonald (1975) found the peak value of cells responding to lung conditioned medium to sediment at 4.4 mm h^{-1}. This is markedly different from the characteristics of cells responding to human urine CSF which have been shown to be more dense (1.081 g cm^{-3}) and sediment at

5.3 mm h^{-1} (Moore & Williams, 1973; Metcalf & MacDonald, 1975). These two cell types probably have a constant volume (225–235 μm^3), the change in sedimentation rate being affected by a change in cell density. Assuming that these two subpopulations of CFU-C are within the same lineage, the data are suggestive of the lower density population being the less differentiated and are in agreement with the ontogenetic studies of Moore, McNeill & Haskill (1970).

A similar heterogeneity of granulocyte-committed stem cells has been reported in human marrow. Clonal techniques have been developed based on the observation that human bone marrow cells can proliferate and differentiate to granulocytes in diffusion chambers implanted into the peritoneum of irradiated mice. Cells forming neutrophil granulocyte colonies in fibrin clot diffusion chambers (CFU-D) have a peak sedimentation rate of 5.2–5.4 mm h^{-1}. This distinguishes them from a rapidly sedimenting CFU-C population (7.0–7.2 mm h^{-1}) which forms colonies in agar after 1 week of incubation and from a CFU-C population of intermediate sedimentation velocity which is primarily responsible for agar colony formation after 14 days (Jacobsen, Broxmeyer & Moore, 1977). Further distinction between CFU-D and day 7 and day 14 CFU-C resides in differences in their cell cycle characteristics, since only 8±16% of CFU-D are in S-phase, in contrast to 33±8% of day 7 CFU-C. There is no evidence to support the view that CFU-D represent multipotential stem cells equivalent to CFU-S in mice, but it is likely that they are intermediary cells, committed to granulopoiesis but more immature than the cells which are detected in the CFU-C assay.

POSITIVE AND NEGATIVE FEEDBACK CONTROL OF CFU-C PROLIFERATION AND DIFFERENTIATION

Recognition that CSF is produced by monocytes and macrophages (Moore & Williams, 1972) and that it acts to promote increased monocyte production and macrophage proliferation (Metcalf & Moore, 1971), introduces the problem of mechanisms designed to counterbalance this positive feedback drive. A number of mechanisms have been revealed in in-vitro studies and many, if not all, may be of physiological significance *in vivo*. The functional heterogeneity of the phagocytic mononuclear cell population must firstly be considered since marked variation in CSF-producing capacity exists. 'Virgin' macrophages developing in agar culture from CFU-C and macrophages generated in continuous marrow culture are not constitutive producers of CSF; however, exposure of these cells to macrophage activating agents such as lipopolysaccharide (LPS) or BCG rapidly induces CSF synthesis and secretion (M. A. S. Moore, unpublished observation). In this sense CSF recruitment of additional monocytes and macrophages would not, *ipso facto*, lead to increased CSF production in the absence of an exogenous source of stimulation such as endotoxaemia due to Gram-negative bacterial infection. Neoplastic monocyte or macrophage cell lines also retain the capacity to produce CSF; however, in some cases the leukaemic cell lines are constitutive producers and in other cases CSF production is observed only after LPS stimulation, suggesting retention of a degree of normal responsiveness by the transformed cells (Ralph, Nakoinz, Broxmeyer & Schrader, 1977). A second feature of monocyte–macrophage CSF production resides in the functional heterogeneity of the activities. Medium conditioned by human monocytes contains two species of CSF. One, of apparent molecular weight 30000, is a true human CSF, stimulating human granulocyte–macrophage colony formation. A high molecular weight (150000) factor is also produced which stimulates mouse marrow colony formation but not human (Shah, Caporale & Moore, 1977). This latter CSF may be identical to the CSF species purified from human urine (Stanley *et al.*, 1976). The physiological significance of these two species of CSF resides in the temporal course of their production. Human-active CSF is produced early in monocyte cultures, but production ceases in 1–2 weeks, whereas mouse-active CSF is produced continuously for many weeks. The paradox of human cells producing a CSF active only in a different species is resolved if the heterogeneity of the responding cell populations in mouse and human marrow culture is considered. The spectrum of CFU-C in the mouse can be defined by biophysical heterogeneity with respect to cell size and buoyant density. This heterogeneity is also reflected in the dose-responsiveness of CFU-C to different species of CSF and in the morphology of the colonies. Human urinary and human macrophage CSF stimulate predominantly macrophage colony formation in mouse marrow culture and appear to act on a more differentiated

CFU-C subset than do other species of CSF which stimulate predominantly granulocytic colony formation. Indeed, these macrophage-stimulating CSFs are capable of promoting macrophage colony formation in agar cultures of thioglycollate-activated mouse peritoneal macrophages (Stanley *et al.*, 1976). Assay systems equivalent to the mouse peritoneal macrophage colony assay have not been developed for human use, but it is of relevance that the high molecular weight species of CSF produced by human monocytes and macrophages stimulates macrophage cluster formation in human bone marrow culture. The apparent species specificity may therefore be related to the arbitrary criteria established for the definition of a colony (> 40 cells) and to differences in the rate of proliferation of human versus mouse cells *in vitro*. The temporal change in production of the two species of CSF by human monocytes and macrophages may thus represent a regulatory mechanism whereby the acute response involves systemic recruitment of additional monocytes and granulocytes by elaboration of a CSF active at the marrow stem cell level and a chronic local response involves continued elaboration of a CSF which exclusively promotes macrophage proliferation.

There is increasing evidence that mature granulocytes and their products participate as one of the regulators of myelopoiesis. Negative feedback control of granulopoiesis has been reported in various systems and the concept of a granulocyte chalone specifically inhibitory of CFU-C (Aardal, Laerum, Paukovits & Maurer, 1977) or to more differentiated myeloid cells (Lord *et al.*, 1974) has received some experimental support. The studies of Broxmeyer, Mendelsohn & Moore (1977*a*) and Broxmeyer, Moore & Ralph (1977*b*) have indicated a more indirect mechanism of granulocyte negative feedback which recognises the implications of the bipotentiality of the granulocyte–monocyte committed stem cell. 'Spontaneous' colony formation in the absence of an exogenous source of CSF is observed in marrow cultures of all species so far investigated when the cells are cultured at a sufficiently high concentration (Moore & Williams, 1972). This spontaneous colony formation is due to endogenous elaboration of CSF by marrow monocytes and macrophages and is considerably enhanced by the removal of mature granulocytes from the cultured cell population. Addition of mature granulocytes, granulocyte extracts or medium conditioned by incubation with granulocytes markedly inhibits spontaneous colony formation (Broxmeyer *et al.*, 1977*b*). This granulocyte-derived colony inhibitory activity (CIA) acts in a non-species specific manner to suppress CSF production by monocytes and macrophages. CIA is not inhibitory to monocyte–macrophage proliferation and is clearly distinct from granulocyte chalone since no inhibition of granulocytic colony formation is observed in the presence of an exogenous source of CSF. A marked and reproducible quantitative defect in CIA content of mature granulocytes has been reported in patients with chronic myeloid leukaemia and the CSF-producing cells in such patients were less sensitive than normal to CIA derived from normal mature granulocytes (Broxmeyer *et al.*, 1977*a*). These negative feedback abnormalities *in vitro* may partially explain the granulocytic hyperplasia associated with chronic myeloblastic leukaemia.

The existence of a negative feedback from the mature granulocyte, acting to modulate CSF production, may explain the steady-state balance of granulocytes and monocytes but does not adequately account for the neutrophil leukocytosis, monocytosis and extensive macrophage proliferation associated with infection, inflammation or immune responses. In this context, granulocyte-derived CIA does not inhibit CSF production by endotoxin-stimulated monocytes and macrophages nor by mitogen-stimulated lymphocytes (Broxmeyer *et al.*, 1977*a*, *b*). The existence of override mechanisms permitting increased CSF production in the face of neutrophil leukocytosis introduces the complexity of additional self-limiting mechanisms for which experimental evidence is available.

Activation of normal or neoplastic B- and T-lymphocytes by an appropriate mitogenic or antigenic stimulus leads to induction of CSF production (Parker & Metcalf, 1974; Ruscetti & Chervenick, 1975). Similar activation protocols also lead to increased production of interferon which has been shown to be profoundly inhibitory to human and murine granulocyte–macrophage colony formation *in vitro* (Greenberg & Mosny, 1977). Thus, interferon may play a physiological role in counteracting the stimulatory activity of lymphocyte-derived CSF.

The regulatory interactions involving diffusible stimulatory and inhibitory activities elaborated by granulocytes, lymphocytes and phagocytic mononuclear cells can clearly involve specific macromolecules or, alternatively, non-specific modulating activities. Pharmacological studies have shown that prostaglandins of the E series (PGE) and other agents capable of elevating intracellular levels of cAMP profoundly inhibit granulopoiesis and macrophage proliferation *in vitro* (Kurland & Moore, 1977*a*). Just as CSF promotes continued replication of the CFU-C and its progeny, PGE limits this effect by an opposing action on the responsiveness of the myeloid stem cell and its proliferative progeny to stimulation by CSF. Kurland & Moore (1977*a*, *b*) have shown that prostaglandin synthesised by phagocytic mononuclear cells may be of central importance in the modulation of haemopoiesis. Measurement of prostaglandin E production by murine macrophages and human monocytes has been performed using a sensitive radioimmunoassay and has shown a linear relationship between the number of phagocytic mononuclear cells and the concentration of PGE in the conditioned medium (Kurland, Bockman, Broxmeyer & Moore, 1977*a*). This observation explains the lack of correlation between the numbers of monocytes and macrophages used to stimulate granulocyte–macrophage colony formation and the incidence of colonies. Titration of varying numbers of adherent macrophages or blood monocytes as a source of stimulus for human or murine marrow CFU-C has clearly shown that colony formation is stimulated by low numbers of phagocytic mononuclear cells (0.05×10^5–2×10^5) and inhibited if higher concentrations are used. Parallel studies using monocytes or macrophages treated with indomethacin, a potent inhibitor of prostaglandin synthesis, have revealed a linear relationship between the number of colonies stimulated and the number of phagocytic mononuclear cells used as the source of CSF (Kurland *et al.*, 1977*a*, *b*). These observations point to the unique ability of the macrophage to control the proliferation of its own progenitor cell by elaboration of opposing regulatory influences. Macrophages activated by in-vitro exposure to LPS show marked enhancement of prostaglandin synthesis and a concomitant increased capacity to inhibit granulocytic colony formation (J. Kurland & M. A. S. Moore, unpublished observation). It may appear paradoxical that LPS stimulation, which elicits a marked increase in macrophage CSF production, should also induce increased production of an opposing activity which can effectively neutralise CSF action. This paradox can be resolved if the temporal sequence of events is considered. LPS *in vivo* and *in vitro* induces increased macrophage CSF production very rapidly with significant changes observed within minutes, whereas increased prostaglandin synthesis is delayed for 18–24 hours (J. Kurland & M. A. S. Moore, unpublished observation). Indeed, the stimulus for increased PGE production is not directly due to LPS, but rather to the increased levels of CSF induced which, in turn, activate macrophage prostaglandin synthetase. Incubation of non-activated macrophages with increasing concentrations of CSF in the absence of LPS leads to a proportional increase in prostaglandin synthesis indicating a very direct relationship between CSF levels and induction of an opposing activity (Kurland *et al.*, 1977*a*). These observations point to the macrophage as a surveillance cell which, under steady-state conditions, is elaborating basal levels of CSF and prostaglandin E. The extreme lability of the PGE molecule provides a further physiological control since the CSF elaborated by fixed tissue macrophages can act systemically to stimulate CFU-C proliferation. The local influence of PGE under basal conditions may thus be limited to inhibition of CSF-dependent proliferation of fixed tissue macrophages. The constitutive contribution of CSF to granulopoiesis and monocyte production is rapidly increased under physiologically perturbed circumstances such as infection. Progressive increases in CSF levels would promote recruitment of additional granulocytes and monocytes by an action on the marrow CFU-C population and would also promote local macrophage proliferation. This process would be self-limiting since a progressive increase in CSF beyond a critical concentration within the local milieu of the macrophage is ultimately sensed and serves to stimulate the coincident production and release of PGE which opposes the stimulatory action of CSF.

THE REGULATORY ROLE OF MACROPHAGES IN LYMPHOPOIESIS

The macrophage has been implicated in the elaboration of a spectrum of molecules which alter or modulate the proliferation and differentiation of lymphoid cells. These include lymphocyte activating factor (LAF) which potentiates the mitogenic response of T-lymphocytes to lectin and histocompatibility antigens (Gery & Waksman, 1972), as well as factors which increase the helper function of T-lymphocytes (Wood & Gaul, 1974) and the promotion of soluble mediator production by lymphocytes (Nelson & Leu, 1975). The role of the macrophage in the regulation of humoral immune responses is, however, less well understood since macrophages are required for humoral immune responses *in vitro* (Hoffman & Dutton, 1971), yet mitogen-induced B-lymphocyte proliferative responses are suppressed in the presence of macrophages (Lipsky & Rosenthal, 1976). The recently developed system of in-vitro murine B-lymphocyte cloning has thrown some light on the immunoregulatory role of macrophages. In this system, a functional subpopulation of murine B-lymphocytes proliferate in semi-solid agar culture to form colonies (Metcalf, 1976; Kincade, Ralph & Moore, 1976). This process is dependent upon 2-mercaptoethanol and B-cell mitogens native to laboratory grade agar. Kurland *et al.* (1977*b*) have investigated the influence of diffusible macrophage-derived factors on this clonal proliferation using a two-layer culture system which prevented macrophage–lymphocyte contact and permitted B-cell activation to be critically assessed under conditions of extremely low cell density. Adherent peritoneal macrophages potentiated both the number and size of developing B-cell colonies, particularly when low numbers of spleen or lymph node cells, or macrophage-depleted lymphoid cell suspensions were used. Macrophage-depleted lymph node cells gave virtually no colonies, but colony formation was restored by the presence of an optimal number of macrophages. When the number of macrophages exceeded that required for optimal stimulation, colony formation was suppressed, an effect which was largely prevented by indomethacin. Similar stimulatory and inhibitory activities were also present in media conditioned by varying numbers of peritoneal macrophages. The diffusible inhibitory activity was identified as PGE, which suppresses B-lymphocyte cloning at concentrations as low as 10^{-8}–10^{-10} M (Kurland & Moore, 1977*a*). The stimulatory factor may be similar to the 5×10^4 molecular weight factor described by Namba & Hanaoka (1974) which is produced by adherent, phagocytic cells and stimulates normal and neoplastic B-cell proliferation. In this regard, conditioned medium from a murine myelomonocytic cell line (WEHI-3) which elaborates both CSF and T-lymphocyte activating factor, can not substitute for macrophages in the initiation of B-lymphocyte colony formation, suggesting the non-identity of this B-lymphocyte stimulating activity with other known factors which alter haemopoietic and lymphoid function.

The role of the antigenic or mitogenic stimulus for B-lymphocyte proliferation is more complex than a simple interaction with the responsive B-cell, since agents such as LPS and sheep red blood cells (SRBC) may also modulate the elaboration of immunoregulatory factors by the macrophage. In the presence of low numbers of macrophages, SRBC and LPS facilitate the production of B-lymphocyte stimulating activity, but when added to high concentrations of macrophages, suppressor activity is generated due to induction of prostaglandin synthesis (Kurland *et al.*, 1977*b*).

CONCLUSION

An adequate case may be made for considering the phagocytic mononuclear cell population and, specifically, the resident macrophages within the haemopoietic tissues, as of central importance in controlling the proliferation and differentiation of granulocyte–macrophage committed stem cells and B-lymphocytes. A more general regulatory role of macrophages in haemopoietic differentiation is also suggested by studies implicating production of stimulatory factors influencing other haemopoietic cell lineages. In human marrow culture, eosinophil colony formation is stimulated by factors elaborated by human monocytes and eosinophil colony formation in the mouse is stimulated by conditioned medium produced by WEHI-3, a murine myelomonocytic leukaemic cell line (Metcalf, Parker, Chester & Kincade, 1974). Velocity sedimentation studies indicate that the cells generating eosinophil colonies are separable from cells generating regular granulocytic or macrophage colonies. Furthermore, the factor stimulating eosinophil differentiation is antigenically and electrophoretically distinct from conventional neutrophil–macrophage stimulating factors. Since eosinophil colony formation in human and mouse marrow cultures is inhibited both by synthetic prostaglandin E and by diffusible products of activated macrophages, a dualism between a specific stimulatory macromolecule and prostaglandin would appear to be involved in the regulation of eosinophil production.

Regulation of megakaryopoiesis is also beginning to show a number of parallels with granulopoiesis. A committed unipotential megakaryocytic stem cell (CFU-M) with a frequency of 10–20 per 10^5 nucleated bone marrow cells can proliferate in agar culture in the presence of an appropriate stimulating factor to produce colonies of polyploid, platelet-producing megakaryocytes (Metcalf, MacDonald, Odartchenko & Sordat, 1975; Williams *et al.*, 1977). In-vivo studies have indicated the role of thrombopoietin in the regulation of platelet production and active material has been obtained from a number of sources (McDonald, 1976). The relationship between thrombopoietin and megakaryocyte colony stimulating factor is unclear; however, the latter has been obtained in media conditioned by mitogen-stimulated mouse spleen cells (Metcalf *et al.*, 1975), embryonic kidney cells (Nakeff & Daniels-McQueen, 1976), and the murine myelomonocytic cell line WEHI-3 (Williams *et al.*, 1977). While neoplastic monocytes in the WEHI-3 cell line produce a megakaryocytic CSF distinct from eosinophil or neutrophil CSFs, normal macrophages do not. However, megakaryocyte colony formation in the presence of an optimal concentration of myelomonocytic leukaemic conditioned medium is potentiated two- to threefold by a diffusible enhancing activity produced by low numbers of mouse peritoneal macrophages (M. A. S. Moore, unpublished observation). As in all the other haemopoietic cloning systems, megakaryocytic colony formation is inhibited by prostaglandin elaborated by high numbers of macrophages.

The in-vitro techniques for detection of erythroid stem cells in the presence of erythropoietin have not, so far, implicated macrophages in any regulatory aspect of normal erythropoiesis. We have, however, recently observed spontaneous growth of erythroid colonies in soft agar of cells derived from patients with certain proliferative diseases of the haemopoietic system such as polycythaemia vera (Horland, Wolman, Murphy & Moore, 1977). In serially sectioned colonies, single or multiple macrophages central within the erythroid colony were frequently observed. This observation is reminiscent of the 'erythroblast islands' in bone marrow, where one or more central macrophages are surrounded by concentric rings of differentiating erythroblasts as a distinct anatomical unit.

The intimate interrelationship of macrophages with proliferating myeloid and lymphoid cell populations in marrow, spleen and lymph node suggests that such cells may be uniquely situated to stimulate cell differentiation and modulate proliferation by elaboration of cell-line specific stimulatory macromolecules and an opposing non-specific activity, prostaglandin E. The generality of this concept is based on detection of diffusible activities operating in low density clonal assay systems where intimate cell contact is prevented, a situation quite unlike that pertaining *in vivo*. In the more physiologically relevant system of continuous marrow culture, intimate cell interactions occur between adherent marrow-derived cells, including macrophages, and haemopoietic stem cells (Allen & Dexter, 1976). The absence of detectable levels of granulocyte–macrophage CSF or megakaryocytic CSF in the continuous culture medium despite extensive granulopoiesis, megakaryopoiesis and monocyte production (Dexter, Allen & Lajtha, 1977; Williams *et al.*, 1977) would suggest that macrophage-derived stimulatory and inhibitory activities may normally act very locally and possibly are transferred directly from the macrophage to the responsive haemopoietic stem cell.

This work was supported by grants CA-19052 and CA-17085 from the National Institutes of Health and the Gar Reichman Foundation.

REFERENCES

Aardal, N. P., Laerum, O. D., Paukovits, W. R. & Maurer, H. R. (1977). Inhibition of agar colony formation by partially purified granulocyte extracts. *Virchows Archiv, Series B Cell Pathology*, **24**, 27–39.

Abramson, S., Miller, R. G. & Phillips, R. A. (1977). The identification in adult bone marrow of pluripotent and restricted stem cells of the myeloid and lymphoid systems. *Journal of experimental Medicine*, **145**, 1567–79.

Allen, T. D. & Dexter, T. M. (1976). Cellular interrelationships during *in vitro* granulopoiesis. *Differentiation*, **6**, 191–4.

Bradley, T. R. & Metcalf, D. (1966). The growth of mouse bone marrow cells *in vitro*. *Australian Journal of experimental Biology and medical Science*, **44**, 287–300.

Broxmeyer, H. E., Mendelsohn, N. & Moore, M. A. S. (1977*a*). Abnormal granulocyte feedback regulation of colony forming and colony stimulating activity producing cells from patients with chronic myelogenous leukemia. *Leukemia Research*, **1**, 3–12.

Broxmeyer, H. E., Moore, M. A. S. & Ralph, P. (1977*b*). Cell-free granulocyte colony inhibiting activity derived from human polymorphonuclear neutrophils. *Experimental Hematology*, **5**, 87–102.

Dexter, T. M., Allen, T. D. & Lajtha, L. G. (1977). Conditions controlling the proliferating of haemopoietic stem cells *in vitro*. *Journal of cellular Physiology*, **91**, 335–44.

Dexter, T. M. & Moore, M. A. S. (1977). *In vitro* duplication and cure of haemopoietic defects in genetically anaemic W/W^v and Sl/Sl^d mice. *Nature, London*, **269**, 412–14.

Gery, I. & Waksman, B. H. (1972). Potentiation of T lymphocyte responses to mitogens. II. The cellular source of potentiating mediator(s). *Journal of experimental Medicine*, **136**, 143–55.

Greenberg, P. L. & Mosny, S. A. (1977). Cytotoxic effects of interferon *in vitro* on granulocytic progenitor cells. *Cancer Research*, **37**, 1794–9.

Hoffman, M. & Dutton, R. W. (1971). Immune response restoration with macrophage culture supernatants. *Science*, **172**, 1047–8.

Horland, A. A., Wolman, S. R., Murphy, M. J. & Moore, M. A. S. (1977). Proliferation of erythroid colonies in semi-solid agar. *British Journal of Haematology*, **36**, 477–81.

Jacobsen, H., Broxmeyer, H. & Moore, M. A. S. (1977). Relationship between colony forming cells in diffusion chambers *in vivo* (CFU-D) and colony forming cells in agar in *vitro* (CFU-C): physical separation and kinetics in culture. In *Experimental hematology today*, ed. S. Baum. Berlin: Springer-Verlag.

Kincade, P. W., Ralph, P. & Moore, M. A. S. (1976). Growth of B lymphocyte clones in semi-solid agar is mitogen dependent. *Journal of experimental Medicine*, **143**, 1265–70.

Kurland, J. I., Bockman, R., Broxmeyer, H. & Moore, M. A. S. (1977*a*). Limitation of excessive myelopoiesis by the intrinsic modulation of macrophage-derived prostaglandin E. *Science*, (in press).

Kurland, J. I., Kincade, P. W. & Moore, M. A. S. (1977*b*). Immunoregulation of B lymphocyte clonal proliferation by stimulatory and inhibitory macrophage-derived factors. *Journal of experimental Medicine*, **145**, 1420–35.

Kurland, J. I. & Moore, M. A. S. (1977*a*). Modulation of hemopoiesis by prostaglandin. *Experimental Hematology*, **5**, 357–73.

Kurland, J. I. & Moore, M. A. S. (1977*b*). Regulatory role of the macrophage in normal and neoplastic hemopoiesis. In *Experimental Hematology Today*, ed. S. Baum, pp. 51–61. Berlin: Springer-Verlag.

Lin, H. S. & Stewart, C. C. (1973). Peritoneal exudate cells. I. Growth requirement of cells capable of forming colonies in soft agar. *Journal of cellular Physiology*, **83**, 369–78.

Lipsky, P. E. & Rosenthal, A. S. (1976). The induction and regulation of guinea pig B lymphocyte proliferation *in vitro*. *Journal of Immunology*, **117**, 1594–602.

Lord, B. I., Cercek, L., Cercek, B., Shah, G. P., Dexter, T. M. & Lajtha, L. G. (1974). Inhibitors of haemopoietic cell proliferation? Specificity of action within the haemopoietic system. *British Journal of Cancer*, **29**, 168–75.

McDonald, T. P. (1976). A comparison of platelet size, platelet count and platelet ^{35}S incorporation as assays for thrombopoietin. *British Journal of Haematology*, **34**, 257–67.

Metcalf, D. (1976). Role of mercaptoethanol and endotoxin in stimulating B lymphocyte colony formation *in vitro*. *Journal of Immunology*, **116**, 635–38.

Metcalf, D. & MacDonald, H. R. (1975). Heterogeneity of *in vitro* colony- and cluster-forming cells in mouse marrow: segregation by velocity sedimentation. *Journal of cellular Physiology*, **85**, 643–53.

Metcalf, D., MacDonald, H. R., Odartchenko, N. & Sordat, B. (1975). Growth of mouse megakartocyte colonies *in vitro*. *Proceedings of the National Academy of Sciences, USA*, **72**, 1744–8.

Metcalf, D. & Moore, M. A. S. (1971). In *Haemopoietic cells: their origin, migration and differentiation*. Amsterdam: North-Holland.

Metcalf, D., Parker, J., Chester, H. M. & Kincade, P. W. (1974). Formation of eosinophilic-like granulocytic colonies by mouse bone marrow cells *in vitro*. *Journal of cellular Physiology*, **84**, 275–89.

Moore, M. A. S., McNeill, T. A. & Haskill, J. S. (1970). Density distribution analysis of *in vivo* and *in vitro* colony forming cells in developing foetal liver. *Journal of cellular Physiology*, **75**, 181–92.

Moore, M. A. S. & Williams, N. (1972). Physical separation of colony stimulating cells from *in vitro* colony forming cells in hemopoietic tissue. *Journal of cellular Physiology*, **80**, 195–206.

Moore, M. A. S. & Williams, N. (1973). Functional, morphological and kinetic analysis of the granulocyte-macrophage progenitor cell. In *Proceedings, Second International Workshop on Hemopoiesis in Culture*, ed. W. A. Robinson, pp. 17–27. Washington: US Government Printing Office.

Nakeff, A. & Daniels-McQueen, S. (1976). *In vitro* colony assay for a new class of megakaryocyte precursor: colony forming unit megakaryocyte (CFU-M). *Proceedings of the Society for experimental Biology and Medicine*, **151**, 587–90.

Namba, Y. & Hanoaka, M. (1974). Immunocytology of cultured IgM-forming cells of mouse. II. Purification of phagocytic cell factor and its role in antibody formation. *Cellular Immunology*, **12**, 74–84.

Nelson, R. D. & Leu, R. W. (1975). Macrophage requirement for production of guinea pig migration inhibitory factor (MIF) *in vitro*. *Journal of Immunology*, **114**, 606–9.

Parker, J. W. & Metcalf, D. (1974). Production of colony-stimulating factor in mitogen-stimulated lymphocyte cultures. *Journal of Immunology*, **112**, 502–10.

Ralph, P., Nakoinz, I., Broxmeyer, H. E. & Schrader, S. (1977). Immunological functions and *in vitro* activation of cultured macrophage tumor lines. *Journal of the National Cancer Institute, Monograph* (in press).

Ruscetti, F. W. & Chervenick, P. A. (1975). Regulation of the release of colony-stimulating activity from mitogen-stimulated lymphocytes. *Journal of Immunology*, **114**, 1513–17.

Russell, E. S. & Bernstein, S. E. (1966). In *Biology of the laboratory mouse*, 2nd edn, ed. E. L. Green, p. 351. New York: McGraw-Hill.

Shah, R., Caporale, L. & Moore, M. A. S. (1977). Characterization of colony stimulating activity produced by human monocytes and PHA stimulated lymphocytes. *Blood*, (in press).

Siminovitch, L., McCulloch, E. A. & Till, J. E. (1963). The distribution of colony forming cells among spleen colonies. *Journal of comparative Cell Physiology*, **62**, 327–36.

Stanley, E. R., Cifone, M., Heard, P. M. & Defendi, V. (1976). Factors regulating macrophage production and growth: identity of colony stimulating factor and macrophage growth factor. *Journal of experimental Medicine*, **143**, 631–47.

Till, J. E., McCulloch, E. A. & Siminovitch, L. (1964). A stochastic model of stem cell proliferation, based on the growth of spleen colony forming cells. *Proceedings of the National Academy of Sciences, USA*, **51**, 29–36.

Williams, N. & Jackson, H. (1977). Analysis of populations of macrophage–

granulocyte progenitor cells stimulated by activities in mouse lung conditioned medium. *Experimental Hematology*, **5**, 523–34.

WILLIAMS, N. & VAN DEN ENGH, G. J. (1975). Separation of subpopulations of *in vitro* colony forming cells from mouse marrow by equilibrium density centrifugation. *Journal of cellular Physiology*, **86**, 237–45.

WILLIAMS, N., JACKSON, H., SHERIDAN, A. P. C., MURPHY, M. J., ELSTE, A. & MOORE, M. A. S. (1977). Regulation of megakaryopoiesis in long term murine bone marrow cultures. *Blood* (in press.)

WOLF, N. S. & TRENTIN, J. J. (1968). Hemopoietic colony studies. V. Effect of hemopoietic organ stroma on plutipotential stem cells. *Journal of experimental Medicine*, **127**, 205–14.

WOOD, D. D. & GAUL, S. L. (1974). Enhancement of the humoral response of T cell-depleted murine spleens by a factor derived from human monocytes *in vitro*. *Journal of Immunology*, **113**, 925–33.

The role of proliferation inhibitors in the regulation of haemopoiesis

B. I. LORD, E. G. WRIGHT AND K. J. MORI

Paterson Laboratories, Christie Hospital and
Holt Radium Institute, Withington, Manchester M20 9BX, UK

Evidence has been accumulating in recent years that for a number of tissues, DNA synthesis and/or mitosis are controlled by a process of feedback inhibition brought about by substances (sometimes called chalones) which are produced by, and can be obtained from, the mature cells of the tissue in question (see Bullough, 1975). The greater the population of end cells, the greater is the concentration of inhibitor and hence by limitation of the number of cell cycles during the maturation sequence, the lower the output of end cells. In this way, it is thought that the population size of the functional cell compartment can be maintained appropriate to the tissue's requirements. By definition, it is considered that these inhibitors must be cell line specific, though not species specific, in their effect. In addition their effect must be reversible and not the result of cytotoxicity. For some tissues, for example skin (Bullough, 1975), granulocytes (Rytömaa & Kiviniemi, 1968), proliferation stimulatory substances (antichalones?) have also been suggested, but these appear to have attracted rather less attention. Bjerknes & Iversen (1974), however, have considered their possible role from a theoretical point of view.

The need for proliferation inhibitory processes in the haemopoietic system is not immediately obvious. It has long been recognised that erythropoietin, produced in response to a demand for an increased red cell mass, is a major factor stimulating the transition from committed erythroid precursor cells to maturing erythroid cells (Gurney, Wackman & Filmanowicz, 1961) and thence stimulating their proliferation processes through the maturation compartment (Stohlman *et al.*, 1968; Hodgson, 1970). More recently, evidence has been presented to show that erythropoietin also stimulates the proliferation and amplification of the early committed erythroid precursor cells (Reissmann & Samorapoompichit, 1970; H.-R. von Wangenheim, R. Schofield, S. Kyffin & B. Klein, unpublished data). Erythropoietin, therefore, is instrumental in promoting the erythropoietic process through all the stages of its development. Colony stimulating activity (CSA), a group of factors extractable from a variety of tissues and found to be essential for the development of granulocytic colonies in agar culture, has been postulated to play an equivalent role to erythropoietin in stimulating granulopoiesis *in vivo* (Stanley & Metcalf, 1971). This has not been fully substantiated, however, and conditions of increased granulopoiesis in the absence of increased levels of CSA have been reported (Udupa & Reissmann, 1975). Furthermore, CSA is undetectable in long-term suspension cultures supporting active granulopoiesis (Dexter, Allen & Lajtha, 1977).

In spite of the known effects of erythropoietin and the postulated in-vivo effects of CSA, proliferation inhibitors of both granulocytic (Rytömaa & Kiviniemi, 1968) and erythroid cells (Kivilaakso & Rytömaa, 1971) have been reported. We have, therefore, investigated these factors to determine their role in the control of granulopoiesis and erythropoiesis.

All the above factors relate to control of proliferation of cells committed to specific lines of development. Considerably less is known about the regulation of proliferation in the pluripotent stem cell compartment. These cells, defined and measured as cells (CFU-S) capable of forming colonies of haemopoietic cells in the spleens of irradiated mice (Till & McCulloch, 1961), are known to have a low rate of proliferation under normal conditions ($< 10\%$ in DNA synthesis) but a rapid rate under conditions of regeneration following depletion of their number (30–50% in DNA synthesis). It is clear, from experiments involving partial body irradiation, that proliferation control over CFU-S is operative locally (Croizat, Frindel & Tubiana, 1970; Gidali & Lajtha, 1972). For example, in the experiments of Gidali & Lajtha, in mice which have been given a lethal dose of irradiation to all but one hind limb of the animal, one sees, after an initial transitory phase, that marrow in the protected limb quickly becomes indistinguishable from normal bone marrow both in its cellularity and in the number and proliferative activity of its CFU-S. The rest of the marrow is still totally depleted and the CFU-S proliferating very rapidly. A further example of this local control, and one which will be utilised later, comes from mice treated with phenylhydrazine. In such mice, CFU-S are found to be proliferating rapidly in the bone marrow while those in the spleen are not (Rencricca *et al.*, 1970). It seemed probable, therefore, that the level of DNA synthesis in CFU-S is maintained, permitted or promoted by the presence or lack of endogeneously produced local factors which are capable of inhibiting or stimulating CFU-S proliferation. We have, therefore, been investigating haemopoietic tissue for the presence of such factors and will demonstrate that CFU-S proliferation, like that of the differentiated haemopoietic cells, is probably determined by both stimulatory and inhibitory factors.

METHODS

Production of cell extract

Extracts were made as previously described (Lord *et al.*, 1974*a*) from granulocytes, erythrocytes, normal bone marrow cells and regenerating bone marrow cells (5 days post 450 rads X-rays) by incubating the cells in saline (5×10^6 cells/ml) for 2 hours at 37 °C and then removing the cells by centrifugation. The cell-free supernatants were retained as crude extracts which were then partially purified and concentrated by passing through Amicon Diaflo ultrafiltration membranes to give a series of fractions as follows: I, 500–10000 daltons; II, 10000–30000 daltons; III, 30000–50000 daltons; IV, 50000–100000 daltons; and V, > 100000 daltons. These fractions were finally freeze-dried and stored at −20 °C until use.

Assay of extracts

The effects of granulocyte and erythrocyte extracts (GCE and RCE) were assessed *in vitro* by measurement of the structuredness of the cytoplasmic matrix (SCM) of appropriate haemopoietic cells (Cercek, Cercek & Ockey, 1973; Lord *et al.*, 1974*a*), or *in vivo* by measurement of autoradiographic labelling indices and cell production. Bone marrow extracts were assayed by their effects on DNA synthesis in CFU-S using the tritiated thymidine ([^{3}H]TdR) suicide technique (Becker, McCulloch, Siminovitch & Till, 1965; Lord, Lajtha & Gidali, 1974*c*).

In experiments entailing the mixing of whole cell populations for the assessment of effects on CFU-S proliferation, the population under study (termed the 'Modifier' population) was always irradiated (900 rads γ-rays) in order that CFU-S in that population would not contribute to spleen colonies produced by the CFU-S in the 'Assay' population.

THE MATURING CELL COMPARTMENTS

As pointed out above, an important criterion for the definition of a physiologically active proliferation inhibitor is its cell line specificity and we have maintained all along that it is not a sufficient demonstration of this property to show it is inactive in a totally unrelated tissue. We have, therefore, confined ourselves to measurements only in the related haemopoietic cell lines. Extracts of erythrocytes and granulocytes, assayed against the SCM of proliferative granulocytic cells, foetal liver normoblasts or phytohaemagglutinin (PHA) stimulated lymphocytes do demonstrate this specificity of effect (Table 1). RCE increases the SCM only of normoblasts, and GCE only that of granulocytic cells. This specificity extends in fact to the lymphoid system also, since lymphocyte extracts affect only the PHA-stimulated lymphocytes (Lord *et al.*, 1974*a*).

In addition, the extracts are not toxic to the cells, the increased

Table 1. *SCM (% control) of haemopoietic cells treated with GCE or RCE*

Test extract	Test cell populations: Cultured granulocytic cells	Foetal liver normoblasts	PHA-stimulated lymphocytes
GCE-I	128	99	101
GCE-III	104	104	100
RCE-I	98	149	99
RCE-III	96	146	99

GCE, granulocyte extract; RCE, red cell extract; I, Amicon Diaflo fraction I (molecular weight 500–10 000 daltons); III, Amicon Diaflo fraction III (molecular weight 30 000–50 000 daltons).

The SCM of a cell population is related to the proportion of cells in DNA synthesis (Cercek *et al.*, 1973). An increase in SCM is compatible with a reduction in the proportion of cells in DNA synthesis and consequently a lower state of proliferation.

Table 2. *SCM (% control) of haemopoietic cells treated with GCE or RCE: reversibility of effect*

Test	Test cell populations: Bone marrow from polycythaemic mice[a]	Regenerating bone marrow in irradiated spleens[b]
Control	100	100
Extract[c]	126	138
	Cells washed three times	
Washed control	101	101
Extract[c]	131	147

[a] Principally granulocytic cell proliferation.
[b] Principally erythroid cell proliferation.
[c] GCE-I or RCE-I (see legend to Table 1) used against bone marrow from polycythaemic mice or regenerating bone marrow cells respectively.

values of the SCM being readily abrogated and returned to normal by washing the cells free of extract (Table 2; Lord *et al.*, 1974*b*).

Although the SCM of a cell varies with the phase of the cell cycle and as a corollary, therefore, the SCM of an asynchronous cell population depends, at least partly, on the proliferative activity of the population (Cercek *et al.*, 1973), these changes are not necessarily related directly to changes in proliferative activity. It is clear, however, that there was some basis on which to consider these extracts as possible cell line specific physiological control factors. Furthermore, since a population of cells with a low proliferative activity would have a relatively high SCM value, the increases in SCM produced by the extracts are at least compatible with a reduced proliferative activity. Several workers (e.g. Rytömaa & Kiviniemi, 1968; Kivilaakso & Rytömaa, 1971; Paukovits, 1971; Bateman, 1974) have demonstrated the ability of granulocyte and erythrocyte extracts to decrease [^{3}H]TdR incorporation by bone marrow cells *in vitro*. Our use *in vivo* with subsequent autoradiographic observation has confirmed their cell line specificity and shown that GCE and RCE reduce both the proportion of cells in DNA synthesis and the rate of flow of their respective precursor cells through the cell cycle to about half the normal levels within 1 to 2 hours of injecting the extract (Lord, 1975; Lord, Shah & Lajtha, 1977*a*). It is very difficult, however, to reduce the labelling indices below 50% of control and even with repeated doses of the extract or with continuous dosage from cells in peritoneally implanted diffusion chambers (Bøyum, Løvhaug & Boecker, 1976), cell production still continues, though at a reduced rate. For example, RCE injected every 8 hours for 4 days during erythroid colony development limits the colony to one half its normal cellularity (Lord *et al.*, 1977*a*). Similarly, GCE limits granulocytic colonies to the extent of two population doublings following a similar injection regime (Lord, 1975). The extracts, therefore, do not cause a complete shutdown of proliferation but appear rather to prolong the cell cycle (mainly in the G1 phase) and thus limit the number of cell cycles and, therefore, the cell-number amplification possible within the maturation time of the cells.

A physiological role, for at least the GCE, is indicated by the fact that injected into androgen-stimulated mice, GCE limits the wave of increased granulopoiesis in exactly the same way as does plasma taken from mice at the peak of their elevated blood granulocyte levels and immediately prior to their normalisation (B. I. Lord & P. Milenkovic unpublished data). Furthermore, dose–SCM response measurements have indicated that both GCE and RCE are active at concentrations lower than 10^{-12} M. The cells are thus very sensitive to the extracts and at levels which must be considered physiological rather than pharmacological.

In the tests carried out so far, it should be noted that the GCE and RCE were of rat or pig origin and the lymphoid extracts bovine. By contrast, the test cells were mouse or, in the case of lymphocytes for the SCM test, were human. Thus, although the extracts do exhibit cell line (and cell source) specificity, they do not carry any species specificity.

THE COMMITTED PRECURSOR CELL COMPARTMENTS

Conflicting evidence has appeared in the literature as to whether cell extracts are effective against the committed precursor cells, particularly in respect of GCE and the production of granulocytic colonies in agar culture. Some workers (Laerum & Maurer, 1973; MacVittie & McCarthy, 1975) report gross deleterious effects on the number of colonies produced. We have found that although the labelling index of the colony cells and the overall cellularity of the colonies is somewhat reduced there are no effects of our GCE on either the colony number, or on the reproduction of the committed granulocytic colony forming cells (CFU-C) in the culture (Lord, Testa, Wright & Banerjee, 1977*b*). This observation is supported by Bøyum *et al.* (1976) who also found no effect on the growth of the CFU-C population over a period of 4 days in peritoneally implanted diffusion chambers. The disparity, however, may relate to the different methods of GCE extraction, some of which seem likely to carry a greater risk of collecting more toxic cellular breakdown products.

Likewise, RCE appears to have no effect on the committed erythroid precursor cells (EPC). RCE treatment to these cells causes no loss of red cell production (Lord *et al.*, 1977*a*) and in the case of hypertransfusion-induced polycythaemia, where a high concentration of RCE might be expected, the number and high proliferation rate of the EPC are maintained for long periods of time, if not indefinitely (H.-R. von Wangenheim, R. Schofield, S. Kyffin & B. Klein, unpublished data).

THE PLURIPOTENT STEM CELL COMPARTMENT (CFU-S)

In spite of the cellular heterogeneity of whole bone marrow cell populations it is possible to test extracts against them because the assays used for the precursor cells are dependent only on specific cell types. The spleen colony technique measures the CFU-S while the agar culture measures the CFU-C which are directly descended from and chronologically very closely related to the CFU-S. Thus, in order to test the cell line specificity of a potential CFU-S proliferation inhibitor, the same test cell populations (in this case regenerating bone marrow or spleen) can be used and assayed for both CFU-S and CFU-C. (In the mouse and in all adult mammalian haemopoietic tissue, the CFU-C are always rapidly proliferating so that potential CFU-S proliferation stimulators are more difficult to test directly for cell line specificity.)

When tested against proliferating CFU-S, we have found that normal bone marrow extract, fraction IV (NBME-IV) either stops cells in, or more probably stops cells entering, DNA synthesis (Lord, Mori, Wright & Lajtha, 1976). Over a 5 hour incubation period, the proportion of CFU-S in DNA synthesis falls from 30 %

Table 3. *Effect of bone marrow extracts on the number and proliferative activity of stem cells and committed granulocyte precursor cells*

	Proliferating CFU-S		CFU-C		Resting CFU-S	
Treatment	CFU-S/ 10^6 cells	% killed by [^{3}H]TdR	CFU-C/ plate	% killed by [^{3}H]TdR	CFU-S/ 10^5 cells	% killed by [^{3}H]TdR
Control	20.6	31.3	76.3	49.1	24.7	6.0
RBME-I NBME-I RBME-II NBME-II	20.0	32.0	—	—	20.0	6.8
RBME-III	20.8	33.0	—	—	25.0	36.5
NBME-III	19.3	24.7	—	—	24.6	4.4
RBME-IV	21.0	35.1	—	—	22.5	12.2
NBME-IV	18.7	4.5	70.1	44.2	23.6	2.0

RBME, regenerating bone marrow extract; NBME, normal bone marrow extract; Fractions I–IV by approximate molecular weight (see text).

to ~ 5%. No other NBME fraction is effective and neither is this inhibitor detectable in any fraction from regenerating bone marrow extracts (RBME). NBME-IV, however, has no effect on CFU-C proliferation. It is clear therefore that NBME-IV is capable of inhibiting proliferation specifically in the CFU-S population and since it also has no effect on the overall seeding efficiency of the CFU-S the effect is non-toxic and reversible. These results are summarised in Table 3.

Frindel & Guigon (1977) similarly reported an inhibitor of CFU-S proliferation in bone marrow. However, their extract was of much smaller molecular weight than our fraction IV, appearing in the dialysate of the crude extract. It is difficult to reconcile these two extracts since we found no activity in our corresponding fraction I. It is also a little disconcerting that their extract is obtained from foetal tissue in which one would expect the stem cells to be proliferating rapidly.

By contrast, regenerating bone marrow yields a fraction III (RBME-III) which is capable of stimulating DNA synthesis in normal bone marrow CFU-S (Table 3). It is difficult to test this factor for cellular specificity since the descendants of CFU-S are all proliferating cells. An indication of its cell line specificity is, however, obtained from the fact that, in terms of SCM measurements, it will not stimulate lymphocytes as does PHA nor does it block the stimulation of lymphocytes of PHA as does the extract of equivalent molecular weight from lymph nodes. In addition, since it is not detectable in NBME-III and neither are the equivalent fractions of RCE or GCE stimulatory to CFU-S proliferation, it is at least source specific. Frindel, Croizat & Vassort (1976) similarly found that bone marrow damaged by hydroxyurea stimulates CFU-S proliferation by means of a diffusible factor. It has been known for some time that even light damage to the CFU-S population results, very rapidly, in proliferation of the CFU-S (see review by Bryon, 1975). Thus, while Frindel *et al.* made no attempt to isolate the factor responsible it seems probable that the hydroxyurea produced a regenerating marrow which generated our stimulatory fraction III.

It is probable that these inhibitory and stimulatory factors are produced also *in vivo*. Proliferating CFU-S in bone marrow from a phenylhydrazine-treated mouse mixed with irradiated spleen cells from the same mouse are inhibited by the 'modifier' spleen population, which had originally contained non-proliferating CFU-S.

Table 4. *Interaction of cell populations containing proliferating and resting CFU-S*

'Assay' population	'Modifier' population	% CFU-S killed by $[^3H]$TdR
PHZ[a] bone marrow Regenerating bone marrow Regenerating spleen Foetal liver	—	45
PHZ[a] bone marrow Regenerating bone marrow Regenerating spleen Foetal liver	PHZ[a] spleen Normal bone marrow	< 10
PHZ[a] spleen Normal bone marrow	—	< 10
PHZ[a] spleen Normal bone marrow	PHZ[a] bone marrow Regenerating bone marrow Regenerating spleen Foetal liver	34

[a] PHZ, phenylhydrazine treatment.

Conversely, CFU-S in the spleen are stimulated by use of a 'modifier' bone marrow population. In fact, it appears that any haemopoietic tissue containing non-proliferating CFU-S can be used as a 'modifier' population to inhibit CFU-S in any 'assay' population containing proliferating CFU-S and vice versa (Table 4; Wright & Lord, 1977).

It should be noted that in some experiments, hydroxyurea or cytosine arabinoside have been used as an alternative to [^{3}H]TdR, singly or in combination with [^{3}H]TdR, to eliminate S-phase CFU-S and with identical results (E. G. Wright & B. I. Lord, unpublished data). This means that the effects described above cannot simply be explained as changes in thymidine pool sizes as has sometimes been suggested. Rather it must be considered that the proliferative status of haemopoietic stem cells varies in response to both proliferation inhibitors and stimulators. To this end, therefore, we studied the interaction of our extracted factors and found similarly that the stimulatory effect of RBME-III on normal non-proliferating CFU-S is abrogated by the subsequent addition of the inhibitory NBME-IV (Table 5; Lord, Mori & Wright, 1977). Conversely the inhibitory effect of NBME-IV on regenerating proliferating CFU-S is reversed by RBME-III. Thus, the interaction of cell populations and cell extracts to bring about a reversal in the proliferative status of haemopoietic stem cells means that these cells

Table 5. *Interaction of CFU-S proliferation stimulatory and inhibitory extracts from bone marrow populations*

'Assay' population	Extract treatment	% CFU-S killed by [^{3}H]TdR
Normal bone marrow	—	< 10
Normal bone marrow	RBME-III	39
Normal bone marrow	RBME-III+NBME-VI[a]	< 10
Regenerating bone marrow	—	27
Regenerating bone marrow	NBME-IV	< 10
Regenerating bone marrow	NBME-IV+RBME-III[a]	38

RBME-III, fraction III of regenerating bone marrow extract; NBME-IV, fraction IV of normal bone marrow extract.
[a] Extracts added together or separated by 1–2 hours.

are almost certainly maintained at an appropriate level of proliferation by a precise balance of inhibitory and stimulatory factors. The exact cellular source or sources of these factors is, however, as yet unknown.

CONCLUSION

Our present state of knowledge about the haemopoietic system suggests that the proliferation of both stem cells and maturing cells is under the control of inhibitory and stimulatory influences. In the case of the maturing cells these factors appear first in the circulation and exert their effect in the bone marrow. It seems probable that the stimulator acts principally, in the case of erythropoiesis at least, to put more committed erythroid precursor cells into the maturation compartment and thereafter accelerates their proliferation rate. On the other hand, the inhibitor increases the cycle duration of the maturing cells thus limiting the final output. Stem cell control is governed by factors produced locally within the marrow itself. These factors appear to act in a directly contrary manner to switch stem cell proliferation on or off, though their source and actual mode of action on the cells is still unknown.

REFERENCES

Bateman, A. E. (1974). Cell specificity of chalone-type inhibitors of DNA-synthesis released by blood leucocytes and erythrocytes. *Cell and Tissue Kinetics* **7**, 451–61.

Becker, A. J., McCulloch, E. A., Siminovitch, L. & Till, J. E. (1965). The effect of differing demands for blood cell production on DNA synthesis by haemopoietic colony forming cells of mice. *Blood*, **26**, 296–308.

Bjerknes, R. & Iversen, O. H. (1974). 'Antichalone'. A theoretical treatment of the possible role of antichalone in the growth control system. *Acta pathologica et microbiologica scandinavica, Section A*, Supplement **248**, 33–42.

Bøyum, A., Løvhaug, D. & Boecker, W. (1976). Regulation of bone marrow cell growth in diffusion chambers. *Blood*, **48**, 373–84.

Bullough, W. S. (1975). Mitotic control in adult mammalian tissues. *Biological Reviews*, **50**, 99–130.

Byron, J. W. (1975). Manipulation of the cell cycle of the hemopoietic stem cell. *Experimental Hematology*, **3**, 44–53.

Cercek, L., Cercek, B. & Ockey, C. H. (1973). Structuredness of the cytoplasmic matrix and Michaelis–Menten constants for the hydrolysis of FDA during the cell cycle in Chinese hamster ovary cells. *Biophysik*, **10**, 187–94.

Croizat, H., Frindel, E. & Tubiana, M. (1970). Proliferative activity of the stem cells in the bone marrow of mice after single and multiple irradiations (total or partial body exposure). *International Journal of radiation Biology*, **18**, 347–58.

DEXTER, T. M., ALLEN, T. D. & LAJTHA, L. G. (1977). Conditions controlling the proliferation of haemopoietic stem cells. *Journal of cellular Physiology*, **91**, 335–44.

FRINDEL, E., CROIZAT, H. & VASSORT, F. (1976). Stimulating factors liberated by treated bone marrow: *in vitro* effect on CFU kinetics. *Experimental Hematology*, **4**, 56–61.

FRINDEL, E. & GUIGON, M. (1977). Inhibition of CFU entry into cycle by a bone marrow extract. *Experimental Hematology*, **5**, 74–6.

GIDALI, J. & LAJTHA, L. G. (1972). Regulation of haemopoietic stem cell turnover in partially irradiated mice. *Cell and Tissue Kinetics*, **5**, 147–57.

GURNEY, C. W., WACKMAN, N. & FILMANOWICZ, E. (1961). Studies on erythropoiesis. XVII. Some quantitative aspects of the erythropoietic response to erythropoietin. *Blood*, **17**, 531–46.

HODGSON, G. (1970). Mechanism of action of erythropoietin. In *Regulation of hematopoiesis*, ed. A. S. Gordon, vol. 1, pp. 459–69. New York: Appleton-Century-Crofts.

KIVILAAKSO, E. & RYTÖMAA, T. (1971). Erythrocyte chalone, a tissue specific inhibitor of cell proliferation in the erythron. *Cell and Tissue Kinetics*, **4**, 1–9.

LAERUM, O. D. & MAURER, H. R. (1973). Proliferation kinetics of myelopoietic cells and macrophages in diffusion chambers after treatment with granulocyte extracts (chalone). *Virchows Archiv*, **14B**, 293–305.

LORD, B. I. (1975). Modification of granulocytopoietic cell proliferation by granulocyte extracts. *Bollettino dell'Istituto sieroterapico milanese*, **54**, 187–94.

LORD, B. I., CERCEK, L., CERCEK, B., SHAH, G. P., DEXTER, T. M. & LAJTHA, L. G. (1974*a*). Inhibitors of haemopoietic cell proliferation?: specificity of action within the haemopoietic system. *British Journal of Cancer*, **29**, 168–75.

LORD, B. I., CERCEK, L., CERCEK, B., SHAH, G. P. & LAJTHA, L. G. (1974*b*). Inhibitors of haemopoietic cell proliferation: reversibility of action. *British Journal of Cancer*, **29**, 407–9.

LORD, B. I., LAJTHA, L. G. & GIDALI, J. (1974*c*). Measurement of the kinetic status of bone marrow precursor cells: three cautionary tales. *Cell and Tissue Kinetics*, **7**, 507–15.

LORD, B. I., MORI, K. J. & WRIGHT, E. G. (1977). A stimulator of stem cell proliferation in regenerating bone marrow. *Biomedicine Express*, **27**, 223–6.

LORD, B. I., MORI, K. J., WRIGHT, E. G. & LAJTHA, L. G. (1976). An inhibitor of stem cell proliferation in normal bone marrow. *British Journal of Haematology*, **34**, 441–5.

LORD, B. I., SHAH, G. P. & LAJTHA, L. G. (1977*a*). The effects of red blood cell extracts on the proliferation of erythrocyte precursor cells, *in vivo*. *Cell and Tissue Kinetics*, **10**, 215–22.

LORD, B. I., TESTA, N. G., WRIGHT, E. G. & BANERJEE, R. K. (1977*b*). Lack of an effect of a granulocyte proliferation inhibitor on their committed precursor cells. *Biomedicine*, **26**, 315–20.

MACVITTIE, T. J. & MCCARTHY, K. F. (1975). The influence of a granulocytic inhibitor(s) on hematopoiesis in an *in vivo* culture system. *Cell and Tissue Kinetics*, **8**, 553–60.

Paukovits, W. R. (1971). Control of granulocyte production: separation and chemical identification of a specific inhibitor (chalone). *Cell and Tissue Kinetics*, **4**, 539–47.

Reissmann, K. R. & Samorapoompichit, S. (1970). Effect of erythropoietin on proliferation of erythroid stem cells in the absence of transplantable colony forming units. *Blood*, **36**, 287–96.

Rencricca, N. J., Rizzoli, V., Howard, D., Duffy, P. & Stohlman, F., Jr (1970). Stem cell migration and proliferation during severe anemia. *Blood*, **36**, 764–71.

Rytömaa, T. & Kiviniemi, K. (1968). Control of granulocyte production. I. Chalone and antichalone, two specific humoral regulators. *Cell and Tissue Kinetics*, **1**, 341–50.

Stanley, E. R. & Metcalf, D. (1971). Enzyme treatment of colony stimulating factor: evidence for a peptide component. *Australian Journal of experimental Biology and medical Science*, **49**, 281–90.

Stohlman, F., Jr, Ebbe, S., Morse, B., Howard, D. & Donovan, J. (1968). Regulation of erythropoiesis. XX. Kinetics of red cell production. *Annals of the New York Academy of Sciences*, **140**, 156–62.

Till, J. E. & McCulloch, E. A. (1961). A direct measurement of the radiation sensitivity of normal mouse bone marrow cells. *Radiation Research*, **14**, 213–22.

Udupa, K. B. & Reissmann, K. R. (1975). Stimulation of granulopoiesis by androgens without concomitant increase in the serum level of colony stimulating factor. *Experimental Hematology*, **3**, 26–43.

Wright, E. G. & Lord, B. I. (1977). Regulation of CFU-S proliferation by locally produced endogenous factors. *Biomedicine Express*, **27**, 215–18.

Ultrastructural aspects of in-vitro haemopoiesis

T. D. ALLEN

Department of Ultrastructure, Paterson Laboratories,
Christie Hospital and Holt Radium Institute, Withington,
Manchester M20 9BX, UK

A system of cultured bone marrow producing active proliferation of stem cells or CFU-S (Till & McCulloch, 1961) and committed granulocyte precursors or CFU-C (Bradley & Metcalf, 1966) has been maintained in these laboratories for some time. Proliferation of stem cells in such conditions can be maintained *in vitro* for several months (Dexter & Lajtha, 1976; Dexter & Testa, 1976). This account presents both transmission electron microscope (TEM) and scanning electron microscope (SEM) observations on the various cell types present in the cultures, and their interactions; parallels are drawn with the in-vivo haemopoiesis.

Culture techniques

These have been described in full elsewhere (Dexter & Testa, 1976) and comprise briefly the inoculation of 10^7 femoral mouse bone marrow cells into 4-oz flat-bottomed, screw-capped glass bottles in medium supplemented with 20% horse serum. Half the culture medium (and those cells removed by this gentle agitation) is removed at weekly intervals and replaced by fresh medium and the cells assayed for CFU-S and CFU-C. Over the initial 3-week period an 'adherent' layer of cells develops, and after this time the cultures are 'recharged' by the addition of a further 10^7 cells. The cultures are then 'mature' and produce CFU-S and CFU-C for up to several months.

Electron microscopy

The methods used were as described previously (Allen & Dexter, 1976*a*). Briefly, they are as follows:

For TEM, a pellet spun from the decanted medium was fixed with glutaraldehyde and osmium tetroxide and embedded in Epon–

Araldite. For SEM, cells were fixed, dehydrated and critical-point dried from carbon dioxide either *in situ*, on coverslips or on silver filters; sputter-coated with gold (Allen & Simmens, 1976) and examined in a Cambridge S4-10 scanning electron microscope.

RESULTS AND DISCUSSION

Stem cell and granulocyte morphology

Until an unequivocally pure population of stem cells can be isolated, morphological criteria for stem cell identification will remain a matter for conjecture. Those suggestions as to the morphology of 'candidate' stem cells which have been put forward (Dicke *et al.*, 1973; Rosse, 1973) do, however, generally agree on the undifferentiated nature of the cell and its superficial resemblance in general to a small lymphocyte (Yoffey, 1973). The cell is 5–8 μm in diameter, with a smooth round nuclear profile and a thin rim of marginal condensed chromatin. The cytoplasm is undifferentiated and the cell membrane is also smooth in profile (Fig. 1*a*). This cell type has been readily observed in the cultures (see Allen & Dexter, 1976*a*) and also characterised in the scanning electron microscope as a largely smooth spherical cell (Fig. 1*b*). No attempt has been made to differentiate the cells of this type into CFU-S, CFU-C or early myeloblasts on morphological grounds and they have all been classified as 'undifferentiated' cells.

The various stages of in-vitro granulocyte differentiation show close similarity to those described *in vivo* (Wetzel, 1970). As maturation proceeds, nuclear segmentation becomes more marked, accompanied by an increase in the amount of condensed chromatin. In the promyelocyte–myelocyte stage this is shown in the initial formation of nuclear clefts, and the initial appearance of an obvious Golgi region. At this stage the earliest formation of cell surface

Fig. 1. (*a*) Transmission electron micrograph of an undifferentiated cell, interpreted to represent the morphology of the stem cell to myeloblast stages in the culture. The nucleo-cytoplasmic ratio is relatively high, with the cytoplasm largely devoid of organelles other than ribosomes and mitochondria. The condensed peripheral chromatin is limited to a narrow margin of the nucleus, and the nuclear profile is smooth, as is the cell membrane itself. (*b*) Scanning electron micrograph of the same cell type as in (*a*), showing the overall undifferentiated nature of the surface of this spherical cell. (From Allen & Dexter (1976). Surface morphology and ultrastructure of murine granulocytes and monocytes in long term liquid culture. *Blood cells*, **2**, 591–606.)

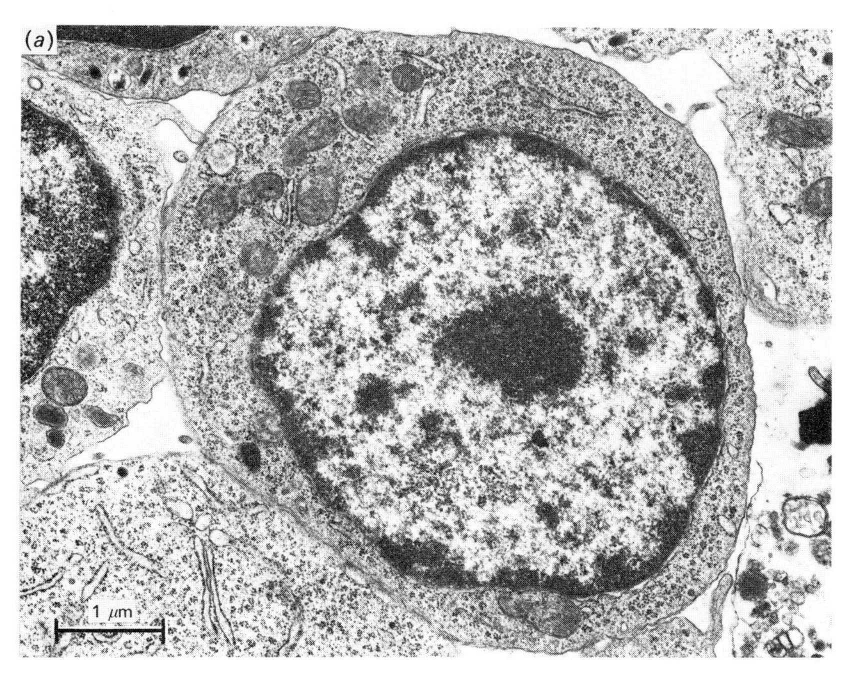
(a)
1 μm

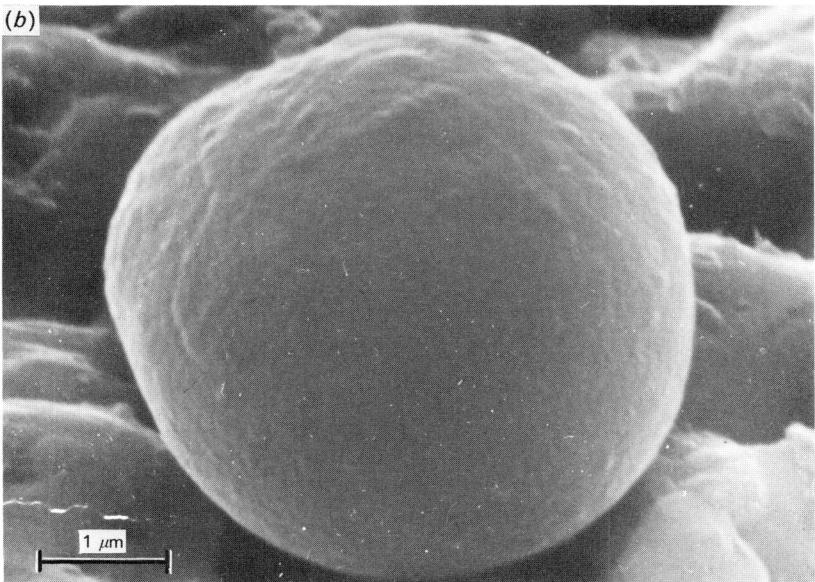
(b)
1 μm

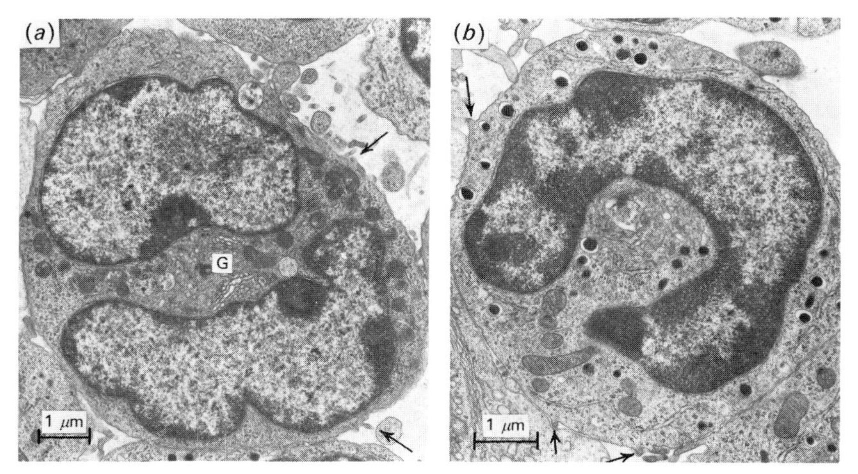

Fig. 2. Transmission micrographs of granulocyte maturation: (*a*) (promyelocyte–myelocyte) shows the initial nuclear cleft and appearance of the Golgi region (G), with some surface projections (arrowed); (*b*) (metamyelocyte) shows a typical 'band cell' nucleus, the appearance of the granule population and also surface projections (arrowed). (From Allen & Dexter (1976). Surface morphology and ultrastructure of murine granulocytes and monocytes in long term liquid culture. *Blood Cells*, **2**, 591–606.)

projections becomes visible (Fig. 2*a*). Nuclear segmentation proceeds further in the metamyelocyte stage, leading to the term 'band cells'. Granule formation becomes established and the cell membrane also shows surface projections (Fig. 2*b*). Scanning electron micrographs of cells in these intermediate stages of differentiation show a spectrum of surface morphology from an almost completely smooth surface to one completely covered with ridges and projections (Fig. 3). Use of differential scorings in parallel light microscope preparations, and the increase in size in intermediate stages of granulocyte maturity (Wetzel, 1970) has enabled a tentative classification on surface morphology of the various stages of granulocyte maturation (see Allen & Dexter, 1976*a*). That increasing maturation brings increased surface elaboration would seem evident from the sections of mature granulocytes which show the nuclear segmentation and granule population characteristic of circulating neutrophils, and also an extremely involved cell surface profile (Figs. 4 and 5*a*). When the mature polymorphonuclear (PMN) granulocytes are settled on a cellular substratum (Lajtha, 1976) they retain their spherical shape and often exhibit a localised region of

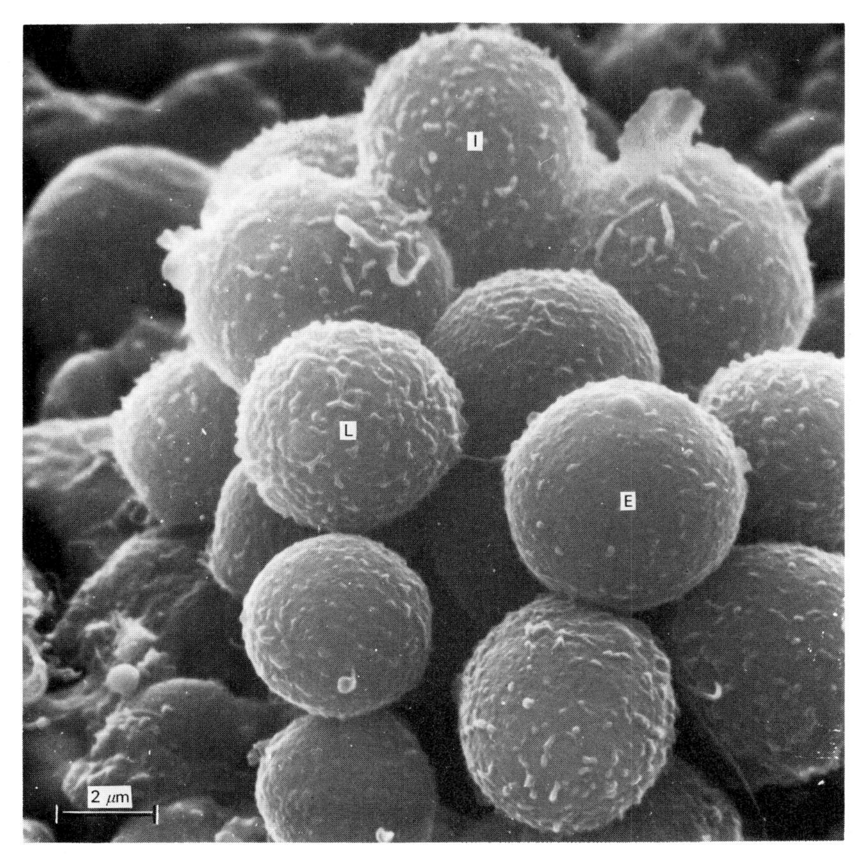

Fig. 3. Scanning micrograph of a clump of granulocytes interpreted to be at different stages of maturation as indicated by their varying surface morphology. The earlier stages are the more smooth, and increasing maturity is indicated by further sophistication of the membrane topography. Size variations are also apparent, as the granulocyte series shows an increase in size at the myelocyte–metamyelocyte stage. Interpretation of early, intermediate and late stages of maturity are indicated by E, I, and L. (From Allen & Dexter (1976). Surface morphology and ultrastructure of murine granulocytes and monocytes in long term liquid culture. *Blood Cells*, **2**, 591–606.)

numerous petal-like projections which may be characteristic for mature granulocytes (Fig. 5*b*).

Monocytes

The morphology of monocytes in this culture system is characterised mainly by their ability to settle and spread on the glass substratum, forming a large proportion of the adherent layer. In this situation

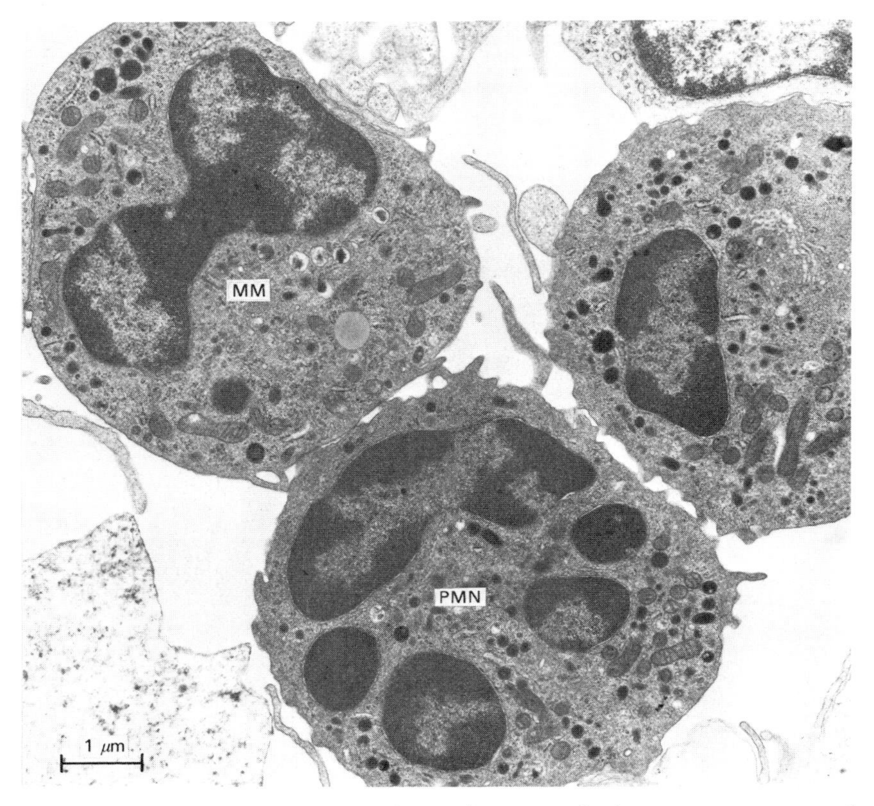

Fig. 4. Transmission micrograph of granulocytes at the late to mature stages of differentiation. The nuclear segmentation of the metamyelocyte (MM) has proceeded to its full extent in the mature granulocyte (PMN) which shows five nuclear profiles with largely condensed chromatin, and a granule population characteristic of a neutrophil. The surface profile of the PMN is almost completely elaborated into ridges and folds. (From Allen & Dexter (1976). Surface morphology and ultrastructure of murine granulocytes and monocytes in long term culture. *Blood Cells*, **2**, 591–606.)

Fig. 5. Scanning micrographs of mature granulocytes, showing (*a*) the ridges and projections that cover the entire surface of the PMN granulocytes, and (*b*) the localised groups of 'petal-like' projections that also occur (arrowed). ((*a*) from Allen & Dexter (1976). Surface morphology and ultrastructure of murine granulocytes and monocytes in long term liquid culture. *Blood Cells*, **2**, 591–606.)

Fig. 6. Scanning micrograph of monocyte cells (M) on the substratum, showing a spread morphology mainly along a single longitudinal axis with a well marked leading edge (L) and trailing retraction fibres (R). (From Dexter, Allen & Lajtha (1977). Conditions controlling the proliferation of haematopoietic stem cells *in vitro*. *Journal of cellular Physiology*, in press.)

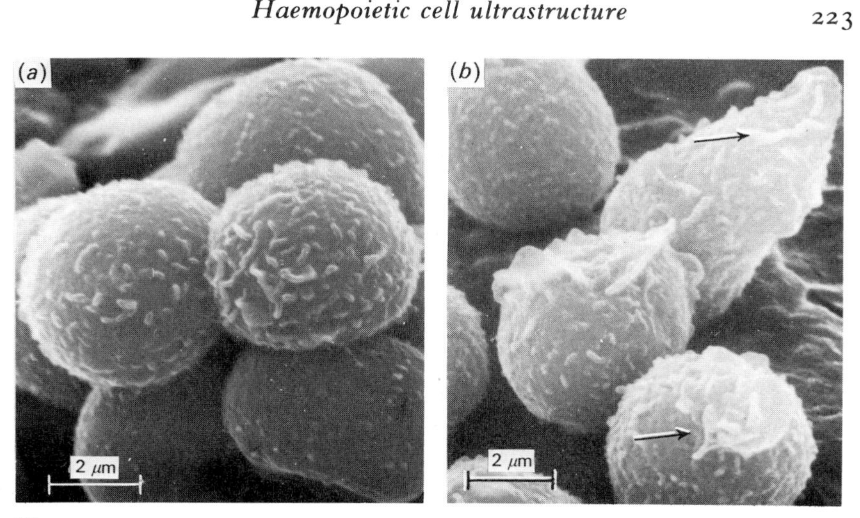

Fig. 5

Fig. 6

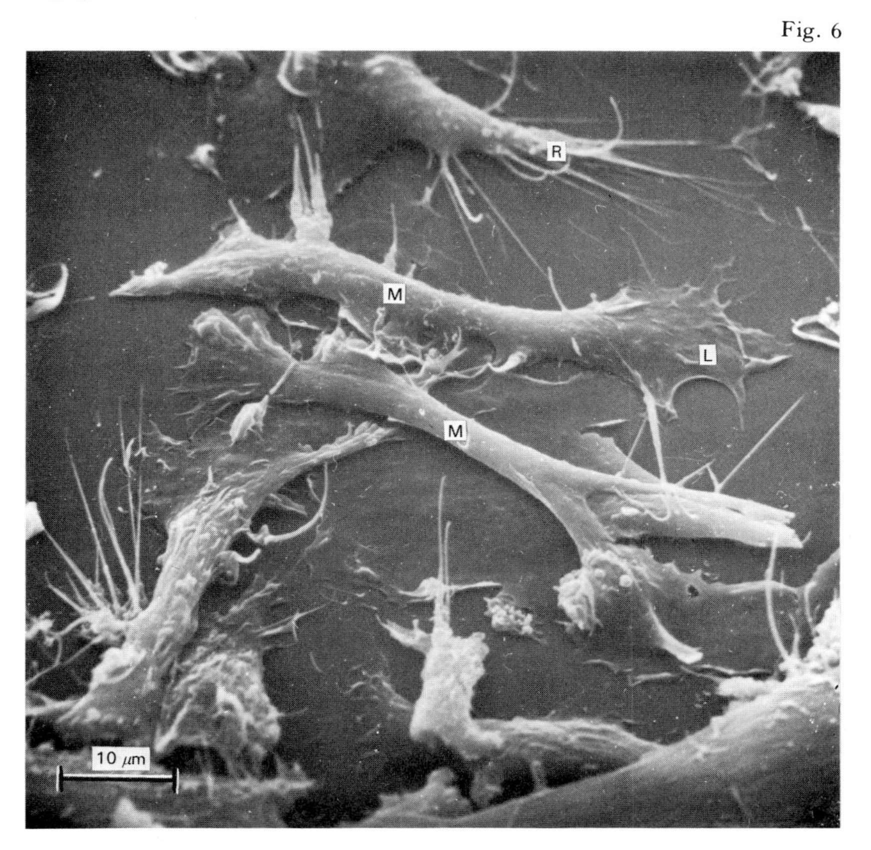

SEM shows the monocytes to spread mainly along a single longitudinal axis, with a well-marked leading edge and trailing retraction fibres (Fig. 6). Their morphology when attached to the surface of the epithelioid members (see below) of the adherent layer tends to be rather less well spread (Fig. 9) and in vertical sections in TEM show obvious phagocytic extensions (Fig. 10*a*). The monocytes are also easily identified in TEM sections by their well-marked cytoplasmic population of phagocytic inclusions (Figs. 16–18). Their distribution within the cultures would appear ubiquitous, occurring at all levels of the multilayered adherent population and also all over the substratum of the culture vessel, with no tendency for aggregation (as in the case of granulocytes and fat cells) nor cell–cell adhesion to form monolayer sheets (as in the case of the epithelioid cells).

Megakaryocytes

Although it has not yet been possible to identify all stages of thrombopoiesis in the cultures, occasional cells of the megakaryocytic series have been observed. These are large cells, up to 30 μm in diameter, and when collected on silver filters from suspension they have an overall spherical morphology (Fig. 7*a*) with the surface characterised by exoplasmic bullae (Bessis, 1973) some 2–4 μm in diameter; they bear a strong resemblance to circulating non-activated platelets in humans (Walsh & Barnhart, 1973) and mouse (T. D. Allen, unpublished data). These 'candidate' platelets are circular to oval and 2–4 μm in diameter (Fig. 7*b*). Their surface is somewhat indented but without any external extensions, in contrast to the remainder of the megakaryocyte surface which is covered with numerous extended microvilli, some of which may still connect with the platelet surface (Fig. 7*b*).

'Epithelioid' cells

Two cell types are characterised by their extreme flattening and consequent 'epithelioid' morphology in the cultures: (*a*) large single cells some 40–60 μm in diameter ('R' cells), with a sparsely microvillous surface (Fig. 8); or (*b*) smaller cells with a smooth surface aggregated into sheets (Fig. 9). The smaller cells ('E' cells) which form monolayer sheets are interpreted to arise from the endothelium of the venous sinuses in the bone marrow. The origin of the larger 'epithelioid' cells (R cells) is uncertain but they may originate from

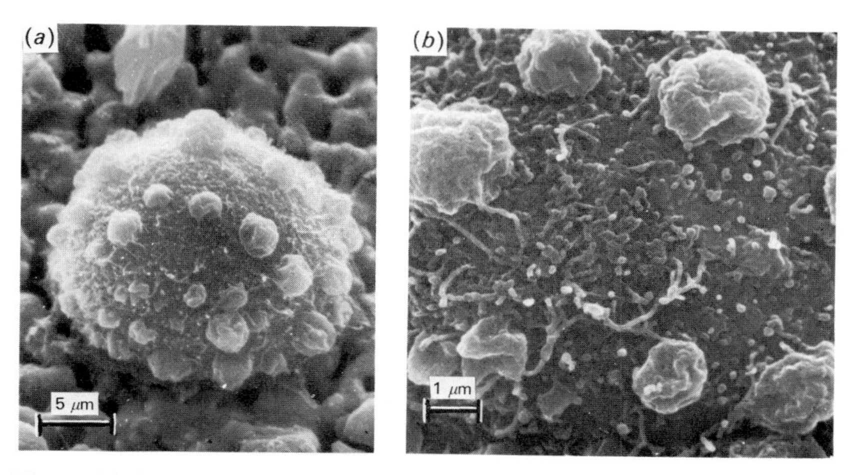

Fig. 7. (*a*) Scanning micrograph of a cell of the megakaryocyte series, showing exoplasmic bullae over the surface some 2–4 μm in diameter. (*b*) Higher power micrograph of a region of the surface of (*a*), showing the individual platelets with their somewhat indented surface and the dense microvillous surface of the remainder of the megakaryocyte.

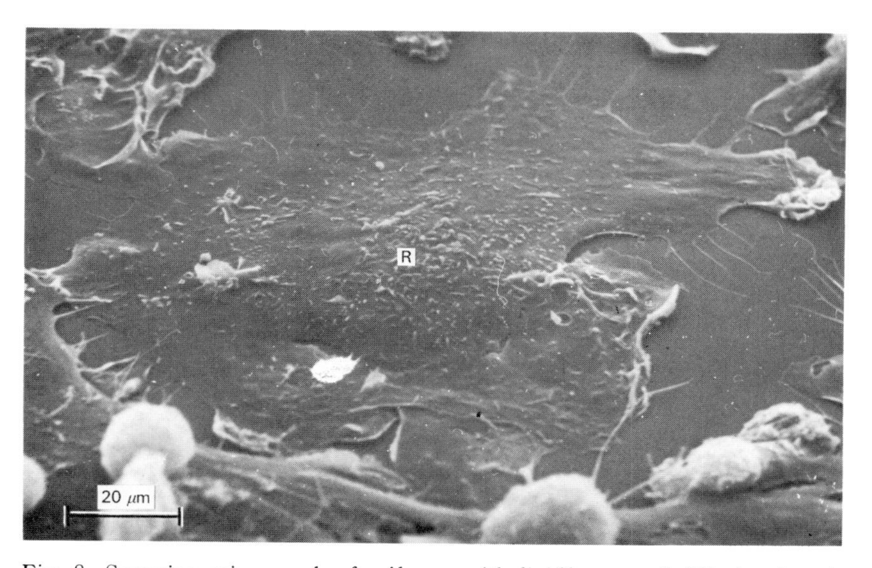

Fig. 8. Scanning micrograph of a 'large epithelioid' type cell (R) showing the extreme flattened morphology and sparse surface microvilli which may be adopted by a bone marrow reticular cell in culture conditions. (From Dexter, Allen & Lajtha (1977). Conditions controlling the proliferation of haematopoietic stem cells *in vitro*. *Journal of cellular Physiology*, in press.)

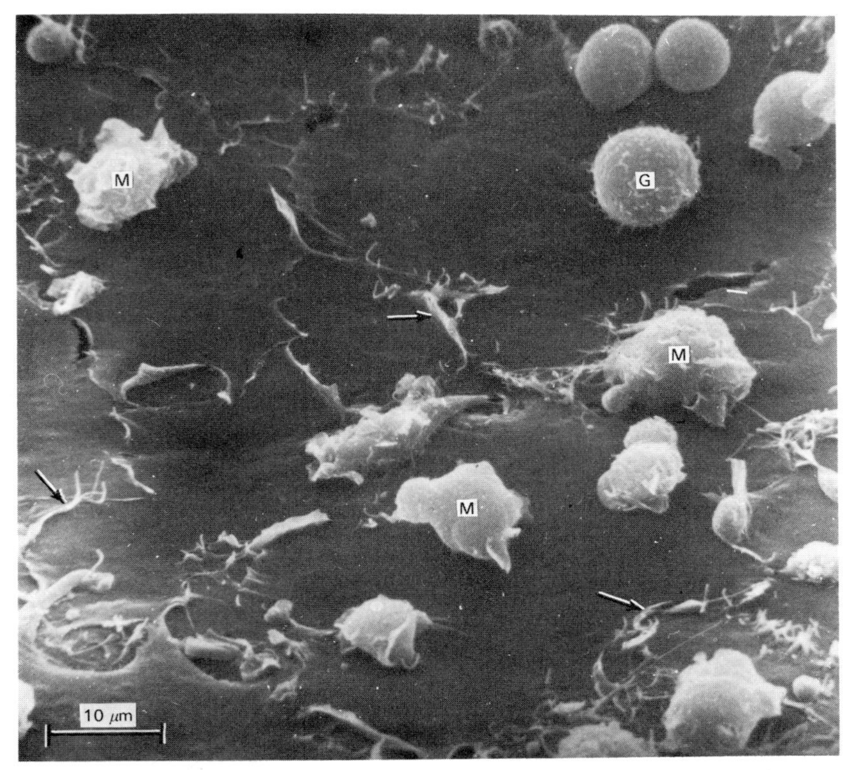

Fig. 9. Scanning micrograph of a cell sheet formed from endothelial-derived cells. This sheet may overly other cells (Fig. 10) and act itself as a base for further cell attachment, in this instance of monocytes (M) and granulocytes (G). The surface of the endothelial cells themselves is smooth apart from the regions of edge overlap (arrowed).

the reticular cells (Weiss, 1976) of the marrow which instead of forming the reticulum of the bone marrow stroma *in vivo*, adopt a flattened morphology when allowed to settle on a glass substratum. Although the flattened morphology of the R cells is usually epi-

---

Fig. 10. (*a*) Transmission micrograph of a section cut at right-angles to the culture vessel growing surface (along the long axis of the page). The adherent layer is two to three cells deep in most places. A monocyte (M, far right) is attached to the upper surface of a flattened epithelioid (endothelial-derived) cell (E) which covers in turn the remainder of the cells present, including an early stage of fat accumulation (F, far right), several other monocytes (M) and an immature granulocyte (IG). (*b*) Scanning micrograph of an area similar to that shown in section in (*a*). The edge of the sheet of epithelioid (endothelial-derived) cells (E) partially covers several monocytes (M).

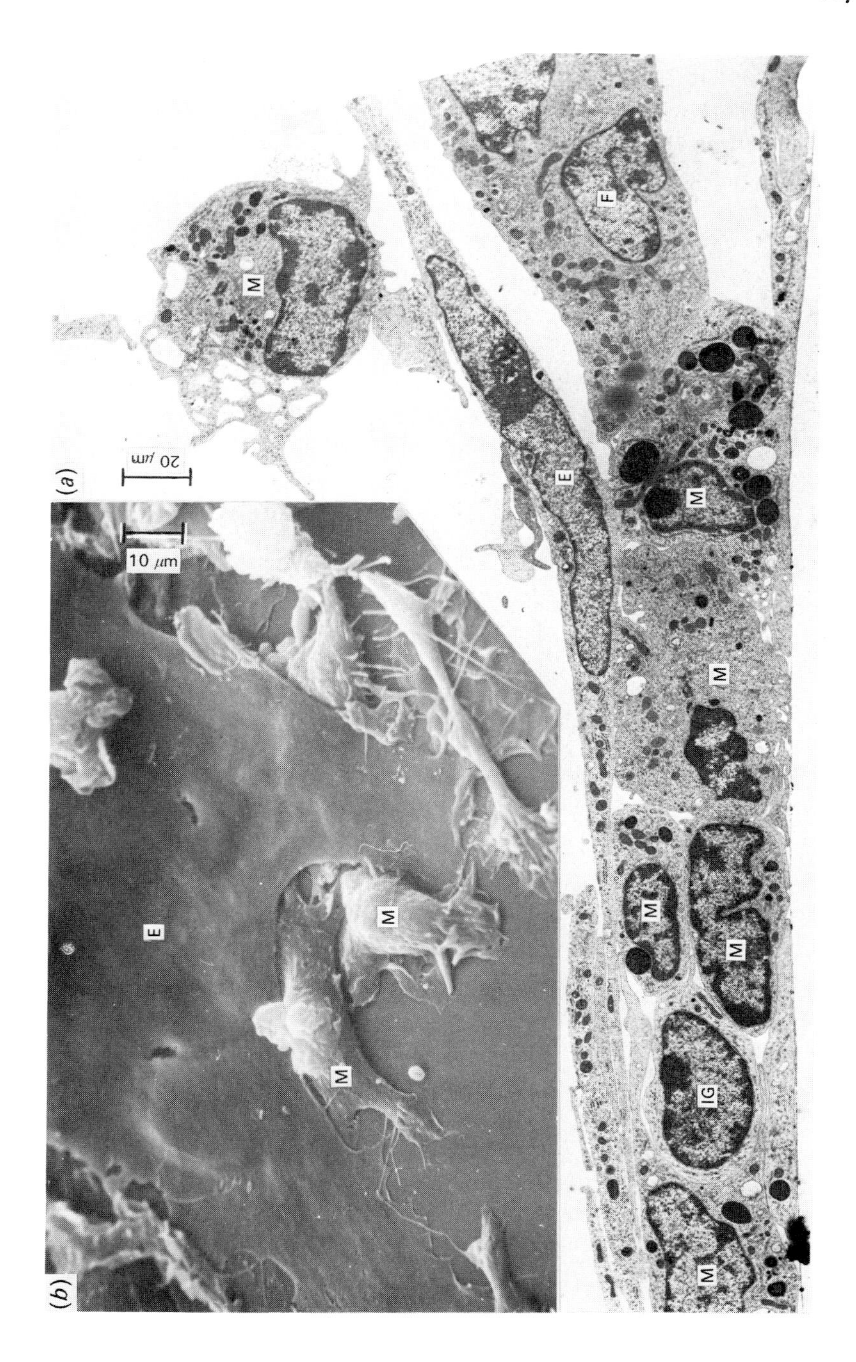
(a)
20 μm
M
F
E
M
M
M
M
IG
M
(b)
10 μm
E
M
M

thelioid it may also be fibroblastic, as illustrated when these cells accumulate fat (see below).

In vertical section, the E cells appear as a flattened sheet overlying possibly two or three other cell thicknesses (Fig. 10*a*, *b*) with the upper surface forming a substratum for further cell attachment (Figs. 16–18). The E cells are easily distinguished in section from the spread monocytes as they contain no phagocytic inclusions and have a more flattened nuclear profile with only a narrow rim of condensed peripheral chromatin (Figs. 10*a*, 16–18). The flattened nuclear profile also causes the nucleolus to traverse the entire depth of the nuclear profile (Fig. 10*a*). The smooth surface of these cells visualised with the SEM is also reflected in their surface profiles in the TEM (Figs. 10*a*, 16–18). The overall ultrastructural similarity of the 'in-vitro' endothelial cell to its possible 'in-vivo' counterpart may be judged from Fig. 20, a section through a venous sinous of normal mouse bone marrow.

Fat cells

The presence of large fat-containing cells in the bone marrow cultures is taken as a reliable indicator of continued production of stem cells (CFU-S) and granulocyte precursors (CFU-C). In contrast, the absence of fat cells indicates that the cultures will quickly convert to monocyte production alone (Dexter, Allen & Lajtha, 1977). Cultures with continuing proliferation of CFU-S and CFU-C are therefore easily recognisable by their large populations of fat cells, which form densely packed regions several square millimetres in area and easily visible on the base of the culture vessel.

The appearance of fat cells in the cultures occurs during the second half of the initial 3-week period of adherent layer establishment, prior to 'recharging' with a further inoculum of 10^7 bone marrow cells. Fat droplets are seen to accumulate in large cells with both an 'epithelioid' or 'fibroblastic' morphology (Fig. 11*a*). In

Fig. 11. (*a*) Scanning micrograph of a region of fat accumulation during the establishment of the adherent layer. Cells with 'epithelioid' (E) or fibroblast (F) morphology can be seen to be distended with spherical fat droplets. (*b*) Scanning micrograph of a region of fat accumulation at a later stage than that shown in (*a*) but before the addition of fresh marrow and consequent stem cell production and granulopoiesis. The overall shape of the cells tends to become spherical as they become 'inflated' by fat droplets which coalesce to form large single vacuoles. This increase in size results in a crowded appearance of the previously spaced cells.

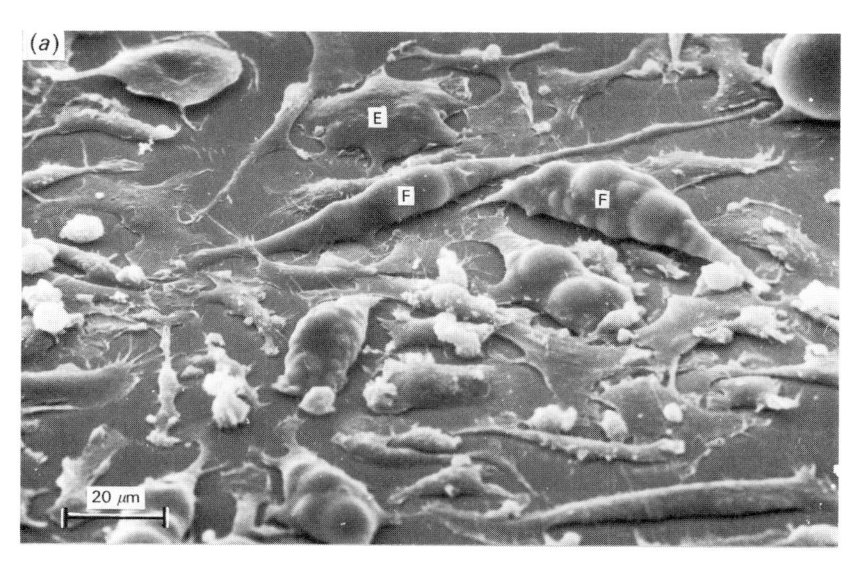
(a)
E
F
F
20 μm

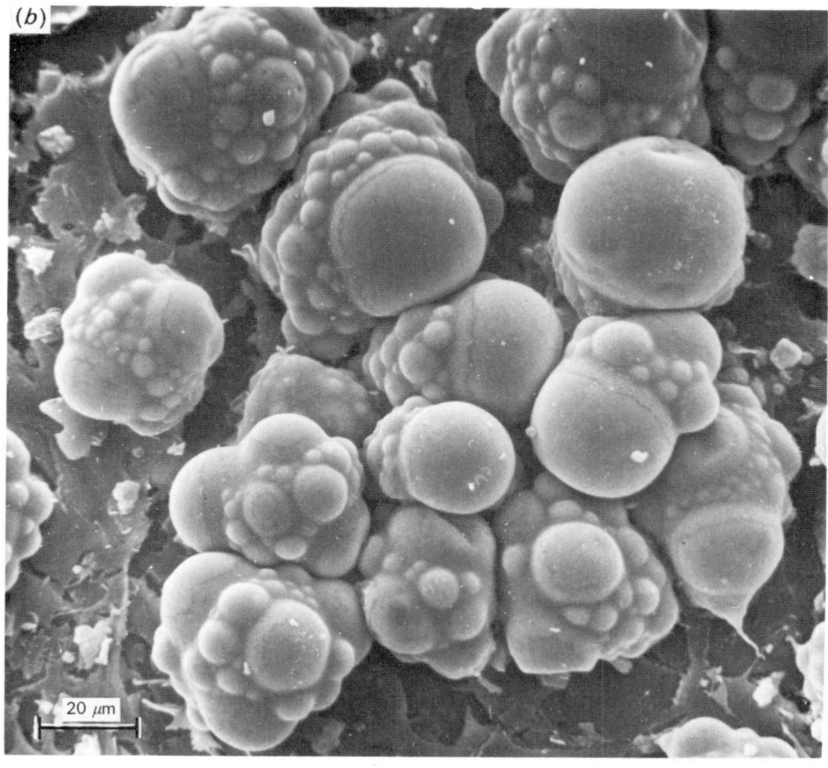
(b)
20 μm

TEM sections cut at right-angles to the growing surface, incipient fat cells (Fig. 16) can be identified by small fat droplets in the cytoplasm, at a stage at which the ultrastructural characteristics are still recognisable before the cell becomes 'inflated' by lipid accumulation. This lipid is synthesised within the cell, as shown by its single vacuolar membrane, and not phagocytosed from the external environment. The nuclear profile of the fat cell is round to oval with localised indentation of the nuclear membrane, and regions of condensed chromatin. The cytoplasm in stages of early fat accumulation is largely undifferentiated except for some extensive cisternae of endoplasmic reticulum, and numerous mitochondria. As fat droplets accumulate, so the mitochondria appear to increase in size and number (Fig. 16). With increasing fat accumulation, the cells assume an overall spherical shape (Figs. 11*b*, 13). Internally the numerous lipid vacuoles coalesce to form droplets 60–80 μm in diameter (Fig. 18). Stretched round these droplets is a cytoplasmic rim so thin that mitochondrial profiles become visible from the surface of the scanning microscope. These profiles may run a sinous, sometimes branched, course over the surface of the fat cell lipid vacuole for 10–15 μm.

On the gross level, the fat cells all tend to arise in a single region so that they form close aggregates of many cells in similar stages of fat accumulation (Fig. 11*b*). With time the aggregates become larger and radiate outwards in foci measuring in excess of 1 cm and containing many thousands of fat cells (Fig. 14).

The normal precursors of fat cells in bone marrow are the adventitial reticular cells (Weiss, 1976; see Fig. 12) and it seems reasonable to assume that these elements of bone marrow stroma have been preserved in the culture along with the other cell types present, as 'R' cells. The slightly more difficult aspect to correlate with the in-vivo situation is their accumulation of lipid, for 'fatty marrow' as such occurs rarely in mouse. As the same reticular cells accumulate fat in rat, rabbit, man, etc. (Weiss, 1976) it is apparent that this is a basic property of reticular cells, and occurs in the prevailing culture conditions *in vitro* in mouse reticular cells. More significant, however, is the fact that in the absence of the fat cells, stem cell and granulocyte production is not maintained and the cultures produce monocytes only (Dexter *et al.*, 1977). Furthermore, there is a well-marked clustering of granulocytes around the areas of fat cells in the cultures (Fig. 15*a*). Comparison of Fig. 11*b* (from a

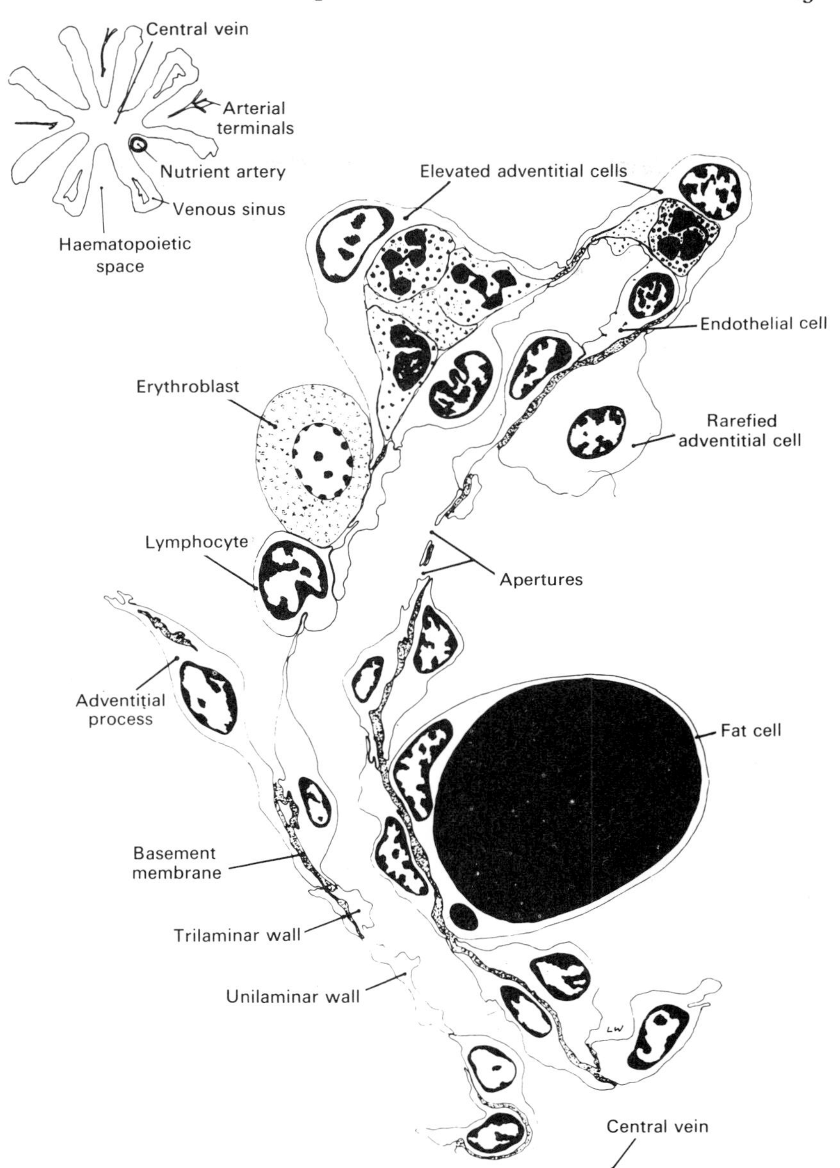

Fig. 12. Diagram of bone marrow histology (from Weiss, 1970). Inset shows the venous sinuses anastomosing into the central vein. Main diagram shows the arrangement of cells around an individual sinus, which is lined with endothelial cells, which are in some regions separated from the adventitial cells by a basement membrane, and in other areas there is direct contact. An adventitial cell which has accumulated fat is also shown. (Reproduced by kind permission of Dr L. Weiss and the Wistar Press.)

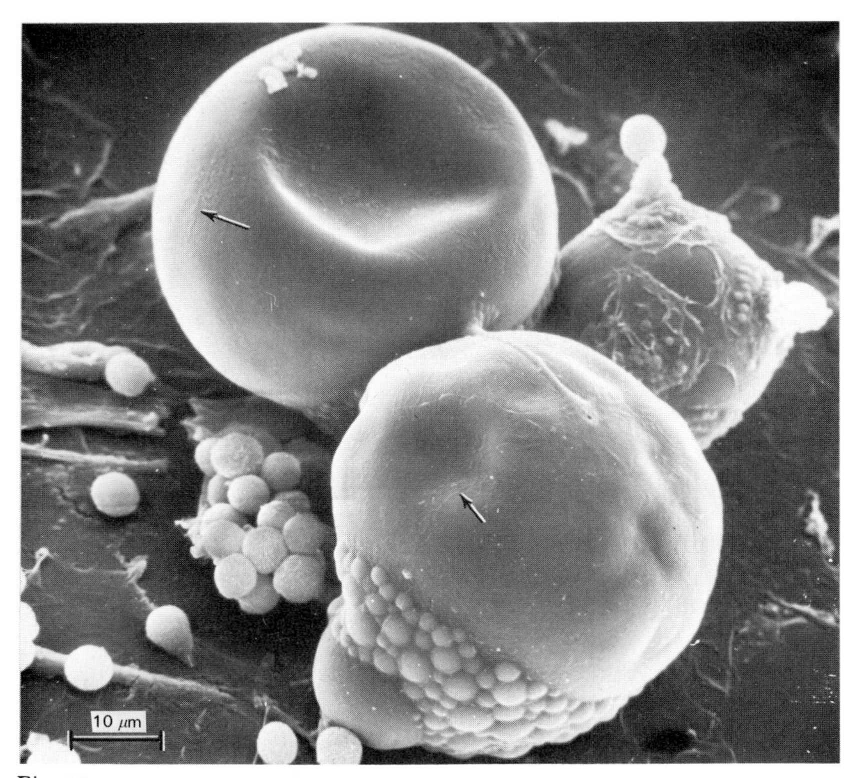

Fig. 13

Fig. 14

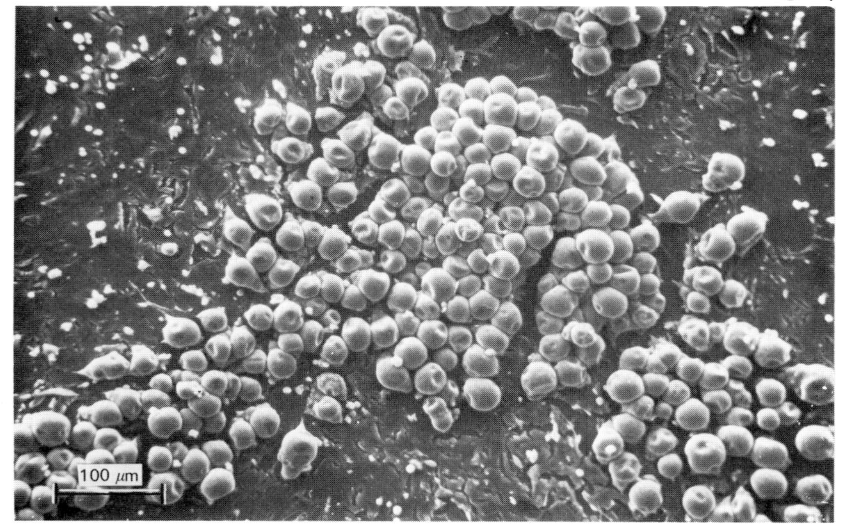

culture at the end of the period of establishment of the adherent layer) with Fig. 15*a* (from a culture soon after the addition of the second inoculum) in full granulocyte production, shows the clustering of approximately 200 granulocytes around a relatively small region of some 20 fat cells. Much larger numbers of granulocytes are concentrated in the regions where the fat cell aggregation numbers hundreds. Closer examination, however, indicates that there is rather more association in the areas of the periphery of the fat cell regions than in the centre, suggesting a preference for association of granulocytes in the area of those cells actively synthesising lipid rather than those which have already accumulated it (Fig. 15*b*).

Cellular interactions in the adherent layer

Sections cut vertical to the growing surface reveal an overall organisation of the adherent layer in the regions where it is composed of several cell layers. The epithelial cell monolayer is not usually observed directly apposed to the growing surface (Figs. 10, 16–18) possibly as a result of a slower growth rate in comparison to the immediate colonisation of the growing surface by the monocyte population. Although granulocytes will also attach directly to a glass surface (Allen & Dexter, 1976*a*) they appear to prefer attachment to a cellular substratum (Lajtha, 1976) and consequently tend to be attached to the surfaces of epithelioid, monocyte or fat cells. Cell division is also often observed in this position (Fig. 17). With respect to their vertical position in the adherent layer, however, the more mature granulocytes tend to be situated at the surface (Figs. 16, 18) whereas the blasts (and possibly stem cells) are often located within the adherent layer, usually between an underlying monocyte and an overlying epithelioid cell (Figs. 10*a*, 18). Furthermore, junctional complexes are observed (Allen & Dexter, 1976*b*) between

---

Fig. 13. Scanning micrograph of two fat cells shown at a stage close to their maximum. The majority of the cell volume is occupied by a single large fat vacuole (measuring 40–50 μm in diameter). In the thin layer of cytoplasm over the surface, mitochondria profiles are visible (arrowed). The relative size of the granulocyte population can be compared with those on the left of the fat cells.

Fig. 14. Low-power scanning micrograph of a region of fat cells forming part of a well-established area of fat accumulation. This region of approximately 200 fat cells forms only a fraction of the total area, which may occupy many square millimetres on the surface of the culture vessel.

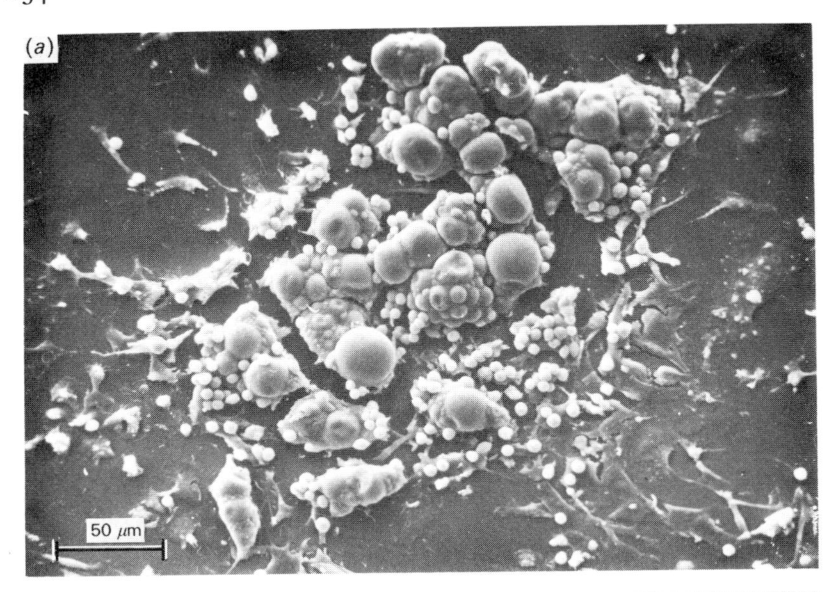
(a)
50 μm

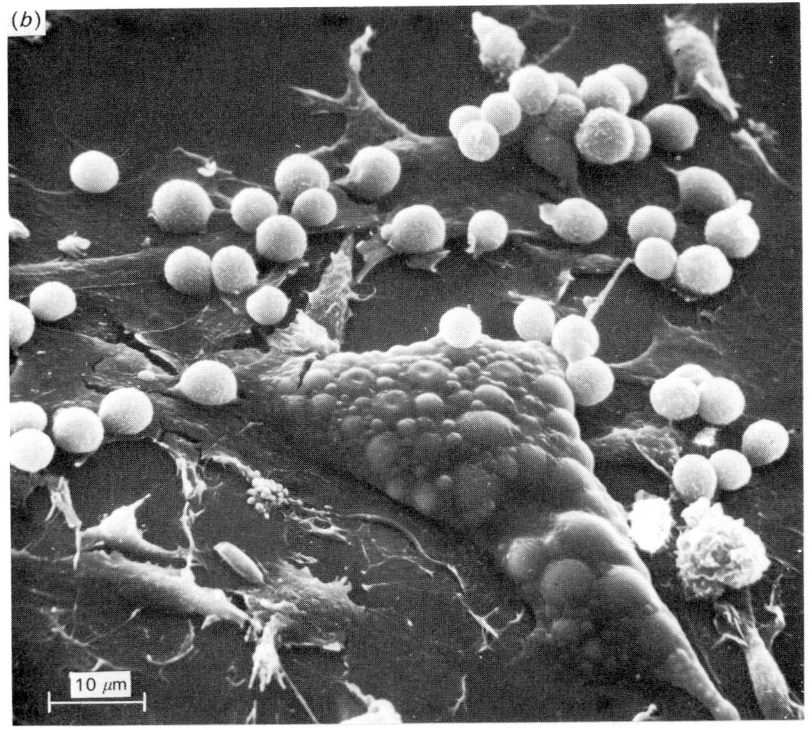
(b)
10 μm

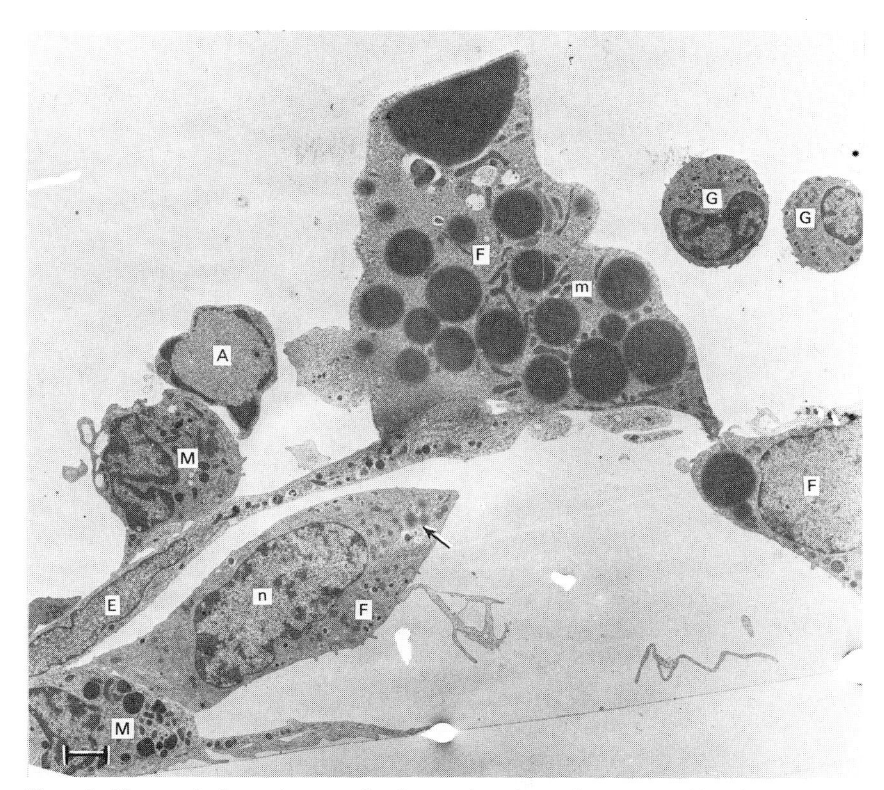

Fig. 16. Transmission micrograph of a section through an area of lipid synthesis. Three fat cells are visible (F), left, right and upper. The left fat cell is in a very early stage of lipid synthesis, with only two small fat droplets apparent in the cytoplasm (arrowed). The nuclear profile in this cell and the cell on the right is typical, showing an unsegmented nucleus (n) with indented profile and discrete regions of condensed peripheral chromatin. The upper fat cell shows an intermediate level of fat accumulation with several obvious lipid vacuoles, but no coalescence into larger droplets. The apparent increase in mitochondrial size and number (m) that occurs concomitantly with lipid synthesis is apparent by comparison with the fat cell on the left. Also visible in this section are an epithelioid (endothelial-derived) cell (E), two monocytes (M), two granulocytes (G) and an apoptotic body (A) which is indicative of the observed low level of cell death that occurs in the cultures. Scale bar, 2 μm.

Fig. 15. (*a*) Scanning micrograph of a small area of fat accumulation shortly after the addition of the second inoculum of bone marrow (1 week post establishment of the adherent layer). The fat cells show various stages of lipid production, and around and amongst them are clustered some 200 granulocytes in various stages of differentiation. (*b*) Scanning micrograph of a peripheral region of the fat cell area. The granulocytes are clustered in the vicinity of, rather than directly attaching to, the cell in lipid production.

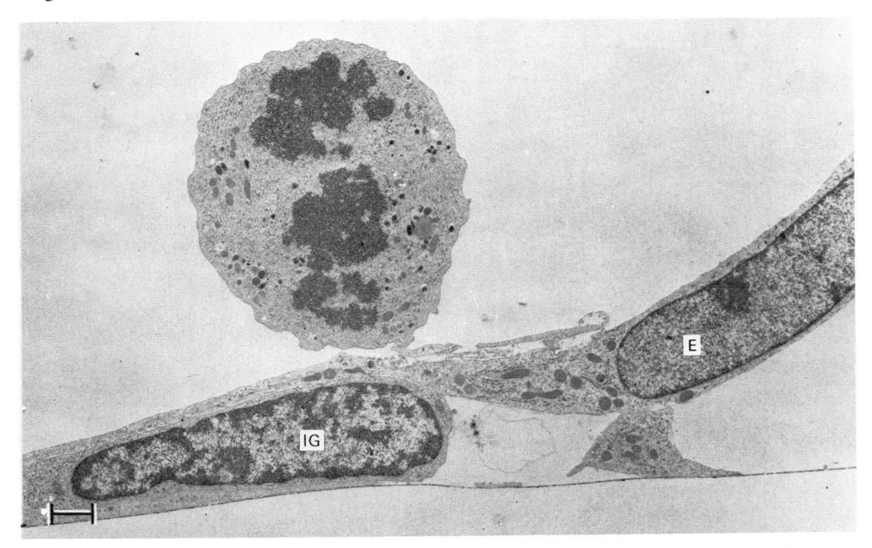

Fig. 17. Transmission micrograph of a section through a dividing metamyelocyte. The dividing cell is positioned on the surface of an epithelial cell (E) which covers a slightly flattened immature granulocyte (IG). Scale bar, 1 μm.

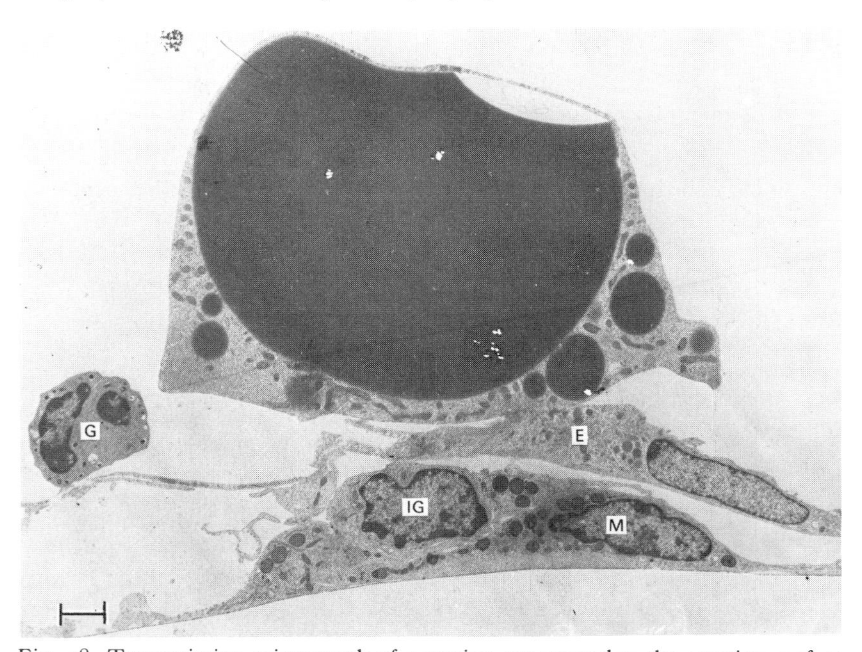

Fig. 18. Transmission micrograph of a section cut normal to the growing surface showing a fat cell with a single large lipid vacuole, and an undifferentiated cell (IG) sandwiched between the overlying epithelioid cell (E) and underlying monocyte (M). A granulocyte (G) is also present adjacent to the fat cell. Scale bar, 2 μm.

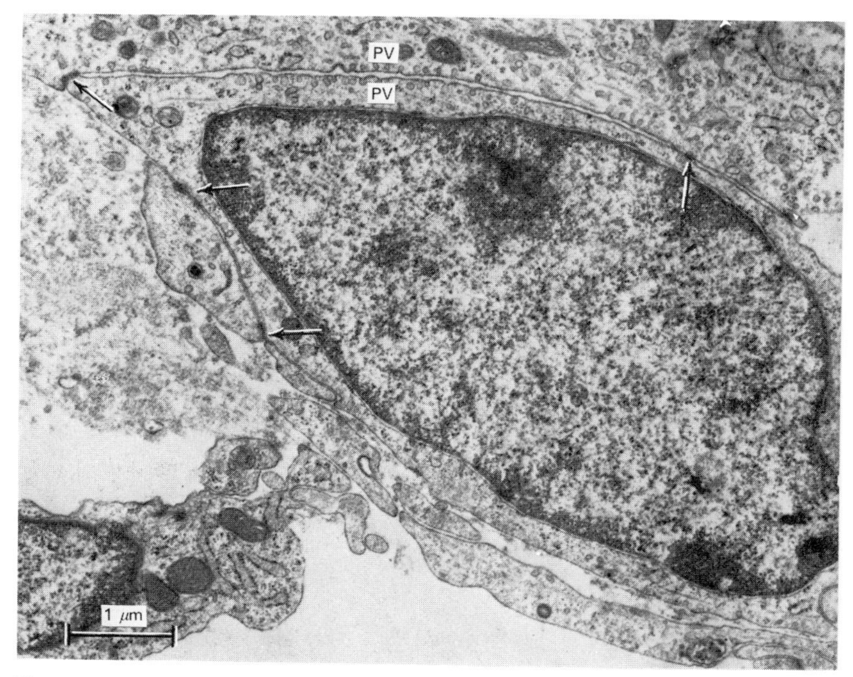

Fig. 19. Transmission micrograph of a section through an undifferentiated cell and the cytoplasmic regions of its immediate neighbours. Junctional complexes are apparent (arrowed) and pinocytotic vesicles (PV) are also present.

the blasts and their immediate neighbours (Fig. 19). These complexes closely resemble gap junctions, which are characteristic of cells interacting metabolically and ionically (Gilula, 1974). Future investigations will be directed to a further elucidation of structure and function of these cellular interactions.

Origin of cell types in culture

Haemopoiesis *in vivo* in bone marrow is directed along the main pathways of erythrocyte, monocyte, granulocyte and platelet production. Consequently freshly isolated bone marrow contains all the precursors for these cell types as well as cells of the vasculature and stroma which support haemopoiesis. No erythropoiesis has been observed in the cultured marrow and thrombopoiesis is also extremely limited. In a mature proliferative culture, therefore, stem cells and the granulocyte series are in an obviously proliferative condition, and it would not be unreasonable to assume that the persistence of cells derived from the vascular endothelium and

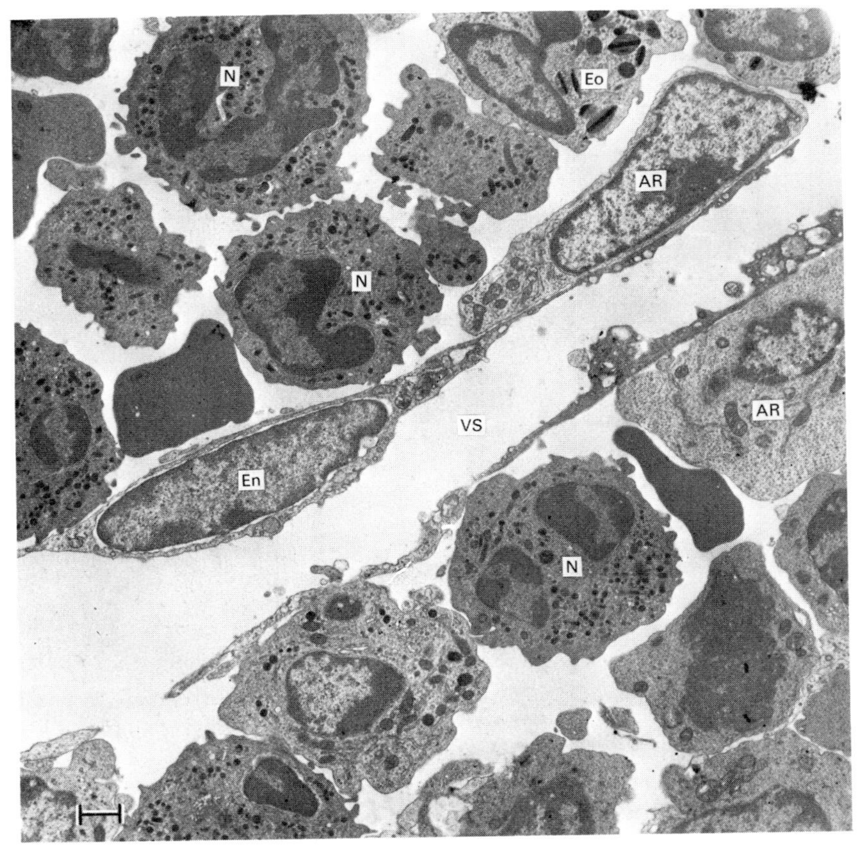

Fig. 20. Transmission micrograph of a section through intact mouse bone marrow, illustrating the in-vivo situation. The central venous sinus (VS) is surrounded by neutrophils (N) and an eosinophil (Eo). The wall of the venous sinus is composed of an inner endothelial cell lining (En) bordered by adventitial reticular cells (AR). Cf. Fig. 12. Scale bar, 1 μm.

stroma may play a role in the support of stem cell production and granulopoiesis.

Investigations into the haemopoietic microenvironment of bone marrow have shown compartments of differentiating haemopoietic cells to be arranged upon a stroma and a vasculature which together compose a system of venous sinuses which carry maturing blood cells into the circulation (Fig. 12; Weiss, 1970). The wall of these sinuses may be of a single endothelial cell layer, or an endothelial cell with a reticular cell on the outside (Fig. 20). The reticular cells are the dominant cells of the marrow stroma, covering the outside of the sinuses and branching into the haemopoietic spaces to form a sponge-like matrix similar to the reticular meshwork of splenic cords (Chen & Weiss, 1972, 1973). The reticular cells have also been postulated to trap stem cells in human foetal marrow (Chen & Weiss, 1975) and exert an inductive effect on the type of haemopoietic pathway of stem cell differentiation (Weiss, 1976). As these elements are necessary for granulopoiesis *in vivo*, their participation in providing the necessary 'in-vitro' microenvironment for granulopoiesis might, therefore, be expected. Reticular cells have been shown to accumulate fat in marrow (Weiss, 1970) and are therefore interpreted as the fat cell precursors *in vitro*, where they are initially visible as large cells with either an epithelioid or fibroblastic morphology (R cells). The second epithelioid cell type in the cultures (E cells) – which is much smaller and forms monolayers – is considered to be derived from the endothelium of the vascular sinus. It is suggested, therefore, that this culture system in which an inductive microenvironment exists for continued production of stem cells and granulopoiesis, may greatly assist the future understanding of the mechanics of granulopoietic and other haemopoietic pathways.

The author would like to acknowledge Dr T. M. Dexter, who initiated the culture procedure, and the expert technical assistance of Mr G. R. Bennion, Mrs Gaynor Johnson and Mr F. Leigh for preparing the micrographs.

REFERENCES

Allen, T. D. & Simmens, S. C. (1976). Conversion of vacuum coating units for sputter coating. *Micron*, **7**, 141–4.

Allen, T. D. & Dexter, T. M. (1976*a*). Surface morphology and ultrastructure of murine granulocytes and monocytes in long term liquid culture. *Blood Cells*, **2**, 591–606.

Allen, T. D. & Dexter, T. M. (1976*b*). Cellular interrelationships during *in vitro* granulopoiesis. *Differentiation*, **6**, 191–4.

Bessis, M. (1973). In *Living blood cells and their ultrastructure*, pp. 367–95. Berlin: Springer-Verlag.

Bradley, T. R. & Metcalf, D. (1966). The growth of mouse bone marrow cells *in vitro*. *Australian Journal of experimental Biology and medical Science*, **44**, 207–300.

Chen, L. T. & Weiss, L. (1972). Electron microscopy of the red pulp of human spleen. *American Journal of Anatomy*, **134**, 425–57.

Chen, L. T. & Weiss, L. (1973). The role of the sinus wall in the passage of erythrocytes through the spleen. *Blood*, **451**, 529–37.

Chen, L. T. & Weiss, L. (1975). The development of vertebral bone marrow of human fetuses. *Blood*, **46**, 389–408.

DICKE, K. A., VAN NOORD, M. J., MAAT, B., SCHAEFER, U. W. & VAN BEKKUM, D. W. (1973). Attempts at morphological identification of the haemopoietic stem cell in primates and rodents. In *Haemopoietic stem cells, CIBA Foundation Symposium 13*, pp. 47–70. Amsterdam: Associated Scientific Publishers.

DEXTER, T. M., ALLEN, T. D. & LAJTHA, L. G. (1977). Conditions controlling the proliferation of haemopoietic stem cells in culture. *Journal of cellular Physiology*, **91**, 335–44.

DEXTER, T. M. & LAJTHA, L. G. (1976). Proliferation of haemopoietic stem cells and development of potentially leukaemic cells *in vitro*. In *Proceedings of the VIIth International Symposium on Comparative Research on Leukaemia and Related Diseases*, Copenhagen, pp. 1–5. Basel: Karger.

DEXTER, T. M. & TESTA, N. G. (1976). Differentiation and proliferation of haemopoietic cells in culture. *Methods in Cell Biology*, **14**, 387–405.

GILULA, N. B. (1974). Junctions between cells. In *Cell communication*, ed. R. P. Cox, pp. 1–29. New York: Wiley.

LAJTHA, L. G. (1976). Influence of surface on white cell movement. *Blood Cells*, **2**, 411–13.

ROSSE, C. (1973). Precursor cells to erythroblastic and to small lymphocytes of the bone marrow. In *Haemopoietic stem cells, CIBA Foundation Symposium 13*, pp. 105–30. Amsterdam: Associated Scientific Publishers.

TILL, J. E. & McCULLOCH, E. A. (1961). A direct measurement of the radiation sensitivity of normal mouse bone marrow cells. *Radiation Research*, **14**, 213–22.

WALSH, R. T. & BARNHART, M. I. (1973). Blood platelet surfaces in 3 dimensions. In *Scanning electron microscopy 1973*, ed. O. Johari & I. Corvin, pp. 481–8. Chicago, USA: ITT Research Institute.

WEISS, L. (1970). The histology of the bone marrow. In *Regulation of hematopoiesis*, vol. 1, ed. A. S. Gordon, pp. 79–92. New York: Appleton-Century-Crofts.

WEISS, L. (1976). The hematopoietic microenvironment of the bone marrow: an ultrastructural study of the stroma in rats. *Anatomical Record*, **186**, 161–84.

WETZEL, B. K. (1970). The fine structure and cytochemistry of developing granulocytes, with special reference to the rabbit. In *Regulation of haematopoiesis*, vol. 2, ed. A. S. Gordon, pp. 769–817. New York: Appleton-Century-Crofts.

YOFFEY, J. M. (1973). Stem cell role of the lymphocyte transitional cell (LT) compartment. In *Haemopoietic stem cells, CIBA Foundation Symposium 13*, pp. 5–46. Amsterdam: Associated Scientific Publishers.

Molecular aspects of erythroid cell regulation

P. R. HARRISON, D. CONKIE, T. RUTHERFORD AND G. YEOH*

Beatson Institute for Cancer Research Wolfson Laboratory for Molecular Pathology, Garscube Estate, Switchback Road, Bearsden, Glasgow G61 1BD, UK

NORMAL DEVELOPMENTAL SYSTEMS

Erythropoiesis has been recognised for many years as a favourable system for molecular studies of how a variety of functions are co-ordinately regulated during the development of a specific differentiated cell type. One reason for this is the fact that the red blood cell is characterised by a series of well-defined functions or proteins, of which haemoglobin is only the most obvious example. A wide range of other markers has now been characterised and quantitated during erythroid development: for example cellular enzyme changes (Denton, Spencer & Arnstein, 1975), membrane changes such as formation of spectrin (Marchesi, Steers, Marchesi & Tillack, 1970; Hiller & Weber, 1977), glycophorin (Tomita & Marchesi, 1975) and other antigens (Furusawa & Adachi, 1968), loss of H-2 antigen (Klein, 1975) and changes in the fluidity of the plasma membrane (Arndt-Jovin & Jovin, 1976). Many of the characteristic changes in enzyme levels and some membrane components occur at a relatively late stage of maturation, at about the same time as haemoglobin accumulates (Denton *et al.*, 1975); but spectrin and globin mRNA in particular accumulate earlier at about the proerythroblast/basophilic erythroblast stage (Terada *et al.*, 1972; Harrison, Conkie, Affara & Paul, 1974; Conkie, Kleiman, Harrison & Paul, 1975; Ramirez *et al.*, 1975; Chang, Langer & Lodish, 1976).

Possibly some of the most important questions to be answered concern how this differential expression of a variety of specific genes is co-ordinated. What, for example, is the molecular basis for the classical two-stage distinction between commitment and differen-

* Present address: Department of Physiology, University of Western Australia.

tiation during development? Does commitment involve reversible modifications to the DNA bases, changes in the pattern of genes transcribed, or are post-transcriptional controls important? Does the timing of appearance of different functions reflect the sequential activation of the relevant structural genes or are all erythroid-specific genes transcribed *en bloc* very early in development followed by sequential processing of these transcripts to functional mRNA molecules? Is in fact the formation of early markers a necessary prerequisite for the expression of late functions? Questions such as these are perhaps some of the questions which can be investigated using the erythroid cell system which offers the potential for monitoring the expression of a variety of *specific* genes at different molecular levels.

Other features of the erythroid system which render it amenable to experimental analysis are that the ontogeny of red blood cells is well understood (see for example Harrison, 1976, for a recent review) and in particular that stem cells at various stages can be assayed. The mature erythroid cells are known to be derived from intermediate stem cells (ECP) which become committed to the erythroid line in the microenvironment of target tissues (Trentin, 1970; Wolf, 1974); these committed cells proliferate and subsequently become sensitive to the hormone erythropoietin (ERC cells), after which they may mature into the recognisable erythroid series. Multipotential stem cells can now be maintained in culture for some time; but under these conditions their differentiation seems to be almost exclusively restricted to the granulocyte/monocyte pathway (Dexter & Testa, 1976). Both ECPs and ERCs can be assayed *in vitro* by cloning techniques (Stephenson, Axelrad, McLeod & Shreeve, 1971; Axelrad, McLeod, Shreeve & Heath, 1973; Iscove, Sieber & Winterhalter, 1974). These systems may therefore have considerable potential for studying *in vitro* the process of commitment and the tissue interactions which may be involved. However one of the problems at present is that commitment to the erythroid line can only be inferred retrospectively by studying the nature of the progeny of any given cell. Moreover, due to the nature of the experimental systems, large-scale biochemical studies of the processes involved are presently impossible by conventional methods. Thus analysis of the molecular mechanism involved in the process of commitment and differentiation of these early stem cells is not yet feasible. What may prove to be extremely useful in this context are techniques for assaying individual erythroid gene products or their mRNAs at the single level using immunofluorescence techniques for detecting specific proteins or in-situ hybridisation methods for individual mRNAs (Harrison, Conkie, Paul & Jones, 1973; Harrison, Conkie, Affara & Paul, 1974; Conkie, Kleiman, Harrison & Paul, 1975).

ERYTHROID DIFFERENTIATION IN FRIEND CELLS

An alternative approach to elucidating the regulation of erythropoiesis is using Friend cells in culture (Friend, Patuleia & de Harven, 1966). Friend cells seem to be derived from haematopoietic stem cells transformed by the Friend virus complex (Friend, 1957) after commitment to the erythroid line (Tambourin & Wendling, 1971; Fredrickson, Tambourin, Jasmin & Smadja, 1975; Tambourin & Wendling, 1975). This subject has been reviewed recently in some detail (Harrison, 1976, 1977); therefore only the most salient points will be summarised here. One of the primary consequences of Friend virus action is the finding that infected (transformed?) erythroid precursor cells are able to proliferate and differentiate without dependence on the hormone erythropoietin. Cells from spleens of mice infected with Friend virus can be cloned *in vitro* without erythropoietin, unlike their normal counterparts; but otherwise they proliferate and differentiate to produce erythroid colonies indistinguishable in size or stage of maturation from normal erythroid colonies (Liao & Axelrad, 1975). Nevertheless, *in vivo*, erythroblasts infected with Friend virus may have a shorter lifetime than usual (Smadja-Joffe *et al.*, 1976). However, permanent cell lines obtained in culture by explantation of fragments of spleen from leukaemic animals seem to be arrested in differentiation at about the pro-erythroblast stage (Rossi & Friend, 1970). This maturational arrest could be a specific consequence of viral infection (Dube *et al.*, 1975); but it seems simpler to assume that Friend cells arise due to the inevitable selection in culture for rare cells which fail to differentiate terminally and which are thus able to overgrow the differentiating cell population.

Treatment of cultured Friend cells with dimethyl sulphoxide (DMSO) (Friend, Scher, Holland & Sato, 1971) or a range of chemicals induces erythroid maturation as judged by the appearance of many of the markers characteristic of normal erythroid cells referred to previously (reviewed in Harrison, 1977), for example: strain-specific α and β globin chains and globin mRNAs, haem and δ-aminolaevulinic acid synthetase, carbonic anhydrase, changes in the enzymes involved in purine metabolism, various membrane changes such as formation of spectrin, glycophorin and other antigens, loss of H-2 antigens, changes in membrane fluidity and permeability, and finally morphological changes involving chromatin condensation and in some cases nucleus extrusion. Thus although Friend cells are malignant and their regulation is clearly abnormal in its independence of erythropoietin, nevertheless when they are enabled to differentiate (by the addition of DMSO for example) then the co-ordinate expression of erythroid functions seems to proceed relatively normally, at least in a qualitative sense, as far as can be judged at the present time. With these reservations in mind, the Friend cell system can therefore be used as a model for studying the molecular mechanisms whereby erythroid functions are expressed. Indeed it is of considerable importance to elucidate the way in which a truly malignant cell can nevertheless differentiate apparently normally under the appropriate environmental conditions. Since the Friend cell is not pluripotent but is committed to the erythroid line, obviously such studies are directly relevant only to the regulation of the later stages of erythroid cell development. Whether the mechanisms which regulate the co-ordinate expression of gene activity during erythroid maturation are basically similar to those involved in the process of commitment itself is an open question; but there seems to be no *a priori* reason to assume that this should be so.

The timing of Friend cell differentiation

An interesting aspect of Friend cell differentiation is the timing of appearance of the various erythroid markers and the manner in which individual cells become differentiated. This issue is important in assessing whether Friend cell differentiation requires some fundamental change in the state of the cell or whether it involves merely modulation of a pattern of gene expression which basically had already been established.

Haemoglobin production in Friend cells requires treatment with DMSO for a period equivalent to one or more cell generation times (Friend *et al.*, 1971; McClintock & Papaconstantinou, 1974). After this latent period of treatment with inducer, the fraction of cells producing haemoglobin subsequently is not reduced by removal of DMSO (Conkie *et al.*, 1974; Levy, Terada, Rifkind & Marks, 1975). This could mean that treatment of Friend cells with inducer irreversibly commits the cells to differentiate and that thereafter inducer is no longer required. Recent work is most simply interpreted to mean that Friend cells are induced to differentiate stochastically, the probability of which is governed by the environmental conditions (Orkin, Harosi & Leder, 1975; Gusella *et al.*, 1976), A stochastic model has also been postulated for normal erythroid cell development (Till, McCulloch & Siminovitch, 1964; Korn, Henkelman, Ottenstrugen & Till, 1973).

One of the major issues concerning development is whether a change in the state of differentiation of a cell is cell-cycle-dependent at a critical stage (see, for example, Holtzer, this volume). The same question has been investigated in relation to Friend cell differentiation, but the issue is still controversial. Levy *et al.* (1975) have attempted to investigate directly whether DNA synthesis is required for induction of haemoglobin, using Friend cells synchronised by a double exposure to excess thymidine. Apparently DMSO has to be in contact with the cells for 24–30 hours in order to achieve an effective intracellular concentration of DMSO. But in addition, DMSO appears to be required during the S-phase after release of the cell cycle block in order for haemoglobin to accumulate subsequently. In contrast, Leder, Orkin & Leder (1975) have reported that inhibition of DNA synthesis with hydroxyurea or cytosine arabinoside does not prevent butyric-acid-induced haemoglobin production in a different Friend cell line.

We have investigated this question somewhat differently by single cell studies. Friend cells can be cloned in semi-solid methocel medium and their proliferation arrested if necessary by the addition of hydroxyurea, cytosine arabinoside or 5′-fluorodeoxyuridine (FUdR), or by isoleucine deprivation. Formation of haemoglobin can be detected *in situ* by benzidine staining. Our experiments show that Friend cells arrested at the single cell stage in the presence of various inducers fail to accumulate haemoglobin, whereas doublet colonies in the same cultures frequently produce haemoglobin (Fig. 1; Table 1). Most of the single cells arrested by isoleucine deficiency and some of the cells arrested by hydroxyurea are still potentially functional since on reversal of the arresting conditions

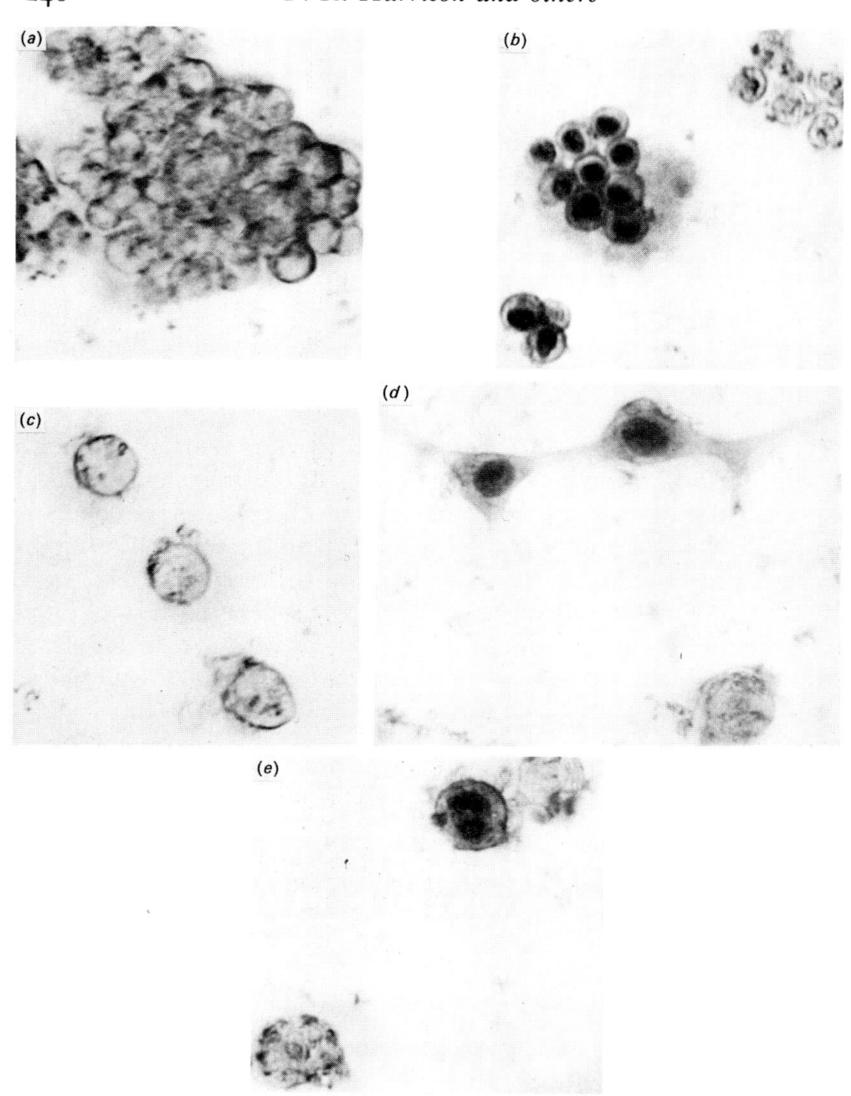

Fig. 1. Effect of inhibition of cell proliferation on the production of haemoglobin in Friend cells. (*a*)–(*d*) Clone 17C cells. Cells were seeded out in semi-solid methocel medium in the presence of (*a*) no additions, (*b*) 1.5% DMSO, (*c*) 1.5% DMSO plus cytosine arabinoside (10^{-6} M) and (*d*) 1.5% DMSO plus 0.05 mM hydroxyurea; (*e*) represents a different Friend cell line (T3C12) treated with 1 mM butyric acid plus 0.05 mM hydroxyurea. All cultures were stained *in situ* with benzidine after 4 days in culture. Haemoglobin is apparent in the nucleus and cytoplasm when stained in this manner, as is found for normal erythroblasts (O'Brien, 1960). Single cells do not produce haemoglobin whereas doublet or binucleate cells do.

Table 1. *Induction of haemoglobin in Friend cell clones*

Clone	Treatment: Inducer	Treatment: Inhibitor	Colonies: Size	Colonies: % benzidine-positive[a]
17C	1.5% DMSO	10^{-6} M cytosine arabinoside	Single cell	3
			Other	65
	1.5 mM butyric acid	0.05 mM hydroxyurea	Single cell	3 (mononucleate)
				12 (binucleate)
			Other	65
	15 mM N-methyl acetamide	Isoleucine	Single cell	2*
			Other	48*
B10/1	1.5% DMSO	0.05 mM hydroxyurea	Single cell	1
			Other	50
	1.5% DMSO	Isoleucine	Single cell	4[a]
			Other	15[a]
T3C12	1 mM butyric acid	Isoleucine	Single cell	3 (mononucleate)
				10 (binucleate)
			Other	15
	1–1.5 mM butyric acid	0.05–0.1 mM hydroxyurea	Single cell	8±3, 8±4[+] (mononucleate)
				20±10, 24±11[b] (binucleate)
			Other	20±8, 55±14[b]

Measurements of colony characteristics refer to 4 days of treatment, but sometimes to 3 days ([a]) or 5 days ([b]). Values are the results of separate experiments (obtained by scoring 200 colonies in at least two separate dishes) or the average±standard deviation of four separate experiments. Haemoglobin formation is assayed by benzidine staining.

they proliferate and produce haemoglobin (Fig. 2). However such studies can never be entirely unequivocal since the methods of arrest used might have undesirable side-effects and affect haemoglobin production independently of the arrest of cell replication *per se*. Nevertheless our experiments are basically consistent with those of Levy *et al.* (1975) which also seem to implicate a cell-cycle-dependence of Friend cell maturation.

Our own studies with the T3C12 Friend cell line may also explain

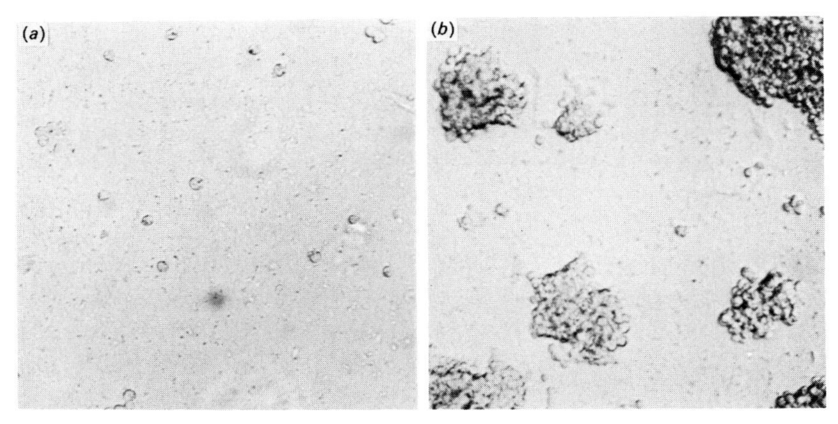

Fig. 2. Proliferation and differentiation of Friend cells after arrest by deprivation of isoleucine. (*a*) Clone 17C cells were arrested without isoleucine for 4 days in the presence of 15 mM N-methyl acetamide. (*b*) Isoleucine was then added to the medium for 4 days whereupon most of the single cells proliferated to form colonies.

why the results of Leder *et al.* (1975) appear to indicate that Friend cell maturation is not dependent on prior DNA synthesis. In agreement with Leder *et al.* (1975) we find that some arrested T3C12 cells do accumulate haemoglobin; but significantly a large proportion of these haemoglobin-containing single cells are in fact binucleate (Fig. 1*a*; Table 1). Thus the critical event for which inducer is required may be associated with, or prior to, nuclear rather than cellular division.

Mechanism of action of Friend cell inducers

Much speculation has arisen concerning the mechanism whereby the wide range of chemical agents is able to induce erythroid maturation in Friend cells. It has generally been assumed that they all act basically by a common mechanism, and two main hypotheses have been proposed. In the first (Tanaka *et al.*, 1975; Preisler & Lyman, 1975) it is suggested that inducers act by destabilising DNA structure at specific genes by analogy with the effects of DMSO and other DNA-denaturing agents on the transcription of bacterial operons (Nakanishi, Adhya, Gottesman & Pastan, 1974). However, the concentrations of inducers which are effective with Friend cells are substantially lower than those required for modifying the transcription of bacterial operons. The main alternative hypothesis assumes that Friend cell inducers act by interfering with membrane function in some way. The fact that various membrane changes correlate with the ability of Friend cells to differentiate after various treatments is consistent with such an interpretation (reviewed in detail in Harrison, 1977).

In order to determine whether indeed all inducers do act by a common mechanism, we have isolated a series of Friend cell variants which are resistant to induction of erythroid maturation by particular inducers. This can be achieved by cloning inducible Friend cells in the inducer after treatment with mutagens (e.g. ethylmethanesulphonate or ICR 191). Under these selective conditions, the majority of cells differentiate terminally leaving the rare non-inducible cells to overgrow. These independent non-inducible variants can then be tested for response to a wide range of inducers (Table 2). The variants are found to give different types of response: some are non-inducible by all the inducers tested, whereas others are inducible by some inducers but not others. These results therefore demonstrate unequivocally that the various inducers of Friend

Table 2. *Haemoglobin production in Friend cell variants isolated by resistance to different inducers*

Resistant line isolated by growth in	Code	Response to					
		DMSO	NMA	NMP	HMBA	HX	Haemin +DMSO
DMSO	R707, TR28D 3BE1AS, 3BE3A6	–	–	–	–		–
	Fw	–	–	–	–		+
	TR25D	–	++	++	++		–
	3BE1A10	–	+	+	–		–
	3BE1A1	–	+	±	–		–
	3BE3B1	–	++	+	–		–
NMA		±	–	+	++	–	
NMP		–	–	–	–	–	
HX		–	±	++	±	–	

Key: DMSO, 1.5 %; NMA, N-methyl acetamide, 15 mM; NMP, N-methyl pyrrolidone, 20 mM; HMBA, hexamethylene *bis*acetamide, 5 mM; HX, hypoxanthine, 2 mM; haemin, 0.1 mM. Treatment was for 5–6 days.

++, 30–60 % benzidine-positive cells; +, 15–30 % benzidine-positive cells; –, < 5 % benzidine-positive cells; ±, slightly positive in some experiments.

cells do not act by an identical mechanism, although it is still possible that they act at different points in a common pathway, possibly in the normal erythropoietin response. This series of non-inducible variants should therefore be useful in establishing more rigorously exactly which membrane changes (for example) are obligatorily involved in induction of erythroid differentiation by various inducers.

Appearance of erythroid functions during differentiation

As argued previously, the mechanism whereby the expression of the entire erythroid programme is regulated is one of the most important issues which might in principle be elucidated using the Friend cell system. One approach to this question is to attempt to elucidate Friend cell mutants in which the coupling of erythroid functions has been relaxed. From such studies one should in principle be able to determine whether groups of functions are regulated separately and whether expression of some functions (perhaps early markers) is obligatory for subsequent appearance of late markers.

Haemoglobin. In the previous section, Friend cell variants were described which were resistant to specific inducers of Friend cell differentiation. On account of their method of isolation these variants fail to differentiate terminally to the non-dividing stage; in addition, these variants are found to be no longer inducible for haemoglobin. This suggests that haemoglobin production and terminal differentiation are two markers which are tightly coupled. On the other hand, variants can be obtained which produce haemoglobin constitutively at a reduced level and yet are capable of continued growth. However this does not necessarily mean that the two markers have been uncoupled since in these variants haemoglobin is probably produced only by a subpopulation of differentiated cells which have ceased dividing.

Globin mRNAs. In all our DMSO-resistant variants which fail to induce haemoglobin with DMSO, the levels of α and β globin mRNAs are not increased above the basal level before DMSO treatment (Harrison, 1977), i.e. 10–50 ppm of the cytoplasmic RNA, as compared to levels of 500–1000 ppm in the parental inducible cells after DMSO treatment. Moreover, in two of the DMSO-resistant variants examined in more detail, the levels of globin mRNAs in the nucleus are also not increased after DMSO treatment above the low but significant levels in untreated cells. In none of the DMSO-resistant variants screened to date is there any evidence for production of only α and β globin mRNA. Since these two mRNAs accumulate asynchronously during DMSO treatment of parent inducible cells (Orkin, Swan & Leder, 1975), it might be expected that it should be possible to uncouple these two markers genetically.

The fact that we have failed to isolate Friend cell variants which accumulate globin mRNAs without the concomitant production of globin chains may suggest that translational control is not important in Friend cell differentiation. Thus the main control may be during transcription or processing of the primary transcript. In order to distinguish between these possibilities direct measurement of the rate of transcription of the globin genes will be necessary and also possibly a detailed study of whether cleavage of the primary transcripts is abnormal in these variants. Recent work has demonstrated the existence of 15S precursor to globin mRNA in both normal erythroid cells (Ross, 1976) and Friend cells (Curtis & Weissmann, 1977). This 15S species seems to be quickly cleaved within the nucleus to a 10S species which is the size of the mature globin mRNA. A very recent paper claims to have detected in addition a larger 27S globin mRNA precursor which is cleaved to the 15S precursor after addition of poly A to the 3′ terminus (Bastos & Aviv, 1977). These interesting findings suggest that processing of globin gene transcripts may be an important regulatory control. This means that it should in principle be possible to obtain mutants in which some aspect of processing is prevented. Such mutants would clearly be of great usefulness in further elucidation of the control mechanisms involved.

The basal levels of globin mRNAs in our DMSO-resistant variants are much higher than in non-erythroid cells (Harrison *et al.*, 1976; Humphries, Windass & Williamson, 1976). It is difficult to be certain whether these levels are solely accounted for by the small proportion of haemoglobin-containing cells (0.1–1 % as judged by benzidine staining). In-situ hybridisation experiments show that the globin mRNA in untreated Friend cell populations is distributed throughout most of the cells in the population (Conkie *et al.*, 1974). Moreover, with the exception of globin mRNAs the overall poly-A-containing mRNA population in Friend cells does not change very significantly after DMSO treatment (Minty, Birnie & Paul, personal communication), as judged by hybridisation of mRNAs to complementary DNAs transcribed from them using reverse transcriptase. In contrast, using the same techniques, the most abundant poly-A-containing mRNAs in various normal tissues differ quite significantly (Young, Birnie & Paul, 1976; Hastie & Bishop, 1977). One interpretation of these collective results is that Friend cells are malignant and therefore the regulation of gene activity may have become relaxed and 'leaky' with the result that erythroid genes are transcribed at low levels even in 'undifferentiated' cells. However the fact that the malignant, but non-erythroid, lymphoma cell contains a much lower level of globin mRNA argues against this interpretation unless it is supposed that a malignant cell allows 'leaky' transcription only of certain, developmentally-related genes. Alternatively the high basal level of globin mRNA in untreated Friend cells may reflect the fact that they are committed erythroid cells. There is evidence that the level of globin gene transcripts in normal non-erythroid cells or in haemopoietic stem cells is very low (Groudine, Holtzer, Scherrer & Therwath, 1974) although probably significant and genuine (Humphries *et al.*, 1976). Thus accumulation of specific erythroid gene transcripts or mRNAs during maturation of Friend cells might then be governed post-transcriptionally, as seems to occur in other systems, for example during transport of sequences from nucleus to cytoplasm (Getz *et al.*, 1975; Hough, Smith, Britten & Davidson, 1975; Chan, 1976; Levy & McCarthy, 1976; Humphries *et al.*, 1976) or by differential mRNA stability (Lodish & Small, 1976; Aviv, Voloch, Bastos & Levy, 1976).

Spectrin. Spectrin is induced early in Friend cell differentiation and therefore it is important to determine whether it is regulated independently of late markers such as haemoglobin. Spectrin formation was assayed in our DMSO-resistant variants by immunofluorescence using rabbit anti-spectrin antiserum (kindly donated by Dr H. Eison of the Institute Pasteur, Paris) and FITC-labelled goat anti-rabbit antiserum. The results (Table 3) reveal clearly that spectrin formation is uncoupled from other markers in two of the variants. This is most simply interpreted to mean that erythroid maturation can be blocked at a point after spectrin and its mRNA is produced but before globin genes are transcribed or at least before globin RNA accumulates in the nucleus or cytoplasm. This

Table 3. *Co-ordination of erythroid functions in Friend cell variants*

	Hb		Globin mRNA		Spectrin		Terminal differentiation	
Code	DMSO	NMP	DMSO	NMP	DMSO	NMP	DMSO	NMP
Inducible	+	+	+	+	+	+	+	+
R707	–	–	–		–		–	–
TR28D	–	–	–	–	–	–	–	–
3BE1A5	–	–	–		–	–	–	–
FwT	–	–	–	–	+	+	–	–
3BE1A1	–	–	–		+?	+	–	–
TR25D	–	+	–	+	–	+	–	+
3BE1A10	–	+	–		–	+	–	+
Lymphoma	–	–	–	–	–	–	–	–

Key: as Table 2.
++, highly inducible; +, partly inducible; –, non-inducible.

interpretation implies that the spectrin and globin genes are not activated at the same stage of development.

SUMMARY

Erythroid precursor cells transformed *in vivo* by infection of susceptible mice with Friend virus differentiate *in vivo* or *in vitro* without dependence on erythropoietin, unlike their normal counterparts. Cell lines obtained by culturing fragments of spleens of infected animals fail to differentiate past the proerythroblast stage, probably because of selection by the culture conditions for variants which fail to differentiate terminally. However, after treatment with various 'inducers', such as dimethyl sulphoxide, Friend cells undergo erythroid maturation as judged by a variety of erythrocyte markers. Single cell cloning experiments are consistent with the interpretation that haemoglobin production in Friend cells is cell-cycle-dependent. Genetical analysis shows that different inducers do not act by a common mechanism: for Friend cell variants which are non-inducible by a particular inducer may nevertheless be inducible by other inducers. All Friend cell variants isolated which fail to induce haemoglobin also fail to induce globin mRNA. This argues against translational control being important during Friend cell differentiation. However, some Friend cell variants which fail to accumulate haemoglobin or globin gene transcripts may nevertheless induce another early erythrocyte marker, spectrin. This suggests that different erythroid genes may be activated sequentially rather than as a battery.

This work has been supported by the Medical Research Council and the Cancer Research Campaign. The technical assistance of Jimmy Sommerville, Paul Hissey and Janis Fleming is gratefully acknowledged. We are indebted to Dr H. Eisen (Institute Pasteur) for a gift of antispectrin antiserum.

REFERENCES

Arndt-Jovin, D. J. & Jovin, T. M. (1976). Cell separation using fluorescence emission anisotrophy. *Journal of supra-molecular Structure*, **5**, Suppl. 1, in press.

Aviv, H., Voloch, Z., Bastos, R. & Levy, S. (1976). Biosynthesis and stability of globin mRNA in cultured erythroleukemic cells. *Cell*, **8**, 495–503.

Axelrad, A. A., McLeod, D. L., Shreeve, M. M. & Heath, D. S. (1973). Properties of cells that produce erythrocytic colonies *in vitro*. In *Proceedings of Second International Workshop on Hemapoiesis in Culture*, ed. W. A. Robinson, pp. 226–37. DHEW Publication No. WIH 74-205.

Bastos, R. W. & Aviv, H. (1977). Globin RNA precursor molecules: biosynthesis and processing in erythroid cells. *Cell*, in press.

Chan, L.-W. L. (1976). Transport of globin mRNA from nucleus into cytoplasm in differentiating embryonic red blood cells. *Nature, London*, **261**, 157–9.

Chang, H., Langer, P. J. & Lodish, H. F. (1976). Asynchronous synthesis of erythrocyte membrane proteins. *Proceedings of the National Academy of Sciences, USA*, **73**, 3206–10.

Conkie, D., Affara, N., Harrison, P. R., Paul, J. & Jones, K. (1974). *In situ* localisation of globin messenger RNA formation. II. After treatment of Friend virus-transformed mouse cells with dimethylsulphoxide. *Journal of cell Biology*, **63**, 414–19.

Conkie, D., Kleiman, L., Harrison, P. R. & Paul, J. (1975). Increase in the accumulation of globin mRNA in immature erythroblasts in response to erythropoietin *in vivo* or *in vitro*. *Experimental cell Research*, **93**, 315–24.

Curtis, P. J. & Weissmann, C. (1977). Purification of globin mRNA from dimethylsulphoxide-induced Friend cells and detection of a putative globin mRNA precursor. *Journal of molecular Biology*, **106**, 1061–75.

Denton, M. J., Spencer, N. & Arnstein, H. R. V. (1975). Biochemical and enzymic changes during erythrocyte differentiation. *Biochemical Journal*, **146**, 205–11.

Dexter, T. M. & Testa, N. G. (1976). Differentiation and proliferation of haemopoietic cells in culture. In *Methods in cell biology*, ed. D. M. Prescott, vol. 14, pp. 389–405. New York & London: Academic Press.

DUBE, S. K., PRAGNELL, I. B., KLUGE, N., GAEDICKE, G., STEINHEIDER, G. & OSTERTAG, W. (1975). Induction of endogenous and spleen focus forming virus (SFFV) during DMSO induced differentiation of SFFV transformed mouse erythroleukemia cells. *Proceedings of the National Academy of Sciences, USA*, **72**, 1863–7.

FREDRICKSON, T., TAMBOURIN, P., JASMIN, C. & SMADJA, F. (1975). Target cell of the polycythemia-inducing Friend virus: studies with myleran. *Journal of the National Cancer Institute*, **55**, 443–6.

FRIEND, C. (1957). Cell-free transmission in adult swiss mice of a disease having the character of a leukemia. *Journal of experimental Medicine*, **105**, 307–18.

FRIEND, C., PATULEIA, M. C. & DE HARVEN, E. (1966). Erythrocytic maturation *in vitro* of murine (Friend) virus-induced leukemic cells. *National Cancer Institute Monograph*, **22**, 505–14.

FRIEND, C., SCHER, W., HOLLAND, J. C. & SATO, T. (1971). Hemoglobin synthesis in murine virus-induced leukemic cells *in vitro*: stimulation of erythroid differentiation by dimethyl sulphoxide. *Proceedings of the National Academy of Sciences, USA*, **68**, 378–82.

FURUSAWA, M. & ADACHI, H. (1968). Immunological analyses of the structural molecules of erythrocyte membrane in mice. II. Staining of erythroid cells with labelled antibody. *Experimental Cell Research*, **50**, 497–504.

GETZ, M. J., BIRNIE, G. D., YOUNG, B. D., MACPHAIL, E. & PAUL, J. (1975). Nuclear poly(A)-containing RNA: estimation of base sequence complexity and sequence homology with messenger RNA in mouse Friend cells. *Cell*, **4**, 121–9.

GROUDINE, M., HOLTZER, H., SCHERRER, K. & THERWATH, A. (1974). Lineage dependent transcription of globin genes. *Cell*, **3**, 243–7.

GUSELLA, J., GELLER, R., CLARKE, B., WILKS, V. & HOUSMAN, D. (1976). Commitment to erythroid differentiation by Friend erythroleukemia cells: a stochastic analysis. *Cell*, **9**, 221–9.

HARRISON, P. R. (1976). Analysis of erythropoiesis at the molecular level. *Nature, London*, **262**, 353–6.

HARRISON, P. R. (1977). The biology of the Friend cell. In *Biochemistry of cell differentiation*, vol. 11, ed. J. Paul. Lancaster: Medical & Technical Publishing Co. Ltd (in press).

HARRISON, P. R., AFFARA, N., CONKIE, D., RUTHERFORD, T., SOMMERVILLE, J. & PAUL, J. (1976). Regulation of erythroid differentiation in Friend erythroleukemic cells. In *Progress in differentiation research*, ed. N. Muller-Berat, pp. 135–46. Amsterdam: North-Holland.

HARRISON, P. R., CONKIE, D., AFFARA, N. & PAUL, J. (1974). *In situ* localisation of globin messenger RNA formation. I. During mouse foetal liver development. *Journal of cell Biology*, **63**, 402–13.

HARRISON, P. R., CONKIE, D., PAUL, J. & JONES, K. (1973). Localisation of cellular globin messenger RNA by *in situ* hybridisation to complementary DNA. *FEBS Letters*, **32**, 109–12.

HASTIE, N. D. & BISHOP, J. O. (1977). The expression of three abundance classes of messenger RNA in mouse tissues. *Cell*, in press.

HILLER, G. & WEBER, K. (1977). Spectrin is absent in various tissue culture cells. *Nature, London*, **266**, 181–3.

HOUGH, B., SMITH, M. J., BRITTEN, R. J. & DAVIDSON, E. H. (1975). Sequence complexity of heterogeneous nuclear RNA in sea urchin embryos. *Cell*, **5**, 291–9.

HUMPHRIES, S., WINDASS, J. & WILLIAMSON, R. (1976). Mouse globin gene expression in erythroid and non-erythroid tissues. *Cell*, **7**, 267–77.

ISCOVE, N. W., SIEBER, F. & WINTERHALTER, K. H. (1974). Erythroid colony formation in cultures of mouse and human bone marrow: analysis of the requirement for erythropoietin by gel filtration and affinity chromatography on agarose–concanavalin A. *Journal of cellular Physiology*, **83**, 309–20.

KLEIN, J. (1975). *Biology of the mouse histo-compatibility – 2 complex*, p. 620. Berlin: Springer-Verlag.

KORN, A. P., HENKELMAN, R. M., OTTENSMEYER, F. P. & TILL, J. E. (1973). Investigations of a stochastic model of haemopoiesis. *Experimental Haematology*, **1**, 362–75.

LIAO, S. K. & AXELRAD, A. A. (1975). Erythropoietin-independent erythroid colony formation *in vitro* by hemopoietic cells of mice infected with Friend virus. *International Journal of Cancer*, **15**, 467–82.

LEDER, A., ORKIN, S. & LEDER, P. (1975). Differentiation of erythroleukemic cells in the presence of inhibitors of DNA synthesis. *Science*, **190**, 893–4.

LEVY, B. & MCCARTHY, B. J. (1976). Relationship between nuclear and cytoplasmic RNA in *Drosophila* cells. *Biochemistry*, **15**, 2415–19.

LEVY, J., TERADA, M., RIFKIND, R. A. & MARKS, P. A. (1975). Induction of erythroid differentiation by dimethylsulphoxide in cells infected with Friend virus: relationship to the cell cycle. *Proceedings of the National Academy of Sciences, USA*, **72**, 28–32.

LODISH, H. & SMALL, B. (1976). Different lifetimes of reticulocyte messenger RNA. *Cell*, **7**, 59–65.

MCCLINTOCK, P. R. & PAPACONSTANTINOU, J. (1974). Regulation of hemoglobin synthesis in a murine erythroblastic leukemic cell: the requirement for replication to induce hemoglobin synthesis. *Proceedings of the National Academy of Sciences, USA*, **71**, 4551–5.

MARCHESI, S. L., STEERS, E., MARCHESI, V. T. & TILLACK, T. W. (1970). Physical and chemical properties of a protein isolated from red cell membranes. *Biochemistry*, **9**, 50–7.

NAKANISHI, S., ADHYA, S., GOTTESMAN, M. & PASTAN, I. (1974). Activation of transcription at specific promoters by glycerol. *Journal of biological Chemistry*, **249**, 4050–6.

O'BRIEN, B. R. A. (1960). The presence of hemoglobin within the nucleus of the embryonic chick erythroblast. *Experimental cell Research*, **21**, 226–8.

ORKIN, S. H., HAROSI, F. I., LEDER, P. (1975). Differentiation in erythroleukemic cells and their somatic hybrids. *Proceedings of the National Academy of Sciences, USA*, **72**, 98–102.

ORKIN, S. H., SWAN, D. & LEDER, P. (1975). Differential expression of α- and β-globin genes during differentiation of cultured erythroleukemic cells. *Journal of biological Chemistry*, **250**, 8753–60.

PREISLER, H. D. & LYMAN, G. (1975). Differentiation of erythroleukemia cells *in vitro*; properties of chemical inducers. *Cell Differentiation*, **4**, 179–85.

RAMIREZ, F., GAMBINO, R., MANIATIS, G. M., RIFKIND, R. A., MARKS, P. & BANK, A. (1975). Changes in globin mRNA content during erythoid cell differentiation. *Journal of biological Chemistry*, **250**, 6054–8.

ROSS, J. (1976). A precursor of globin messenger RNA. *Journal of molecular Biology*, **106**, 403–20.

ROSSI, G. B. & FRIEND, C. (1970). Further studies on the biological properties of Friend virus-induced leukemic cells differentiating along the erythrocytic pathway. *Journal of cellular Physiology*, **76**, 159–66.

SMADJA-JOFFE, F., KLEIN, C., KERDILES, C., FEINENDEGEN, L. & JASMIN, C. (1976). Study of cell death in Friend leukemia. *Cell and Tissue Kinetics*, **9**, 131–45.

STEPHENSON, J. A., AXELRAD, A. A., MCLEOD, D. L. & SHREEVE, M. (1971). Induction of colonies of hemoglobin-synthesising cells by erythropoietin *in vitro*. *Proceedings of the National Academy of Sciences, USA*, **68**, 1542–6.

TAMBOURIN, P. E. & WENDLING, F. (1971). Malignant transformation and erythroid differentiation by polycythemia-inducing Friend virus. *Nature, New Biology*, **234**, 230–3.

TAMBOURIN, P. E. & WENDLING, F. (1975). Target cell for oncogenic action of polycythemia-inducing Friend virus. *Nature, London*, **256**, 320–2.

TANAKA, M., LEVY, J., TERADA, M., BRESLOW, R., RIFKIND, P. A. & MARKS, P. A. (1975). Induction of erythroid differentiation in murine virus infected erythroleukemia cells by highly polar compounds. *Proceedings of the National Academy of Sciences, USA*, **72**, 1003–6.

TERADA, M., CANTOR, L., METAFORA, S., RIFKIND, R. A., BANK, A. & MARKS, P. A. (1972). Globin messenger RNA activity in erythroid precursor cells and the effect of erythropoietin. *Proceedings of the National Academy of Sciences, USA*, **69**, 3575–9.

TILL, J. E., MCCULLOCH, E. A. & SIMINOVITCH, L. (1964). A stochastic model of stem cell proliferation based on the growth of spleen colony-forming cells. *Proceedings of the National Academy of Sciences*, **51**, 29–36.

TOMITA, M. & MARCHESI, V. T. (1975). Amino-acid sequence and oligosaccharide attachment sites of human erythrocyte glycophorin. *Proceedings of the National Academy of Sciences, USA*, **72**, 2964–8.

TRENTIN, J. J. (1970). Influence of haematopoietic organ stroma (haematopoietic inductive microenvironments) on stem cell differentiation. In *Regulation of haematopoiesis*, ed. A. S. Gordon, vol. 1, pp. 161–86. New York: Appleton-Century-Crofts.

WOLF, W. S. (1974). Dissecting the haematopoietic microenvironment. I. Stem cell lodgement and commitment and the proliferation and differentiation of erythropoietic descendants in the Sl/Sl^d mouse. *Cell and Tissue Kinetics*, **7**, 89–98.

YOUNG, B. D., BIRNIE, G. D. & PAUL, J. (1976). Complexity and specificity of polysomal poly (A^+) RNA in mouse tissues. *Biochemistry*, **15**, 2823–9.

Haemopoietic stem cells and murine viral leukaemogenesis

P. E. TAMBOURIN

Unité de Physiologie Cellulaire (No. 22) de l'INSERM,
Institut de Radium – Bât. 110, 91405 Orsay, France

Virus-induced murine leukaemias have provided an active ground of research for the past decade. Murine leukaemia viruses have, however, not been so widely used by virologists as their avian counterparts. Therefore the genetics of murine viruses or even the exact knowledge of the virus mixture composition required to induce the several typical leukaemias are most often poorly, if at all, known. Virology has acquired most of its recent dramatic results with viruses whose in-vivo effects are far from well documented, while an understanding of the leukaemic process *in vivo* was derived from studies with murine viruses whose biology is still poorly understood. One main reason for this state of things is the alleged non-transforming ability of murine leukaemogenic viruses which were then considered as unsuitable tools for experimental tumorigenesis *in vitro*. Today interesting research developments arise. On one hand the genetics of murine leukaemia viruses is beginning to be better understood (Steeves, 1975) and cloned viruses are more and more often used. Typical works in this field are those published recently by Troxler, Parks, Vass & Scolnick (1977). On the other hand recent research has been done *in vivo* and *in vitro* on the physiopathology of avian leukaemia (Graf, Fink, Beug & Royer-Pokora, 1977).

Knowledge of the physiopathology of murine leukaemia and in particular of the role of the haemopoietic stem cells in this neoplastic process is of more than academic interest. These murine leukaemias appear to be good models for human leukaemias, although the infectious properties of viruses could be a nuisance in extrapolating from the murine system to human. Furthermore, the haemopoietic system in which a pluripotent stem cell (i.e. a cell which can effectively choose between different genetic programmes) gives rise to differentiated committed precursors and then to at least six lines of blood cells, offers a good model to study the interaction between the virus genome and haemopoietic cells at different stages of differentiation. In-vitro culture methods open the way to fruitful research in that direction. I shall try to show how Friend and Rauscher viruses, which have been the most extensively studied, provided a way to revise some concepts in the field of virology, because transformation, followed by a malignant erythroid differentiation, can be observed at about 30 hours after infection *in vivo* and *in vitro* and this transformation process does not give rise directly to true tumour cells. Such tumour cells appeared later in the disease probably as the result of a 'progression' in leukaemia development (Foulds, 1969; Shabad, 1973).

ONCOVIRUSES

'Oncogenesis is defined as the origin and development of a malignant neoplasm, i.e., a new growth in which the cells multiply, progressively invade and destroy normal tissues, sometimes metastasize, and, if unchecked, ultimately kill the affected animal' (Sanford, 1974). One way to study the fundamental problem is to use oncogenic viruses, the so-called oncoviruses (Fenner, 1976), *in vivo* or *in vitro*. Much effort has been devoted to the study of these viruses for the valid reason that identification of a specific oncogenic virus may lead to the development of preventive vaccination or to immunotherapy of the induced tumour. Much of the information on the action of oncogenic viruses had been obtained by means of tissue culture, 'transforming' established cell lines which are not only aneuploid but also heteroploid, and which were sometimes obtained originally by culturing a tumour. The relevance of such techniques to carcinogenesis is unclear. Morphological in-vitro 'transformation' is now widely considered as representative of 'malignant' transformation. In-vivo experiments, in particular those based on leukaemia viruses, undoubtedly show that this extrapolation must be used cautiously (Foulds, 1969; Sanford, 1974; Frei, 1976).

RNA tumour viruses are enveloped animal viruses whose virion contains an RNA genome and an RNA-dependent DNA polymerase. Some RNA tumour viruses are the apparent cause of much 'natural' cancer or leukaemia in chickens and in mice. Some of them, such as the strongly transforming Rous sarcoma virus, seem to qualify amongst the most efficient carcinogenic agents known in animal or in cell culture. In-vitro transformation means a stable, heritable change including alterations in cell morphology, growth properties, cell permeability to sugar, appearance of new antigens, enhanced transplantability, agglutinability by plant lectins, sensitivity to activate macrophages and lymphocytes, malignant conversion and emergence of established cell lines with the potentiality to be subcultured indefinitely *in vitro* (Fedoroff, 1967).

The genome of RNA tumour viruses is composed of two 35S single-stranded RNAs. Each 35S RNA has a molecular weight of approximately 3×10^6 daltons (Temin, 1974). It seems now well established that each 35S RNA subunit is composed of the same sequence (Lai, Duesberg, Horst & Vogt, 1973) and that their non-transforming virus derivatives have a smaller genome that also includes the same sequences. Studies on virus genetics (Wyke, 1975) showed that the viral genome contains at least four genes, or groups of genes, three of which code for RNA-dependent DNA polymerase, for internal proteins and for envelope glycoproteins. The fourth gene, called '*sarc*', is responsible for the in-vitro maintenance of the transformed state of the host cells. Most of these experiments have been performed with the avian sarcoma viruses (Martin, 1970; Vogt, 1971; Martin & Duesberg, 1972; Lai *et al.*, 1973; Stehelin, Guntaka, Varmus & Bishop, 1976*a*; Stehelin, Varmus, Bishop & Vogt, 1976*b*). But Scolnick, Parks and colleagues (1973, 1974) have detected sarcoma-specific sequences in different strains of murine sarcoma viruses. The biological significance of the *sarc* gene remains unclear (neither more nor less than the definition of the 'transformed' state). Viruses that are typically leukaemogenic *in vivo* but can also transform fibroblasts *in vitro*, such as the murine Abelson virus and the avian MC29 virus, do not seem to carry *sarc* sequences (Scolnick *et al.*, 1975), and neither are *sarc* sequences detectable in the cellular RNA of fibroblasts transformed by the Abelson virus.

C-type RNA tumour viruses are commonly divided into two classes according to their 'oncogenic' potential *in vitro*: transforming viruses including the sarcoma viruses, and non-transforming viruses including most of the leukaemia viruses. It seems, at least to me, that this classification is today somewhat artificial and might be erroneous. It must be remembered that this classification is based on in-vitro studies which utilised fibroblasts exclusively since this cell type was, in the past, the easiest to obtain in culture. A fibroblast is not an undifferentiated or multipotent cell but a specialised cell with at least two characteristics: a long in-vivo and in-vitro lifespan and the capacity to grow in culture conditions. This could be not so negligible. Moreover, recent experiments, in particular those of Graf, Royer-Pokora, Schubert & Bekg (1976) in Tübingen, suggest that cloned avian erythroblastosis leukaemia viruses are able to transform fibroblasts in an in-vitro assay. Other chicken (Graf, 1973) or mouse leukaemia viruses have also been described as potential inducers of other neoplasms such as sarcoma. As already mentioned Abelson virus, which induces *in vivo* a 'B'-type lymphoid leukaemia, has been reported to transform 3T3 fibroblast cells (Scher & Siegler, 1975) in addition to lymphoid cells *in vitro* (Rosenberg, Baltimore & Scher, 1975). Apart from this, mutants of the Rous sarcoma virus which are defective in fibroblast transformation can develop a leukaemogenic potential, as demonstrated by Biggs, Milne, Graf & Bauer (1973). Similarly murine Harvey sarcoma virus induces both erythroblastosis and sarcomas.

In conclusion the oncogenic spectrum observed with sarcoma and leukaemia viruses *in vivo*, where all the target cells for the oncogenic action of the viruses are present, does not necessarily reflect their ability to transform specific types of cells in culture. Conversely transformation or absence of transformation *in vitro* does not necessarily reflect the oncogenic or non-oncogenic capacity of the virus. Experiments with murine leukaemia viruses, in particular with Friend virus, will demonstrate the variety of interactions which could occur between this virus and its haematopoietic target cells according to their stage on the differentiative pathway.

MURINE LEUKAEMIAS

The Friend leukaemia viruses

The anaemia-inducing Friend virus (FVA) was isolated originally from a Swiss mouse by Dr Charlotte Friend in 1956 at the Sloan Kettering Institute in New York. A cell-free extract prepared from the Ehrlich ascites mouse carcinoma unexpectedly induced a curious syndrome which appears to be exceedingly rare under natural life conditions but which could be transmitted serially by inoculating filtrates prepared from spleen extracts. After several passages the clinical picture became constant and readily reproducible. Today the description given by Friend (1956, 1957), Metcalf, Furth & Buffett (1959) and Zajdela (1962) is still valid and may be considered as the typical 'standard' description of the Friend disease.

Inoculation of the anaemia-inducing Friend virus (FVA) into young adult (2-month-old) mice of susceptible strains such as DBA/2, Swiss, BALB/c, C3H, CBA, etc., induces considerable enlargement of spleen and liver after a short latency of only a few weeks, depending on the dose of virus initially injected. This striking pathologic manifestation represents the main macroscopic characteristics of the induced disease. Thymic tumours are never observed and, in sharp contrast to the great majority of the spontaneous forms of mouse leukaemias, the peripheral or internal lymph nodes are not enlarged. It should be emphasised that the disease induced with the Friend virus is apparently uninfluenced by the presence or absence of thymus (Metcalf *et al.*, 1959).

The peripheral blood reveals a characteristic picture. There is evidence of anaemia, thrombocytopenia, lymphocytosis, granulocytosis and erythroblastosis. Blood smears stained with Giemsa or similar staining methods show the presence of numerous, very typical hyperbasophilic cells, so-called 'Friend' cells, which can be related morphologically to proerythroblasts (Dawson, Fieldsteel & Bostick, 1963; Dennis & Brodsky, 1965; Brodsky, Dennis & Kahn, 1966; Brodsky, Ross, Kahn & Petkov, 1968; Parr & Rowson, 1970). Anaemia and thrombocytopenia are partly due to a shortened lifespan of red blood cells and platelets (Brodsky *et al.*, 1966, 1968) on which the virus is budding (Zajdela, Tambourin, Wendling & Pierre, 1968). The hyperbasophilic 'Friend' cells predominate in the spleen and liver but are also present in bone marrow, lymph nodes and other organs at terminal stages of the disease. These cells proliferate very rapidly and have a short lifespan (they die or differentiate) (Smadja-Joffe, Jasmin, Malaise & Bournoutian, 1973; Smadja-Joffe *et al.*, 1976). Usually the disease progresses rapidly and the animal dies within 1 to 3 months after virus inoculation.

In the 1960s a very strange and, as yet, unexplained phenomenon occurred concerning the Friend virus. Mirand, Prentice & Hoffmann (1961) working with a virus extract given by C. Friend, noticed that several filtrates prepared from spleens of infected mice produced a hypervolemic polycythaemic response at 21 days instead of the routine anaemia. This 'new' polycythaemia-inducing Friend virus (FVP) not only induced hypervolemic polycythaemia in infected mice but also mimicked the action of erythropoietin (EPO). EPO is the physiological hormone which normally controls the initiation of erythroid differentiation by acting on committed target cells, the so-called erythropoietin responsive cell (ERC) (Lajtha, 1975; Goldwasser, 1976; Harrison, 1976). In FVP-infected mice, erythropoiesis remained active even at a haematocrit level above 70%. In successive reports, Mirand demonstrated that this erythropoiesis occurred without erythropoietin (Mirand, 1967*a*, *b*, 1976; Mirand *et al.*, 1968*a*, *b*). It is a most surprising fact that at the same time C. Friend sent a lyophilysed sample to Dr Nakahara in Japan and this virus, when injected in DDD mice, also induced the same polycythaemic response as described by Mirand (Kasuga & Oota, 1962; Sassa, Takaku & Nakao, 1968). Neither intraperitoneal injection of isologous red blood cells, which raises the haematocrit so as to suppress normal erythropoiesis (Jacobson, Goldwasser, Plzak & Fried, 1957), nor anti-erythropoietin serum (Schooley, Garcia, Cantor & Havens, 1968) could stop FVP-induced erythroid differentiation.

In our laboratory we have worked with both viruses for about 10 years. The FVA virus was a gift from Dr C. Friend to Dr R. Latarjet in 1958. It has been used widely in the Radium Institute, Paris, in particular by Zajdela (1962) and Chamorro (1962, 1967). This virus is still an anaemia-inducing virus which provokes the typical Friend disease. Over the 10 years we never observed any conversion of FVA into FVP. The FVP virus was brought to our laboratory by Dr Yoshikura in 1968. This virus constantly induces a polycythaemia more or less pronounced according to the strain of mice used, associated with lymphocytosis, granulocytosis and thrombocytopenia as the result of a shortened platelet lifespan (Brown & Axelrad, 1976). This FVP disease is also characterised by a splenomegaly and hepatomegaly; thymus and lymph nodes are never involved. FVA- and FVP-induced diseases are histologically and cytologically indistinguishable. The 'same' typical hyperbasophilic leukaemic Friend cells invade the spleen and the liver, there is the same sensitivity to actinomycin D treatment (Wendling, Tambourin, Barat & Zajdela, 1969; Tambourin & Wendling, 1971) and the same short latency.

However, in contrast with FVP leukaemia, erythropoiesis in FVA-induced disease remains extremely but not completely sensitive to experimental hypertransfusion. This treatment corrects the FVA-induced anaemia and greatly increases the survival of infected animals. Spleen weight increases much more slowly and a slight erythropoiesis becomes detectable when the spleen weight goes beyond 500 mg. Exogenous erythropoietin or any other erythropoietic stimulus reverses this delaying effect (Tambourin, Wendling, Barat & Zajdela, 1969). Quite similar results were obtained by Pluznik, Sachs & Resnitsky (1966) and Seidel (1972*a*) with the Rauscher virus, which induces a leukaemia not very different from FVA disease.

In conclusion it seems that both FVA and FVP viruses induce an abnormal erythropoiesis but FVA erythropoiesis appears to be *almost* completely dependent on erythropoietin control while FVP erythropoiesis seems completely autonomous from EPO. Other minor differences in red cell production and blood volume during the FVA and FVP leukaemic development have been reported (Tambourin, Gallien-Lartigue, Wendling & Huaulme, 1973). The explanation for such a difference between FVA and FVP diseases remains an open and interesting problem.

The Rauscher leukaemia virus

The Rauscher virus was isolated by F. J. Rauscher at the National Cancer Institute in Bethesda (1962). The origin of the virus strain is not clear. According to the original report of Rauscher, his initial studies were carried out on a 'virus-induced leukaemia' of adult Swiss mice previously reported by Schoolman and his associates (Schoolman, Spurrier, Schwartz & Szanto, 1957). After successive transplantation of the ascites tumour cells, first in Swiss mice and then in weanling mice of the BALB/c strain, unfiltered extracts were inoculated into BALB/c and Swiss mice. After successive passages through suckling BALB/c mice, leading to a gradual increase in virus potency, the incidence of leukaemia exceeded 95 %.

In susceptible strains, the Rauscher virus was able to induce a two-step type of disease. The first phase consisted of the characteristic erythroblastosis-like syndrome of hepatosplenomegaly just described above. In surviving mice, erythrocytopoiesis was followed by the development of lymphocytic leukaemia 30 to 45 days after viral inoculation. In our hands Rauscher virus never induced lymphocytic leukaemia. Mice developed a typical disease comparable to FVA-induced syndrome with minor but reproducible differences. For example, our strain of Rauscher virus is more potent in male than in female mice, in contradiction to the results obtained with Friend virus (Gillepsie & Rowson, 1968).

The physiopathology of the disease induced by Rauscher virus has also been extensively studied (Boiron, Levy, Lasneret & Oppenheim, 1965; Brodsky *et al.*, 1967; Brommer & Bentvelzen, 1970; Brommer, 1972; Ebert, Maestri & Chirigos, 1972; Seidel, 1972*a*; Cox & Keast, 1973; Knyszynski & Danon, 1976) and the role of the haematopoietic stem cell in the leukaemia development has been considered in detail.

Kirsten and Harvey viruses

Two other erythroblastosis-producing viruses have been isolated. Few studies have been done as yet on their physiopathology.

In 1967, Kirsten and his colleagues at the Department of Pathology and Pediatrics, University of Chicago, isolated a virus unexpectedly recovered in the course of serial cell-free passage of thymic lymphomas which arose spontaneously in old mice of the C3H strain. This virus was passaged through newborn W/Fu rats which developed malignant lymphomas after 6 to 7 months. One cell-free passage induced erythroblastosis in less than 20% of the injected rats, which died within 4 weeks of inoculation. A serial cell-free passage was then initiated from this erythroblastosis and all rats subsequently died with erythroblastosis. In addition multiple sarcomas and polyostotic osteolytic lesions were found in most rats after such erythroblastosis passages. The same filtrates of rat erythroblastosis induced erythroblastosis and sarcomas in C3H mice. Therefore, as previously mentioned, association between sarcoma and erythroblastosis occurred. However, these viruses were not cloned (Kirsten & Mayer, 1967; Kirsten, Mayer, Wollmann & Pierce, 1967). But recently Scher, Scolnick & Siegler (1975) using cloned Kirsten and Harvey viruses also demonstrated that these cloned transforming viruses caused erythroid leukaemia in mice. Some of the animals inoculated developed solid tumours. These tumours (usually rhabdomyosarcomas or fibrosarcomas) grew slowly and did not kill the host. The erythroid leukaemia is characterised by a splenomegaly, anaemia and thrombocytopenia (Wollmann, Pang, Evans & Kirsten, 1970).

In the course of routine passage of the mouse Moloney leukaemia virus in rats, Dr Jennifer Harvey observed that reinjection of the passaged virus into newborn BALB/c induced, after a short latency of only 1 month, pleomorphic sarcomas at or near the injection site instead of leukaemia. Later it was recognised that stock (Chesterman, Harvey, Dourmashkin & Salaman, 1966; Bassin, Simons, Chesterman & Harvey, 1968) or cloned (Scher *et al.*, 1975) Harvey virus induced erythroblastic leukaemia in mice.

The other leukaemia viruses

Numerous other leukaemia viruses have been isolated since the fundamental experiment of Ludwig Gross in 1951. But little work has been done in the field of physiopathology, in connection with the haemopoietic stem cells. Abelson virus also transformed 3T3 cells *in vitro* and caused a lymphoid leukaemia of 'B' or 'null' cell origin (Abelson & Rabstein, 1969, 1970; Siegler, Zajdel & Lane, 1972; Scher & Siegler, 1975; Sklar, Shevach, Green & Potter, 1975; Rosenberg *et al.*, 1975; Rosenberg & Baltimore, 1976). The other lymphoid leukaemia viruses induced 'T' leukaemias in most strains into which they were injected (Gross, 1951, 1959, 1963; Moloney, 1960; Dunn, Moloney, Green & Arnold, 1961; Decleve *et al.*, 1975; Chazan & Haran-Ghera, 1976). Other murine viruses induced myeloid leukaemia but have been rarely used for the study of haemopoietic stem cell behaviour (Graffi, Fey & Schramm, 1966; Fey, 1969; Soule & Arnold, 1970).

PROBLEMS ASSOCIATED WITH THE USE OF MURINE LEUKAEMIA VIRUSES

From time to time a group working with murine leukaemia viruses gives evidence of results quite different from works previous published by other groups. Sometimes some dispute occurs. Most often both groups are right. It rapidly becomes clear that the two diseases in question are not at all the same. The virus may have changed, which is not really a rare event. The transformation of FVA into FVP leukaemia probably represents the most typical example. The group of Professor A. Gordon in New York works with a very unusual variant of Rauscher virus that they called variant 'a' (Broxmeyer *et al.*, 1975). This virus, referred to as RLV/a, caused a profound and usually fatal murine anaemia in BALB/c mice with no splenic destruction. Infected BALB/c mice did not respond to the anaemia with increased levels of plasma erythropoietin (Camiscoli *et al.*, 1972) and phenylhydrazine-induced haemolytic anaemia resulted in greatly decreased incidence of RLV/a disease (Weitz-Hamburger *et al.*, 1973). But these infected animals responded to erythropoietin administration by a wave of erythroid maturation (Weitz-Hamburger *et al.*, 1975). Indeed comparison of results between two hypothetical laboratories, one working with the usual Rauscher virus and the other with RLV/a, would, if these differences were not taken into account, probably lead to some conflicting situations. On the other hand, the physiopathological study of the leukaemias induced in the same strain of mice by these two viruses would be a rather interesting problem.

In fact, the use of different strains of mice is another cause of variation. It is now well known that the host's genetics play a major role in the development of leukaemia (Fieldsteel, Dawson & Bostick, 1961; Ludwig, Bostick & Epling, 1964; Steeves, Mirand, Bulba & Trudel, 1970). When non-inbred mice were used different types of leukaemia were frequently observed after infection with stocks of Friend or Rauscher leukaemia viruses (Chamorro, 1962, 1967). The same results have been found with the Graffi virus (Fiore-Donati & Chieco-Bianchi, 1964). This influence of host genetics has been particularly well documented in terms of susceptibility and resistance (see review by Lilly & Pincus, 1973).

A third cause of variation comes from the virus itself. It is now well established that Friend and Rauscher viruses are rather complex mixtures in which at least two viral entities exist. Firstly, the so-called Spleen Focus-Forming Virus (SFFV) induces macroscopic foci of primitive erythroid cells in the spleen red pulp (Pluznik & Sachs, 1964; Axelrad & Steeves, 1964; Ikawa, Sugano & Oota, 1967; Mirand *et al.*, 1968*a*). The second component is the more typical lymphoid murine leukaemia virus referred to as MuLV. The SFFV is defective and depends for its replication on the presence of its associated MuLV helper for some still poorly understood function(s) (see reviews by Steeves, 1975; Troxler, Parks, Vass & Scolnick, 1977).

Other leukaemia-inducing or non-leukaemia-inducing viruses have been shown to be present as true components or as contaminants in the Friend or Rauscher mixture.

The lactate dehydrogenase virus (LDV) is a very common contaminant of tumours and leukaemia virus stocks (Riley, 1968). Friend virus preparations have also been shown to contain the Rowson–Parr Virus (RPV) that can be easily separated by end-point dilution. In several respects RPV appears to be similar to MuLV (Rowson & Parr, 1970). From a chloroleukaemic C57Bl mouse injected neonatally with the MuLV-containing helper component of the Friend virus complex, a myelogenous leukaemia-inducing virus (MyLV) has been isolated by McGarry *et al.* (1974). Whether this MyLV has arisen by selection of a pre-existing virus in the Friend virus (FV) complex or by another mechanism remains unknown.

Another source of variation in the properties of leukaemia-inducing viruses results from long-term passage *in vitro* (Sinkovics, Bertin & Howe, 1966; Yoshikura, Hirokawa, Ikawa & Sugano, 1969; Barski, Barbieri, Koo Youn & Hue, 1973; Buchhagen, Pincus, Stutman & Fleissner, 1976). The leukaemogenicity of viruses produced by persistently infected cells gradually decreases. There are no satisfactory explanations accounting for the existence of such attenuated viruses.

In conclusion, it must be emphasised that virus variation or contamination is a very common risk in prolonged experiments which also produced some variations in the process of leukaemia. *Therefore, each group working with murine leukaemia viruses ought to work with freshly cloned virus, and specify carefully what kind of leukaemia this particular virus induces.*

EARLY EVENTS

Today the physiopathology of Friend (FVA and FVP) and Rauscher syndromes appears to be well understood. Three aspects have been particularly investigated: (1) early events, (2) behaviour of haemopoietic stem cells after virus infection, (3) identity of target cell(s).

In-vivo studies

The study of early events has first been done *in vivo*, mainly in our laboratory, using large doses ($> 10^5$ FFU) of FVP injected intravenously in hypertransfused plethoric (HP) mice. For 30 hours nothing could be detected in such animals. After that time hyperbasophilic cells, the so-called Friend cells, appeared massively in the spleen red pulp only (not in the lymphoid areas) outside the granulopoietic and megacaryocytic foci.

It has been successively demonstrated that these cells result from the morphological transformation of unidentified cells present *in situ* in the spleen red pulp and in the bone marrow. These Friend cells multiply, and some of them differentiate immediately along the erythrocytic pathways without erythropoietin, as previously shown by Mirand *et al.* (1967*a*, *b*) and Sassa *et al.* (1968) in the situation of well developed leukaemia. Haemoglobin synthesis becomes measurable in the spleen between 28 and 36 hours after infection. The number of Friend cells increases very rapidly as the result of cell multiplication and of the continuation of the transformation process always bringing new hyperbasophilic leukaemic cells. Cell transformation, multiplication and differentiation provoke a rapid growth of spleen (Tambourin & Wendling, 1971).

The transformed hyperbasophilic Friend cell is probably an abnormal proerythroblast. A [^{59}Fe]haem incorporation kinetic study in spleen suggests that these cells, at least some of them, would synthesise haemoglobin (Tambourin & Wendling, 1971). In 1966 Yokoro & Thorell, employing microspectrophotometric methods, determined the haemoglobin concentration in such blastic cells, on impression smears of the hepatic lesions of Rauscher disease. The same kinds of experiments were done by Sundelin & Adam in 1967. These two separate studies showed that most of these blastic cells contained a variable amount of haemoglobin in their cytoplasm. Ikawa, Furusawa & Sugano (1973) demonstrated that the in-vivo transformed hyperbasophilic Friend cells reacted to a *specific* antibody against normal erythrocytic membrane. It is also very interesting to notice that these hyperbasophilic cells are rapidly lost (Smadja-Joffe *et al.*, 1973, 1976), which is to be connected with the characteristic of a normal proerythroblast. Taking advantage of the transforming ability of FVP Hankins & Krantz (1974) devised a rapid in-vivo assay for FVP which fits well with the virus's focus-forming activity.

Hankins & Krantz (1975) have reproduced our experiments using a quite different method. Ex-hypoxic HP mice were infected with a high dose of FVP. At different times after virus infection spleen cells were harvested, cultivated and their ability to synthesise haemoglobin measured at once or 48 hours after in-vitro incubation without erythropoietin in the medium. A similar experiment acted as a control, using EPO *in vivo* in the place of FVP. This experiment showed that, although the rate of haem synthesis had not yet been altered *in vivo* within 16 to 24 hours after infection of the mice the spleen cells harvested at this time and incubated were able to start haemoglobin synthesis which was detectable 48 hours later *without* any EPO in the medium.

In-vitro experiments

Beautiful in-vitro experiments have been done recently by Axelrad's group in Toronto which reproduced the in-vivo phenomena. Using the plasma clot method devised by Stephenson, Axelrad, McLeod & Shreeve (1971), which allows the in-vitro study of colony forming units that respond to EPO (referred to as CFU-E) by giving small erythroid colonies, Clarke, Axelrad, Shreeve & McLeod (1975) obtained erythroid colonies in cultures seeded with *normal* bone marrow cells that had been previously exposed to culture medium from a cell line which continuously produces Friend virus (FVP) *in vitro*, without the addition of erythropoietin. This system appears to be very fruitful for investigation of Friend virus/host cell interaction and could contribute to our knowledge of murine leukaemia virus genetics.

At the same time Nooter & Ghio (1975) at the Radiobiological Institute TNO, The Netherlands, reported that bone marrow cells from BALB/c mice infected *in vivo* with Rauscher virus produced erythroid colonies *in vitro* in the absence of erythropoietin. A similar experiment has also been performed independently by Liao & Axelrad (1975), who found that after in-vivo infection by FVP the normal CFU-E was rapidly replaced by a colony forming unit capable of producing erythroid colonies *in vitro* in the absence of EPO. Exactly at the same date, Horoszewicz, Leong & Carter (1975) published the same results with FVP virus.

In 1976 Nooter & Bentzelven reproduced experiments of Clarke *et al.* (1975). Incubation of normal mouse bone marrow cells with purified Rauscher virus led to transformation *in vitro*. The same year Nooter, Van Den Berg, De Vries & Bentvelzen made a transfection experiment using DNA isolated from Rauscher-virus-induced leukaemic spleen. Mouse bone marrow cells from normal mice could be transfected by Rauscher virus proviral DNA. Successful transfection was manifested by virus production and transformation, i.e. formation of erythroid colonies *in vitro* without addition of erythropoietin to the culture medium.

MALIGNANT ERYTHROPOIESIS

For a few years after the discovery of Friend leukaemia virus (FVA) most investigators considered the leukaemia as a typical reticulum cell sarcoma associated with an intense erythropoiesis, a physiological reaction of the haemopoietic system to compensate for peripheral anaemia. So, until 1965, the disease was classified as undifferentiated reticulum cell sarcoma associated with a normal compensatory erythropoiesis (Friend, 1957; Metcalf *et al.*, 1959; Graffi & Bielka, 1963; Dawson *et al.*, 1963; Brodsky *et al.*, 1966). This idea was supported by the morphological character of the transplantable tumour derived from the splenic and hepatic lesions of Friend disease (Buffett & Furth, 1959; Friend & Haddad, 1960).

Zajdela (1962) was the only author to report that Friend virus (FVA) probably induced an erythroid neoplasia. The isolation of FVP by Mirand and the Japanese groups demonstrated unequivocally that erythropoiesis in FVP leukaemia is a process which fits with the definition of oncogenicity (it takes place in conditions where normal erythropoietic precursors are completely blocked in their differentiation pathway), so that FVP was considered as quite a different virus from the typical Friend virus and Zajdela's conclusion forgotten.

In 1966 Friend, Patuleia & De Harven isolated a Friend leukaemic cell line which differentiated along the erythrocytic cell line *in vitro* and *in vivo* (Rossi & Friend, 1967). It is not explained whether this cell line was derived from a tumour isolated from a FVP or FVA leukaemia (Friend & Haddad, 1960) but this result strengthened Zajdela's hypothesis. Friend, Scher, Holland & Sato further reported, in 1971, a spectacular enhancement of erythrocytic maturation of the same cell line by adding 1–2% of dimethyl sulphoxide (DMSO) to the culture medium. The finding that even in leukaemias induced by FVA (Tambourin *et al.*, 1969) or Rauscher (Seidel, 1972*a*) virus, erythropoiesis was not completely blocked by hypertransfusion was another argument to prove that in Friend (FVA or FVP) or Rauscher leukaemia erythropoiesis belongs to the malignant process. The in-vitro experiments reported in the previous section demonstrated that conclusion definitely.

THE TCFU

The spleen colony assay of Till & McCulloch (1961) using irradiated or non-irradiated recipients has been widely used to quantify or to demonstrate the existence of a tumour cell population (Bruce & Van Der Gaag, 1963; Wodinsky, Swiniarski & Kensler, 1967; Bergsagel & Valeriote, 1968; Tanaka & Lajtha, 1969). Such a population, if present in the spleen of Friend or Rauscher leukaemia or in other murine leukaemia, could interfere with experiments on haemopoietic stem cells. But, in contrast to the lymphoid leukaemia induced by several viruses, the initial erythroblastic disease induced by Friend or Rauscher viruses cannot be easily transplanted using the classical method, i.e. by subcutaneous graft. If fragments or cell suspensions of spleen or enlarged liver of a leukaemic animal are grafted subcutaneously into uninfected syngeneic mice, typical leukaemia ensues which is induced by the virus present in, and liberated by, the graft (Brommer, 1972). No local growth can be seen except in very rare cases which occurred most often when the donor was at the terminal stage of the disease (Friend & Haddad, 1959, 1960; Buffett & Furth, 1959; Dawson *et al.*, 1963; Dawson & Fieldsteel, 1974).

This unusual result could be due to the inability of this drastic method to demonstrate the existence of tumour cells already present in the early phases of infection but incapable of growth under the skin. It could also be due to an absence of true tumour cells, i.e. cells with an infinite lifespan. In 1968 Thomson & Axelrad devised a new method based on the spleen colony assay in unirradiated mice. To eliminate foci induced by the virus (Axelrad & Steeves, 1964; Pluznik & Sachs, 1964; Ikawa & Sugano, 1967; Mirand *et al.*, 1968*a*) Thomson used (C3H×C57Bl) F_1 hybrids which were about 100 times less susceptible than C3H to the focus-forming activity of the Friend or Rauscher virus that might have been brought in with the hypothetical tumour cells to be assayed. These animals are able to accept C3H cell transplants. When injected into such hosts, spleen cells from an *infected* donor regularly induced the formation of macroscopically visible colonies. Injections of normal spleen cell suspension were ineffective as expected. Thus, evidence was presented which seemed to establish the fact that these spleen colonies resulted from the proliferation of a tumour colony forming unit, called TCFU, already present in the early phase of the Friend or Rauscher disease. Since then numerous investigators have employed this method to study the properties of the TCFU or to compare TCFU and haemopoietic CFU (Thomson & Axelrad, 1968; Steeves, Mirand & Thomson, 1969; Thomson, 1969*a*, *b*; Karnjanaprakorn-Yorsook & Thomson, 1973, 1974; Thomson & Shrivanatakul, 1973; Thomson, Shrivanatakul & Mirand, 1974; Axelrad, Cinader, Koh & Van Der Gaag, 1976).

Today we can say that this method is probably based on a wrong concept and that, in most cases, it does detect any type of tumour cells. The importance of that problem and the number of papers published on the subject of the relation between TCFU and the haemopoietic stem cell, demand that one obtains further information. In our laboratory we repeated all the experiments described by Thomson in order to determine if the viruses used in the two laboratories and the behaviour of the mice after infection were similar. The conclusion was quite clear. In all experiments we found exactly the same result. So, we decided to study by immunological methods the nature (donor or host) of the cells which were contained inside the colonies. An antiserum anti-C57Bl/6 was prepared in C3H mice and tested against the cells present in the colonies. These cells were destroyed by the antiserum in all cases (more than 50) studied. In conclusion, the method devised by Thomson & Axelrad would not detect true tumour cells but rather high virus-producing cells, similar to the cells detected by the in-vitro infectious centre method.

Using Rauscher virus and repeating exactly the experiments of Thomson, Bentvelzen (1972) was able to demonstrate that after irradiation of the cell suspension at 3000 R, sufficient to kill all the cells but not to inhibit virus production immediately or to inactivate the virus, the number of colonies obtained in F_1 hybrid semi-resistant mice was half the number obtained with the unirradiated cells. Bentvelzen concluded that most of the colonies observed (at least those which remain after irradiation) were not due to the proliferation of donor cells.

In 1971 Takada, Takada & Ambrus also concluded that leukaemic Friend cells carrying the CBA/HT6T6 chromosome marker were unable to proliferate in CBA/H syngeneic hosts.

A more refined genetic method, similar in principle to the method of Thomson & Axelrad, was proposed in 1974 by Kumar, Bennett & Eckner. This method gave somewhat paradoxical results. The demonstration of the existence of transplantable transformed Friend cells is based upon the fact that $Fv\text{-}2^r$ gene is such that mice homozygous for that gene ($Fv\text{-}2^{rr}$) are *absolutely* resistant to focus formation by FV complex. Therefore, observation of foci on the surface of a B10D2 mouse spleen which is $Fv\text{-}2^{rr}$ implies that such foci, which can be observed only after the intravenous injection of FV-infected cells, result from the proliferation of the infused cells, since B10D2 host cells are supposed to be absolutely resistant.

Unfortunately, the conclusion of the paper, derived directly from this hypothesis, was that B10D2 cells can be transformed by the FV complex and are transplantable in some circumstances! Then the assumptions which allow the demonstration are no longer valid and the cells could not be transplantable. In such a situation the observed foci should result from the transformation *in situ*, by high virus-producing donor cells, of the host cells which gave rise to foci.

A recent paper of Blank, Steeves & Lilly (1976) supports this conclusion. They showed that in a new congenic DBA/2 strain, which is $Fv\text{-}2^{rr}$, FV complex replicated and that such mice when infected with 3000 FFU developed mild to moderate splenomegaly and elevated haematocrit levels (70–80 %) and most of them died within 2 months. Thus, despite the potent resistance of DBA/2 ($Fv\text{-}2^{rr}$) mice, neoplastic erythropoiesis did progress to a lethal termination in adult mice infected with a sufficiently high virus dose. Blank *et al.* made the same conceptual misinterpretation in their paper. To demonstrate the existence of TCFU in DBA/2 ($Fv\text{-}2^{rr}$) mice and follow their growth kinetics, Blank *et al.* injected infected spleen DBA/2 ($Fv\text{-}2^{rr}$) in DBA/2 ($Fv\text{-}2^{ss}$–$Fv\text{-}1^{bb}$). As DBA/2 ($Fv\text{-}2^{rr}$) mice were $Fv\text{-}1^{nn}$, these $Fv\text{-}1^{bb}$ recipients were supposed to be resistant to the virus. Therefore spleen foci observed in DBA/2 ($Fv\text{-}2^{ss}$–$Fv\text{-}1^{bb}$) hosts 9 days after injection of infected spleen cells should have been the result of donor cell proliferation. Blank *et al.* did not check the validity of their assumption so as to be sure that they also were not detecting highly infectious centres.

In conclusion, it is more than likely that the TCFU methods do not detect true tumour cells but rather infected cells able to produce locally sufficient amounts of virus to overcome the genetic resistance of the target cells. The identity of the TCFU remains unknown. A study of Thomson *et al.* (1974) and a more recent paper of McGarry & Mirand (1977) demonstrated that TCFU and haemopoietic CFU do not constitute identical populations although haemopoietic stem cells were, in my opinion, a more likely candidate than others cells. Indeed the TCFU must be an infected cell, with a radiation survival curve comparable to the typical radiation survival curve of haemopoietic cells ($D_0 = 90$ rads; Thomson & Axelrad, 1968). They probably have a relatively long lifespan since sufficient amounts of virus must be produced. They are present in a proportion of about 3–4 TCFU per 10^5 cells in the spleen (Thomson &

Axelrad, 1968; Bentvelzen, 1972) and 20 TCFU per 10^5 cells in the bone marrow (P. Tambourin & F. Wendling, unpublished data).

In another series of experiments, Rossi, Cudckowicz & Friend (1970, 1973) also studied in-vivo neoplastic transformation of haemopoietic cells infected with Friend virus. This malignant transformation, which was described as occurring as early as 3–4 hours after infection of DBA/2 mice with FLV, was characterised by two parameters: (1) the acquisition of autonomous growth potential on intravenous transplantation into non-irradiated histocompatible mice, (2) the serial transplantability in subcutaneous as well as haemopoietic sites. The ability of the grafted cells to proliferate was assessed 7 days after transplantation by measurement of ^{125}IuDR uptake in recipients, i.e. by an indirect method inadequate for discriminating between the proliferation of host cells and that of donor cells. The controls for background uptake of IuDR were inadequate since, from experiments on TCFU, it is quite clear that sonically disrupted or frozen–thawed cells are not at all comparable to intact cells in terms of virus production, as they should be. The intravenous serial transplantability of these cells was also studied by this same method and, therefore, the possibility of studying endogenous cell proliferation could not be excluded. The subcutaneous graft gave rise to small nodules. These experiments were repeated in our laboratory. Small subcutaneous nodules were obtained 21 days later. Histological examination of nodule sections showed undoubtedly that they were normal-appearing ectopic spleens subsequently infiltrated by leukaemic cells. Curiously beyond day 7 after infection the ability of infected spleens to generate such ectopic grafts disappeared (P. Tambourin & P. Bucau, unpublished data). Recently, Levy *et al.* (Levy, Rubenstein & Friend, 1976*a*; Levy, Rubenstein & Tavassoli, 1976*b*) published similar results. It seems reasonable to predict that repetition of these experiments under more controlled conditions would lead to the conclusion that Friend leukaemic cells are not serially transplantable, at least during the first 2 weeks after infection.

Furthermore, the problem of neoplastic cell transformation can be approached from a quite different point of view by listing a catalogue of the respective properties of the hyperbasophilic leukaemic cells and of the tumour cells while keeping constantly in mind that some properties of the tumour cells could be acquired later than the true neoplastic event(s) along the numerous successive in-vivo or in-vitro passages. At the moment in our laboratory we are studying the properties of tumour cells in their initial passage. Preliminary results indicate that most of the typical characteristics of a tumour cell already exist at this time. Tumour cells have an infinite lifespan while Friend leukaemic spleen cells have a very limited capacity for proliferation (Smadja-Joffe *et al.*, 1973, 1976). Ikawa *et al.* (1973) demonstrated that unlike hyperbasophilic cells present in the primary splenic lesions, none of the transplantable tumour cell lines reacted with the labelled erythrocyte membrane-specific antibody. It has been also shown that several chromosomal rearrangements exist in tumour cell lines when no such specific rearrangements where detected in the infected animals (Ostertag *et al.*, 1972; Matioli, 1974; Majumdar & Bilenker, 1975; F. Wendling, unpublished data). Histones extracted from Friends leukaemic spleen or erythroid normal spleen do not show the increase in f2a2 seen in DMSO-responsive Friend cells lines, although haemoglobin synthesis occurs (Blankstein & Levy, 1976). In an irradiated syngeneic host, Matioli (1974) studied the metastasis and growth of Friend tumour cells compared with normal 'leukaemic' stem cells. The most interesting differences were the high metastatic activity, the lack of differentiation, the deterministic growth and the independence from the spleen microenvironment experienced by the tumour cells in contrast to the normal and Friend-leukaemia-derived stem cells.

HAEMOPOIETIC STEM CELLS

The haemopoietic pluripotent stem cell and committed stem cell compartments have been widely investigated in different types of murine leukaemia. If the interpretations and therefore the conclusions of the numerous laboratories involved do not appear to be necessarily concordant, the results themselves are surprisingly coherent. In Friend leukaemia (FVA and FVP) an extensive study of the haemopoietic stem cell compartment has been done by Wendling *et al.* in our laboratory (1973*a*, *b*). In Rauscher leukaemia, a similar study has been reported by Brommer in the Netherlands (1972), Seidel in Germany (1973*a*, *b*, *c*, *d*) and by Okunewick and co-workers in the USA (Okunewick & Phillips, 1973, 1974; Okunewick, Phillips & Brozovich, 1976*b*).

Increased radioresistance of virus-infected mice

The first indirect study of the involvement of the haemopoietic stem cell compartment during the Friend leukaemia development was done by Dr Gallien-Lartigue in our laboratory in 1968. It was an unexpected finding that irradiation of Swiss mice infected with Friend virus (FVA) at the enormously supralethal dose of 950 rads (about 250 rads above the dose which killed 50% of the uninfected irradiated control mice during the 30 days following X-irradiation, LD50/30) allowed the survival of 40% of the mice in the groups infected previous to X-ray exposure (Gallien-Lartigue, Tambourin, Wendling & Zajdela, 1968).

Usually the irradiation of lymphoid leukaemic animals provoked their rapid death, an effect attributed to the disappearance of immunological barriers. Two other facts also seemed surprising at this time:

(1) It was necessary to wait for a period of time after infection (i.e. a minimal splenic enlargement) to obtain enhanced survival. It seemed that the more the animal was leukaemic, the stronger was its resistance to lethal X-irradiation. As long as the spleen weight stayed below 500 mg, the FVA-infected mice died within the same time interval as their controls.

(2) At day 9 after irradiation, no tumour colony was ever observed among the macroscopic surface spleen nodules at least by histological and cytological criteria.

Quite similar results were reported later by Brommer (1972), Okunewick, Fulton & Markoe (1972) and Markoe & Okunewick (1973*a*, *b*) in Rauscher leukaemia, and by Rife, Sizemore & Matioli (1975) in FVP-induced leukaemia. From these different experiments it seemed that the increased radioresistance was mainly due to the increase in the number of splenic haemopoietic stem cells, and secondarily to a slightly increased radioresistance of the stem cells at these high doses of irradiation (Rife *et al.*, 1975).

The spleen colony assay

The method used to study the haemopoietic stem cell compartment in normal as in leukaemic mice is the spleen colony assay of Till & McCulloch (1961). This technique is very simple. A cell suspension prepared from the organ to be studied (bone marrow, spleen, blood, etc.) is injected intravenously into a lethally irradiated normal mouse. If the cell concentration is well adjusted, small but distinct nodules (1 to 3 mm in diameter) can be observed at the surface of the spleen 7 to 10 days after the graft, the number of which is directly proportional to the number of cells injected. Of the injected cells only a proportion will seed into the spleen, but this proportion can be assessed and varies between 10 and 20% (Till, McCulloch & Siminovitch, 1964; Playfair & Cole, 1965). Each colony has been shown to originate from one cell, CFU-S (Becker, McCulloch & Till, 1963). Knowing the seeding efficiency, the 'f' factor, the absolute number of colony forming cells in the sample can be calculated.

As it has already been said this method must be applied with some caution when leukaemic tissue is grafted. As a matter of fact it must be recalled that the spleen colony assay has been used to quantify lymphoma (Bruce & Van Der Gaag, 1963), plasma (Bergsagel & Valeriote, 1968), myeloid (Tanaka & Lajtha, 1969) and L1210 (Wodinsky, Swiniarski & Kensler, 1967) tumour cell populations which produce spleen colonies in either irradiated or non-irradiated syngeneic hosts. Therefore in the irradiated syngeneic recipients the spleen colony assay could detect tumour colony forming cells as well as haemopoietic colony forming cells. It must be remembered, however, that the results obtained after the irradiation of leukaemic animals and reviewed in the previous section strongly suggest the non-existence of such tumour cells. In Friend and Rauscher leukaemia, cytological and histological investigations of the colonies developing in the spleen of irradiated recipients grafted with spleen or bone marrow cells from leukaemic donors, indicated that the pattern of differentiation along erythrocytic, myelocytic (either neutrophilic or eosinophilic) and megakaryocytic lines appeared comparable with the pattern of differentiation observed in spleen colonies derived from the graft of normal haemopoietic cells (Wendling, Tambourin & Jullien, 1972; Brommer, 1972; Brommer & Bentvelzen, 1974). The spleen colonies originating from a leukaemic cell graft showed only a slightly greater percentage of erythroid colonies and a slight decrease in the percentage of the megakaryocytic colonies below control values. No tumour colony was found in most cases except in 1 experiment out of 100 in which a donor infected 27 days previously was used. In that particular case the number of colonies observed on the recipient spleens was at least tenfold greater than usual. These colonies were macroscopically larger and whiter than the typical haemopoietic colonies. Histologically large cells with faintly basophilic cytoplasm and a few cells morphologically related to maturing erythroblasts could be seen. These colonies were indefinitely transplantable in either irradiated or non-irradiated hosts by the subcutaneous or intravenous route (F. Wendling & P. Tambourin, 1972, unpublished data). Other studies on the patterns of differentiation along the haemopoietic cell lines have been reported by Van Griensven, Van Beek & Van't Hull (1974). They found a significant decrease in the percentage of erythroid colonies and an increase in the percentage of the myeloid colonies after the graft of bone marrow from Rauscher-infected BALB/c mice. Van Beek, Van't Hull & Van Griensven (1976) noticed that the erythroid colonies derived from both bone marrow and spleen of Rauscher-infected BALB/c mice exhibited a retarded maturation, the spleen-derived erythroid colonies showing the most severe changes. But, from these different studies using different viruses and mouse strains, and from the exception reported, it can be concluded that the spleen colony assay in syngeneic irradiated mice does detect haemopoietic cells, and that no tumour cells interfere with this assay, at least in the above-mentioned experimental conditions and especially if donor mice infected for a relatively short time (< 25 days) are used. But are these haemopoietic cells true haemopoietic stem cells? The definition of a stem cell implies two main characteristics: (1) self-renewal, and (2) ability to differentiate. In the case of a haemopoietic stem cell, multipotence must be added to these two fundamental properties. From the above experiments it can be seen that the spleen colony forming cells in the spleen collected from leukaemic animals are able to differentiate towards the several haemopoietic cell lines.

Self-renewal and multipotence

To my knowledge, the self-renewal capacity of the haemopoietic stem cells from infected mice has been evaluated only by Hajdik, Niewisch & Matioli (1971) (during the erythroid phase induced by the Kirsten virus), by Matioli (1973) and by F. Wendling in our laboratory.

The capacity of haemopoietic stem cells to renew their own numbers, or self-replication, is important not only for the replacement of cells lost by the process of differentiation, but also in the recovery of the haemopoietic system following injury. The availability of assay methods for CFU-S makes it possible to study the kinetics of stem cell self-renewal. Normal or leukaemic spleen or bone marrow cells are injected into heavily irradiated isologous mice. Thereafter the spleens (or bone marrow) of these animals are taken at intervals of time and cell suspensions prepared and assayed for their CFU-S content in a secondary irradiated recipient. These experiments clearly demonstrated that, as expected, CFU-S from leukaemic animals were able to repopulate their own compartment with a doubling time of 20 to 25 hours, which is quite similar to the values obtained with normal cells (Siminovitch, McCulloch & Till, 1963; McCulloch & Till, 1964). In 1971, Hajdik *et al.* studied the stem cell compartment in mice infected by Kirsten virus. Upon retransplantation of these stem cells into intermediate syngeneic irradiated host mice, the doubling time, self-renewal and extinction probability (Vogel, Niewisch & Matioli, 1968), which seemed to be abnormal when measured by techniques which avoided the step of intermediate irradiated recipients appears to revert towards normal values when measured by the usual two-step method. The interpretation of these results is somewhat questionable. Indeed this study measured on the one hand the growth of stem cells in unirradiated infected newborn mice by transplanting them directly into irradiated mice and counting spleen colonies 9 days later, and on the other hand the growth of some 'infected' stem cells after injection into intermediate irradiated recipients by subsequent serial retransplantation. It is clear that these two situations differed at least by two parameters; normal or leukaemic microenvironment and use of *unirradiated* and *irradiated* mice. It seems difficult to take advantage of these results, but the finding of normal growth of 'leukaemic' CFU-S after injection into normal intermediate irradiated recipients is interesting. This result is in agreement with those of Okunewick, Markoe, Erhard & Phillips (1973). Similar experiments were repeated by Matioli (1973) using Friend virus; similar results were found.

During this research retransplantation experiments were performed, using individual colonies picked at day 7 after the first cell graft. A single colony reinjected in irradiated syngeneic hosts gave rise to secondary colonies. Histological studies of these secondary colonies showed that the different haemopoietic cell lines could all be retrieved from a single colony. Although the karyological method of Becker *et al.* (1963) was not used, it may reasonably be concluded that the CFU-S remain multipotent after Friend virus infection.

Size of the CFU-S compartment after infection

It was first demonstrated that the relation between the number of colonies counted in the spleen and the number of cells injected remained linear after infection. Evidence was thus provided that each colony is derived from a single entity and that over the range of cell doses used each colony developed independently (Brommer, 1972; Wendling *et al.*, 1972; Okunewick, Phillips & Erhard, 1972).

Spleen. After injection of Friend or Rauscher leukaemia viruses, an exponential increase in spleen weight occurred. The number of CFU-S per 10^5 spleen cells remained nearly the same whether the grafted cells came from a normal or a leukaemic spleen. Consequently, the total number of CFU-S in the infected spleen was increased by a factor (15- to 30-fold) depending on the spleen weight (Wendling *et al.*, 1972; Brommer, 1972; Markoe, Okunewick, Erhard & Phillips, 1973; Brommer & Bentvelzen, 1974; Okunewick, Phillips & Erhard, 1972; Seidel, 1973*b*, *c*).

Bone marrow. Results obtained using bone marrow of Rauscher- or Friend-infected mice are somewhat different. A first group found that after infection the number of CFU-S in the femoral bone marrow remained nearly constant (about 3000 CFU-S per femur, despite a more or less pronounced cytopenia (Wendling *et al.*, 1972; Seidel, 1972*b*, 1973*c*, *d*; Iturriza & Seidel, 1974). Okunewick & Phillips (1973) observed a pronounced fall in the number of bone marrow CFU-S of Rauscher-infected SJL/J mice but a slight increase occurred when the same mouse strain was infected with FVP. Van Griensven *et al.* (1974) found that the total number of bone marrow CFU-S was increased by a factor of 2.7 in Rauscher-infected BALB/c mice.

Blood. The mean number of circulating CFU-S was increased 500–1000-fold after FVP virus infection, and the total number of circulating CFU-S detected in the blood of well-developed leukaemic mice has been assessed at 40000 CFU-S per mouse, as compared to 45 in normal mice (Wendling *et al.*, 1972; Seidel, 1973*d*).

No correlation was found between the total number of nucleated circulating cells and the number of circulating CFU. Seidel reported that the increase in CFU-S occurred at about the time of increase in the number of peripheral lymphoid cells.

'Seeding' or 'dilution' factor

The spleen colony assay detects only those injected CFU that reach the spleen after an inevitable dilution in the irradiated hosts. This dilution factor, usually called 'f' factor (Siminovitch *et al.*, 1963), was studied by Wendling *et al.* (1972) after FVP infection and by Okunewick *et al.* (1973) after Friend and Rauscher infection. It appeared that both viruses slightly depressed this factor in the marrow as compared to the controls but to a much lesser degree than in the spleen (about 50% of the normal value). The determination of the 'f' factor and of the size of the stem cell compartment in the different organs allow an approximation of the total number of CFU-S in the mouse body after virus infection. Assuming that one colony was derived from one haemopoietic CFU-S it was calculated that this number was only increased by a factor of 10 (Wendling *et al.*, 1972). This is not so surprising. In normal animals the spleen represents only 6 to 10% of the haemopoietic tissue. A large increase in size of the CFU compartment in blood or spleen does not represent a large increase when the haemopoietic system is considered *in toto*. Nevertheless, the size of the stem cell compartment is greatly modified by Friend, Rauscher and Kirsten viruses.

Effect of anti-virus antiserum on CFU-S

Attempts were made in three different laboratories (Tambourin, 1970; Brommer, 1972; Okunewick, Phillips & Brozovich, 1974, 1976*a*) to demonstrate the presence of viral or viral-associated antigen upon the surface of the colony forming cells. Effective antiserum against a mixture of murine viral leukaemia cells and virus was prepared in syngeneic mice or in rabbits. This antiserum proved effective against leukaemia virus prior to injection both *in vitro* and *in vivo*. Conversely, this antiserum was without effect against normal CFU-S, as was the normal serum. Prior to transplantation the cell suspensions prepared from infected spleens, or from normal spleens as controls, were incubated *in vitro* in syngeneic antiserum. No effect was observed on CFU-S from the control spleens whereas incubation in this antiserum of splenic CFU-S from animals infected with Rauscher virus, FVA or FVP reduced their colony forming ability by up to 60% (Okunewick *et al.* found 50%). The effectiveness of the antiserum was evident at different times after infection according to the virus dose initially injected. With a very high dose, the sensitivity of CFU-S was detectable about 2 days after infection. This point remains to be investigated more extensively. These results showed that a 'specific' antiserum against a mixture of murine viral leukaemia cells and virus was effective against the haemopoietic stem cells of infected animals. It is tempting to conclude that the antigenic changes indicate the presence of the viral genome within the pluripotent colony forming cells. But these results must be considered as quite preliminary and will require further experiments. For example, the haemopoietic stem cells which escaped the antiserum treatment did not seem to be free of virus since the grafted animals subsequently developed a typical Friend leukaemia. The use of antiserum against purified viral glycoprotein would improve these results. Lastly a 'non-specific' effect cannot be excluded at the moment.

Suicide experiments

In-vivo administration of, or in-vitro incubation with, high specific activity tritiated thymidine ([^{3}H]TdR) in very high doses will kill cells that incorporate it (i.e. cells in S-phase at the time of administration of the thymidine). The thymidine suicide technique has been used to determine what proportion of CFU-S is in DNA synthesis. Normal bone marrow or spleen CFU-S are not affected (value less than 10 %) by an exposure to [^{3}H]TdR for 20–30 minutes *in vitro* (Becker, McCulloch, Siminovitch & Till, 1965). During recovery from irradiation or cytotoxic drug damage this can rise to 50 % (Lajtha, Pozzi, Schofield & Fox, 1969).

In Rauscher leukaemia, in-vivo suicide tests with [^{3}H]TdR indicated the presence of a large number of dormant G_0 cells among the new CFU-S produced subsequent to the leukaemia virus infection (Okunewick & Phillips, 1974). Studies of these CFU-S which were in DNA synthesis were carried out using the in-vitro suicide technique. It was found that within 4 hours after infection with the Rauscher virus the number of spleen CFU-S killed by [^{3}H]TdR exposure increased to 20 % from a base line of 10 % in controls. For the next 2 weeks, the proportion of CFU-S in DNA synthesis remained at approximately 20 %. In the third week after virus infection, the proportion of CFU-S killed by the [^{3}H]TdR decreased to the point of being no longer detectable, even though the total number of CFU-S was increased several-fold over normal. This very interesting work was also done by Okunewick, Phillips & Brozovich (1976*b*) who suggested that the virus is not, of itself, an initiator of DNA synthesis. It would, rather, be dependent to a certain extent upon the cell's controls for DNA synthesis.

Repopulating capacity

In Friend, Rauscher and Kirsten leukaemia the splenic or bone marrow stem cells not only can form colonies of different histological types in the spleens of irradiated mice but also are capable of repopulating the depleted haemopoietic organs in these mice, building up a new blood-forming system. Two kinds of assay can be used to test the repopulating efficiency of these leukaemic spleen or bone marrow cell suspensions: the radioprotection assay and the study of blood recovery in irradiated mice grafted with normal or infected cells.

Radioprotection assay. The protection of supralethally irradiated mice by injected bone marrow or spleen cell suspensions depends on the number and efficiency of stem cells capable of repopulating the recipient's haemopoietic system (Urso & Congdon, 1957; McCulloch & Till, 1960; Smith & Vos, 1962). It was found that in FVP-infected (Wendling, Jullien & Tambourin, 1973) and Rauscher-infected mice (Brommer, 1972; Okunewick & Erhard, 1973; Brommer & Bentvelzen, 1974) haemopoietic CFU-S allowed the survival of syngeneic X-irradiated recipients. However, in FVP leukaemia, the splenic CFU-S were about eight-fold less efficient than CFU-S from normal mice, whereas the efficiency of bone marrow CFU-S was only slightly altered. However, in the study of Wendling *et al.* it must be pointed out that CFU-S from normal spleen were exceptionally efficient, since only nine CFU-S were able to protect 50 % of irradiated recipients from acute haemopoietic death. In the Rauscher leukaemia the radioprotective ability of the spleen cells was at least equivalent to normal (LD50/30 = 17 CFU-S). These results are another confirmation that haemopoietic stem cells in the leukaemic spleens not only maintain their relative numbers, but also retain their differentiation efficiency.

Assessment of differentiation ability. In the radiation dose range for haemopoietic death, survival of lethally irradiated recipients depends on transplantability, continued proliferation and potential for differentiation of the injected CFU-S. Wendling, Jullien & Tambourin (1973), postulated that the decreased efficiency of the CFU-S derived from leukaemic spleen was due, in part, to a failure to provide a normal supply of mature end cells and to the production of functionally deficient mature cells. It was successively found

(Wendling *et al.*, 1973*a*, *b*) that CFU-S derived from Friend leukaemic *spleens* were able to provide, as expected, reticulocytes, leucocytes and platelets in the peripheral blood. However, these spleen-derived CFU-S had very poor ability to produce thrombocyte progeny, and erythrocyte production was lowered 7.5-fold, but they were able to produce a normal amount of granulocytes. Conversely CFU-S from bone marrow of FVP-infected mice behaved like their normal counterparts as regards the erythropoietic and granulopoietic functions. The production of peripheral platelets seemed to be quantitatively reduced although the time of platelet emergence was not delayed, in contrast with CFU from infected spleens. It was concluded that the observed therapeutic deficiency of spleen-derived CFU-S could be partly related to the impairment of the thrombopoietic recovery. On the other hand Okunewick, Markoe, Erhard & Phillips (1973) studied ^{59}Fe uptake following transplantation of splenic CFU-S from normal and RLV-infected mice. They observed that although the progeny of control or infected CFU-S were initially growing at the same rate, the erythropoietic activity of the RLV progeny was much lower than the control. These results do not imply that the leukaemic CFU-S by itself was altered. On the contrary, these alterations seem to affect the more mature compartment since it was observed that the initial cell proliferation rate of RLV grafted cells did not differ from the normal.

'Decline' of CFU-S ability after successive proliferation

Spleen colonies may be obtained by transplantation of cells derived from a single colony. However, when such colonies are analysed for CFU-S content, none or very few are found (Simonovitch, Till & McCulloch, 1964). This decrease in capacity for self-renewal with repeated transfer has been termed 'decline'. The phenomenon is not only observed when cloned populations are used. After repeated passages of large numbers of normal bone marrow cells at 14-day intervals, the capacity for self-renewal of colony forming cells was reduced until it eventually became undetectable. If, however, the interval between two passages is increased, CFU-S can be propagated considerably longer than originally reported (Till, 1966). By stretching the passage intervals to 1 year, several marrow lines were maintained for up to 40 months and one line was carried through seven passages over a period of 5 years (Micklem & Loutit, 1966). In connection with hypothetic neoplastic transformation it was of interest to study whether infection of haemopoietic stem cells by a leukaemogenic virus increased the capacity to be transplanted on successive assays. Wendling, Tambourin & Bacau-Varlet (1974) have studied that question after infection of C3H mice by FVP. It was demonstrated that CFU-S from infected donors had a greatly restricted capacity for self-renewal since with infected spleen cells it was quite impossible to get more than three passages, spaced by 2 weeks. As reported by Matioli (1974) the spleen CFU-S from leukaemic animals grew like normal CFU-S in primary irradiated, syngeneic hosts, with, however, a slightly diminished probability of self-renewal. On second transfer, Wendling *et al.* (1974) observed that although the number of stem cells transplanted was equivalent to the number of CFU-S injected on the first passage, the number of CFU-S recovered 14 days after the second graft was drastically reduced. No explanation of this result has been given.

In the same experiment some colonies were tested for the ability to induce subcutaneous tumour. All these assays were unsuccessful.

The in-vitro colony forming cells in Friend and Rauscher leukaemia

Two committed precursor compartments have been investigated by in-vitro methods: the CFU-E already considered and the CFU-C, widely accepted as a committed precursor cell of the granulopoietic and monocytic cell lines (different from the multipotent CFU-S), which has been revealed by the agar colony test system (Bradley & Metcalf, 1966; Dicke, Platenburg & Van Bekkum, 1971).

The CFU-C compartment size from CBA/J mice with Rauscher leukaemia initially decreased 2 days after virus injection (as CFU-S) but thereafter increased rapidly up to 30-fold the normal values 2–6 weeks after the initiation of the disease. In contrast the CFU-C compartment size was somewhat below the normal value (as CFU-S). In the experiments of Iturizza & Seidel (1974), CFU-C paralleled the evolution of CFU-S. The thin-layer agar method of Dicke *et al.* (1971) was used by Van Griensven *et al.* (1973, 1975) to demonstrate that in bone marrow of BALB/c mice infected by Rauscher leukaemia virus, a 2.6-fold increase in CFU (CFU-S, CFU-C) occurred on day 12 after infection; thereafter this number returned to the normal value. Golde, Faille, Sullivan & Friend (1976) studied the same problem in DBA/2 mice at the terminal stage of Friend leukaemia. It was concluded that 'growth in liquid culture in a diffusion chamber was dependent on the presence of colony stimulating activity (CSA) and resulted in the generation of normally differentiated granulocytes and macrophages'. Colony formation in agar was also dependent on CSA and the cloning efficiency of leukaemic spleen cells was found to be approximately 10 times normal. The colonies formed were composed of leukocytes which appeared morphologically normal. Total in-vitro colony forming units per leukaemic spleen exceeded the normal by more than 300-fold but cells elaborating CSA were decreased. Although it is uncertain whether the stem cells stimulated by CSA are normal or leukaemic, it is clear that Friend and Rauscher leukaemia have a profound effect on the proliferation and differentiation of non-erythroid stem cells, the meaning of which will be discussed later.

The effect of Friend and Rauscher virus on CFU-E has been extensively reviewed and it was clear that CFU-E had become independent from erythropoietin.

Haemopoietic stem cells in other murine leukaemias

Relatively little work has been done with other murine leukaemias compared with the numerous experiments just reported on Friend, Rauscher or Kirsten leukaemia. Using the anti-θ antiserum to eliminate leukaemic cells, Frindel & Chevalier (1975) studied the number of CFU-S in the bone marrow of AKR mice with spontaneous leukaemia. It was found that the number of CFU decreased by a factor of only 0.7. No data were given on the size of the haemopoietic stem cell pool in blood or spleen.

Similar experiments were done in mice grafted with tumour or transplantable leukaemic cells. Lajtha, Tanaka & Testa (1972) used RF mice, which spontaneously develop a myeloid leukaemia (Cole & Furth, 1941; Tanaka & Craig, 1970), and injected 10^5 RF leukaemia spleen cells into these mice. Between the sixth and seventh days after injection of leukaemic cells a decline of all normal parameters, involving erythropoiesis, pluripotent stem cells and spleen CFU-C developed. The CFU-C, however, showed an interesting significant rise between days 3 and 6. Using the same experimental model, Husseini, Fried, Knospe & Trobaugh (1976) observed an increase in the haemopoietic stem cells in both spleen and bone marrow. As the disease progressed the number of CFU declined in both organs. Clerici, Mocarelli, Villa & Natale (1971) demonstrated that in mice injected intraperitoneally with Ehrlich ascites carcinoma cells the number of endogenous splenic haemopoietic colonies observed after the irradiation of the tumour-bearing mice decreased. The number of exogenous haemopoietic colonies obtained in lethally X-irradiated mice after the graft of bone marrow cells from donors previously grafted with Ehrlich carcinoma cells was decreased by a factor of 5, 8 days after the injection of the donor with tumour cells. The graft of NCTC 2472 to C3H mice (Frindel & Tubiana, 1967) modified the balance of the pluripotent stem cell compartment and of the differentiated compartments in the bone marrow. The number of CFU-S per femur decreased to 60% of the control 48 hours after the tumour cell injection, but at 11 days the number of CFU-S increased and slightly overshot the normal value. After an initial decrease, the fraction of CFU-S in S phase increased progressively (Croizat & Frindel, 1972). In 1973, Chevalier & Frindel showed that the number of bone marrow stem cells remained constant during the 9 days following the injection of tumour cells, the mean survival time of injected mice being 10 days.

Similar studies have been performed *in vitro* on the behaviour of the CFU-C compartment in preleukaemic and leukaemic AK mice infected at birth with Gross murine leukaemia virus or grafted with leukaemic cells (Hays, Craddock, Haskett & Newell, 1976). The preleukaemic state as well as marrow replacement by leukaemic lymphoblast resulted in increased marrow CFU-C. No evidence for enhanced local production of CSA by lymphoblasts or by the marrow microenvironment was found to account for this rise. Opposite results were obtained by Chevalier, Gaillard & Frindel (1974).

Conclusion

It appears from these numerous studies that a general agreement may be reached on the point that the spleen colony assay in irradiated syngeneic recipients does detect only pluripotent haemopoietic stem cells except in extremely rare cases where true tumour cell colonies can easily be identified macroscopically. If true tumour cells exist during the early phases of infection with Friend or Rauscher viruses these cells are not able to multiply in irradiated mice, which would be surprising.

The injected 'leukaemic' spleen or bone marrow CFU-S retain the main characteristics of the normal haemopoietic stem cells: self-renewal, differentiation and probably multipotentiality. The number of stem cells in leukaemic animals appears to be increased, which could explain the radioresistance of leukaemic animals. These stem cells are probably infected by the virus and the differentiation processes along the haemopoietic cell line are more or less affected, the immunological process, the thrombopoietic activity and to a lesser extent the erythropoietic activity being quantitatively reduced. Whether these results mean that the haemopoietic stem cells are a target will now be discussed.

TARGET CELL(S) AND HAEMOPOIETIC STEM CELLS

The evidence from the previous sections suggests that at least one conclusion will be unanimously accepted: virus-induced murine erythroblastoses are complex syndromes in the course of which infectious and oncogenic viral functions become intermingled with 'physiological' reactions of the haemopoietic system and with the leukaemic process in such a way that the target cell problem constitutes a big puzzle. Even though the research for probable target cells for murine leukaemia viruses has been a very active field these last 10 years, it is not surprising if the question is still under discussion. I think the main cause of dispute is probably the lack of a clear definition of *what is a target cell.* The search for target cells has no purpose as long as the concept of target cell remains so indefinite.

A target cell for a virus can be a cell in which it multiplies *in vivo* or *in vitro*, or a cell for in-vivo oncogenic function, or a cell for in-vitro morphological transformation, or a cell for appearance of new antigens, etc. Results obtained from viral genetic studies show clearly that virus multiplication and oncogenic virus expression are two genetically independent entities (see first section and Martin, 1970). In-vivo morphological studies clearly showed that quite different murine leukaemia viruses such as Gross, Rauscher or Friend virus, can be observed as budding particles on megakaryocytes, granulocytes, lymphocytes, epithelial cells, etc. (De Harven & Friend, 1958, 1966; Feldman & Gross, 1964, 1966, 1967; Silvestre, Levy, Leclerc & Boiron, 1966; Pennelli, Chieco-Bianchi, Tridente & Fiore-Donati, 1968; Forteza-Vila, Seidel & Calvo, 1972; Brown & Axelrad, 1976). This is equally true *in vitro*, each virus strain having a specific host cell range (Peries, Levy, Boiron & Bernard, 1964; Wright & Lasfargues, 1965; Yoshikura, Hirokawa, Yamada &

Sugano, 1967; Eckner, 1975; Clarke, Axelrad & Housman, 1976). A virus can multiply without neoplastic transformation, but this multiplication may affect the cell functions and in some instances even kill the cells. Therefore, if we are looking for an in-vivo virus target cell for oncogenicity, the finding that the virus multiplies inside a cell will not prove that the cell is a good candidate for a target cell, although it is likely, but not compulsory, that the true target cell(s) has to be found among the cells in which the virus enters and probably multiplies. In conclusion, a clear distinction must be made between modifications of the haemopoietic system due only to the virus replication and changes due to the expression of the viral oncogenic function.

Another difficulty in approaching the target cell problem in leukaemia results from the physiological characteristics of the haemopoietic system which can react strongly to the presence of a tumour. This phenomenon, the so-called leukaemoid reaction, has been well documented. It has been defined as a situation in which an examination of the blood suggests that leukaemia might be present. Actually, the changes in the leucocytes, platelets or reticulocytes are subject to particular conditions, most often secondary to a carcinoma, and have been observed in concomitance with a wide variety of murine tumours which sometimes exert drastic effects on blood. The benign character of this reaction can be easily demonstrated when it is a solid tumour: surgical excision results in prompt reversal of the leukaemoid reaction (Bateman, 1950; Mohri, 1962; Lappat & Cawein, 1964; Delmonte & Liebelt, 1965; Delmonte, Liebelt & Liebelt, 1966; Grezes & Salomon, 1972; Milas & Basic, 1972; Sendo, Kodama & Kobayashi, 1973; Boggs *et al.*, 1977; Hacker, Roberts & Jackson, 1977). In the case of spontaneous or virus-induced leukaemia such a reaction probably occurs and somewhat obscures the true leukaemic process. This reaction itself can also interfere with the leukaemic process. The Friend and Rauscher leukaemias appear to be very good examples of such a complexity. As already mentioned, almost all the haemopoietic cell lines appear to be profoundly affected by the virus, including the immunological system. One way to clear up this rather complex situation is to answer the four following questions. (1) What is the nature of the cells which the virus enters and which then become hyperbasophilic, multiply, invade the organism, in particular the liver, and finally kill the animal? (2) Is this cell, which is undoubtedly one of the target cells for oncogenicity, the only cell required in order to observe the typical Friend or Rauscher leukaemia or are there other target cell(s) for another oncogenic expression of the oncogene? (3) What is the role of the haemopoietic stem cell in the leukaemic process? (4) What is the relationship between these cells and the tumour cells isolated by numerous investigators that can be grown as cell lines and respond to DMSO by differentiating along the erythroid line?

First question: the target cell for in-vivo hyperbasophilic transformation

Today it seems quite clearly demonstrated that the target cell needed for expression of this viral function is the erythropoietin responsive cell or a closely related precursor. *In vivo*, it has been successively demonstrated that:

(1) These cells belong to the haemopoietic system (Chirigos & March, 1966; Odaka, 1969; Rossi, De Harven, Haddad & Friend, 1971).

(2) The susceptibility is genetically controlled by the *W* and *Sl* loci (Bennett *et al.*, 1968; Steeves, Bennett, Mirand & Cudkowicz, 1968).

(3) The viral oncogenic action does not require the presence of pluripotent haemopoietic stem cells (Fredrickson *et al.*, 1975; Nasrallah & McGarry, 1976).

(4) It requires *only* the erythroid pathway of differentiation (Fredrickson *et al.*, 1975; Tambourin & Wendling, 1975).

(5) The Rauscher- and FVA-induced erythropoiesis is also a neoplastic process (Pluznick *et al.*, 1966; Tambourin *et al.*, 1969; Seidel, 1972*a*).

(6) The transformation event does not require the presence of proerythroblasts or more mature erythroblasts (Tambourin & Wendling, 1971).

(7) It is totally inhibited by treatment with actinomycin D at a dose which *selectively* blocks the erythroid differentiation at the level of erythropoietin responsive cells (Reissmann & Ito, 1966) but which does not inhibit the thrombocytopenic action of the virus or its replication (Tambourin & Wendling, 1971).

(8) Hyperbasophilic cells differentiate *only* along the erythroid cell line without erythropoietin (Mirand, 1967*a*, *b*; Sassa *et al.*, 1968; Tambourin & Wendling, 1971, 1975).

(9) Hyperbasophilic cells carry specific erythrocyte antigens (Ikawa *et al.*, 1973).

(10) EPO and FVP interact in their respective capacities to induce erythroid differentiation (Tambourin & Wendling, 1971; McGarry & Mirand, 1973).

(11) Transformation occurs at a maturation stage related to the erythropoietin responsive cell population (Tambourin & Wendling, 1975).

From these in-vivo experiments it can easily be inferred that (1) the multipotent haemopoietic stem cell is not a direct target cell for the morphological transformation process (its presence is not required), and (2) the viral 'gene' responsible for hyperbasophilic transformation is expressed *only* in an erythroid-committed cell (probably the erythropoietin responsive cell or a closely related precursor). From the recent in-vitro experiments reported above, which come to strengthen this conclusion, it appears that the target cell can be transformed either *in vivo* or *in vitro*.

Second question: other target cells

Mice infected with Friend or Rauscher leukaemia virus die from erythroblastosis. As demonstrated recently by Graf *et al.* (1977) for avian erythroblastosis virus, the rapid development of erythroleukaemia could obscure the appearance of other leukaemic events or the development of tumours. It must be remembered that in his original description Rauscher said that the virus, which was not a cloned virus, was able to induce a dual type of leukaemia.

Pluripotent stem cell as target cell. Several laboratories have provided conclusive evidence of the CFU-S as viral target cells. The first argument is based on the above-mentioned response of CFU-S to antiserum prepared against the virus and the leukaemic cells. As it has also been recalled several times, the ability of a virus to replicate is not at all related to its oncogenic potency and unfortunately this result only shows that the CFU-S is a cell which the virus enters and in which it probably replicates.

One relevant experiment suggesting that haemopoietic stem cells might be target cells for viral oncogenic action comes from the study of the CFU-S compartment size after infection. The initial effect of the Rauscher virus is a depletion (Seidel, 1972*b*; Okunewick *et al.*,

1976*a*) followed by a large increase in the spleen and blood. However, three sets of results suggest that these effects might only be a consequence of the oncogenic process and belong to the leukaemoid reaction.

The haemopoietic stem cell concentration in spleen and bone marrow remains remarkably constant all through the leukaemia process. This increase may only be a consequence of normal regulatory mechanisms since such a 'local' proliferation control seems to exist for normal CFU-S (see review by Lajtha, 1975). The 'signal' is the decrease in CFU-S concentration resulting from cell death or removal for differentiation.

A treatment of Friend or Rauscher leukaemic animals with actinomycin D, at a daily dose of 60–80 μg per kg body weight, completely stops the transformation process. The spleen weight and the white blood cell count return to normal values after several days. A similar treatment does not affect the AK lymphoid leukaemia development (Wendling *et al.*, 1969). Such a treatment has no detectable effect on normal haemopoietic stem cells. In FV-leukaemic animals so treated, the stem cell compartment, which was greatly enlarged before the treatment, also recovers promptly suggesting a reversal due to the arrest of the transformation process.

Experiments with Myleran (Fredrickson *et al.*, 1975; Nasrallah & McGarry, 1976), a drug which preferentially kills haemopoietic stem cells (Reissmann & Samorapoompchit, 1970; Reissmann & Udupa, 1972; Udupa, Okamura & Reissmann, 1972; Blackett & Millard, 1973), also suggest strongly that CFU-S are not required for the leukaemic process at least for a few days.

These experiments neither prove nor disprove that the CFU-S is a target cell for some unknown oncogenic function of the virus but they might equally well fit with the idea that CFU-S are not directly implicated in the leukaemogenic process.

Another set of facts suggests such a conclusion. A very interesting comparative study of the stem cell compartment after Rauscher and Friend virus infection of SJL/J was reported in 1973 by Okunewick & Phillips, who concluded that the two viruses produced qualitatively similar but quantitatively different effects. Wendling, Tambourin, Gallien-Lartigue & Charon (1974) reported a similar comparative study between FVA- and FVP-induced leukaemias which in spite of considerable physiopathological differences produced similar changes in the CFU-S compartment.

Committed cells as target cells. As regards the committed cell types, they are undoubtedly affected by Friend and Rauscher viruses. Infection with these viruses results in immunodepression (Ceglowski & Friedman, 1968; Häyry, Rago & Defendi, 1970; Friedman, 1974; Ceglowski, Campbell & Friedman, 1975; Eckner, 1975; Siegel & Morton, 1973, 1976). These viruses replicate actively in all recognisable haemopoietic cells, in particular in megakaryocytes, which results in a thrombocytopenia. It must be pointed out that this thrombocytopenic action results partly from a selective decrease in megakaryocytes of high nuclear DNA content while those of low nuclear DNA content remain relatively unaffected (Brown & Axelrad, 1976). But thrombocytopenia is not specific to the action of erythroblastosis-inducing viruses. It is a very common consequence of infection by a wide variety of viruses: influenza (Danon, Jerushalmy & De Vries, 1959), Gross and similar lymphoid viruses (Dalton, Law, Moloney & Manaker, 1961; Brodsky & Dimitrov, 1973), myxoviruses (Jerushalmy, Kohn & De Vries, 1961), Newcastle disease virus (Jerushalmy, Kaminsky, Kohn & De Vries, 1963), rubella (Adkins & Fernbach, 1965), measles (Oski & Naiman, 1966); (see review by Maupin, 1966). Tumour cells also influence thrombopoiesis within a few hours after inoculation (Hacker, Roberts & Jackson, 1977).

Similarly lymphocytosis and leucocytosis are rather common features of numerous diseases unrelated to erythroblastosis. A simple actinomycin D treatment (see above) is able to reduce this disorder.

As for CFU-S, the virus-induced changes already reported do not prove whether this cell population, or at least part of it, allows or does not allow the expression of Friend virus oncogenic function. They would just as well support the idea that these perturbations belong to a benign reaction against the virus and the erythroid oncogenic process itself.

Third question: role of the stem cell

Although the pluripotent stem cells do not seem to be a target cell for the viral oncogenic function, they probably play an indirect but major role in the development of erythroblastosis, probably also in other leukaemias, by rapidly supplying committed precursors in which oncogenic or infectious virus functions occur. In experiments already reported (Fredrickson *et al.*, 1975) which show that early

Table 1. *Comparison between normal haemopoietic multipotent stem cells, hyperbasophilic Friend cells and tumour cells present in Friend leukaemia*

In-vivo property	Normal haemopoietic stem cell	'Infected' haemopoietic stem cell	Hyperbasophilic transformed Friend cells	Friend tumour cells
Time after infection it is detectable	—	Before and after	About 30 hours in HP mice	About 25–30 days
Lifespan	Long but limited	Probably more limited	About 15–30 hours	Infinite
Ability to produce colony in unirradiated host	No	No	No	Yes
Ability to produce colony in irradiated host	Yes	Yes	No	Yes
Multipotence	Yes	Yes	No	No
Differentiation ability	Yes	Yes	Yes (erythroid line)	Only few erythroid cells
Sel-renewal	Yes	Yes	No	Yes
Evidence of 'decline	Yes after seven passages	Yes after three passages	Yes	No
Radioprotective ability	Efficient	More or less efficient	No	No
Viral antigen	No	Yes	Yes	Yes
Specific erythrocyte antigen	No	No	Yes	No
Erythropoietin sensitivity	Indirect	Indirect	Yes	No
Karyotype	Unknown	Unknown	Normal	Often metacentric chromosome
Actinomycin D sensitivity	No	No	Yes	Little
Ability to produce subcutaneous tumour	No	No	No	Yes
Ability to grow *in vitro*	Yes (committed precursors)	Yes (committed precursors)	No	Yes
In-vitro lifespan	Limited	Limited	About 10 hours	Infinite
In-vitro differentiation ability	Yes (high)	Yes (high)	Yes (erythroid line)	Low; high after DMSO
In-vitro self-replication	Unknown	Unknown	No	Yes

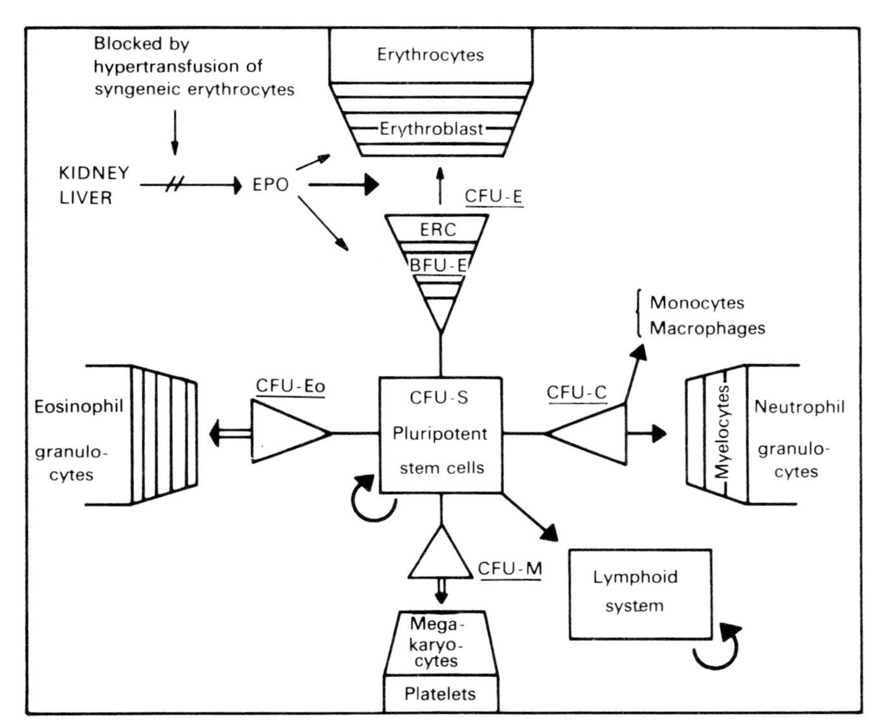

Fig. 1. Simplified schema of the murine haemopoietic system. Normal steady state: the pluripotent haemopoietic stem cell compartment, CFU-S, is a self-maintaining population with a large majority of its cells in G_0. A 'first step' differentiation event transforms the CFU-S into committed precursor cells. Such committed cells have been most often identified and studied by in-vitro culture methods (CFU-E, erythroid colony forming unit; BFU-E, erythroid burst forming unit; CFU-M, megakaryocytic colony forming unit; CFU-Eo, eosinophilic granulocyte colony forming unit; CFU-C, neutrophilic granulocyte and macrophage colony forming unit). These transit populations undergo a number of cell cycles during their maturation but without the capacity for prolonged self-maintenance. A 'second step' differentiation triggers the late forms of the committed cells and initiates the maturation process of the identifiable haemopoietic cell lines. Erythropoietin (EPO) has been demonstrated to stimulate erythropoietin responsive cell (ERC) proliferation and recruitment, to induce ERC differentiation and to promote the erythropoietic maturation process at the level of ERC.

events occurred even in a situation where normal haemopoietic stem cells were almost totally absent, it is reasonable to speculate that such a permanent depletion of the stem cell compartment would, sooner or later, result in an absence of committed precursors and therefore of target cells for the oncogenic action of the virus.

An unanswered problem which would be interesting to study is

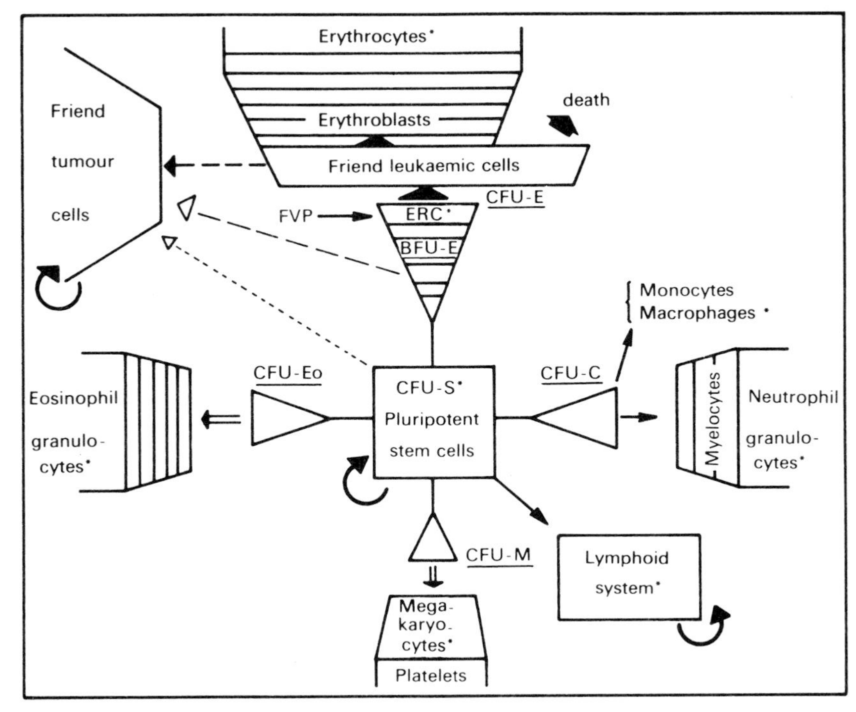

Fig. 2. Schema of leukaemogenesis and of tumorigenesis by Friend leukaemia virus. Oncogenic potency of the virus is expressed only at the late erythroid committed precursor level (probably into the ERC). After viral transformation, which takes about 30 hours, leukaemic hyperbasophilic Friend cells multiply and differentiate along the erythrocytic pathway. Erythropoietin (EPO) is not required to initiate the pathological erythropoietic differentiation. As indicated by kinetic studies, Friend leukaemic cells are not self-maintaining (they die or differentiate). Then the progression of the disease requires a constant recruitment of target cells from the ERC and CFU compartments. The viral replicative ability is not restricted to the erythroblastic cells. The most primitive haemopoietic stem cells and the different haemopoietic cell lines are infected by the virus. Their development remains qualitatively normal but quantitatively more or less affected. Later in the disease the tumour cell population arises from the yet unidentified (broken line for hypothetical pathway) target cell for tumorigenesis. These tumour cells have acquired an infinite lifespan and have retained a differentiative ability towards the erythroid cell line *in vivo* or *in vitro*. The cell populations marked with an asterisk denote populations infected by at least one component of the Friend virus complex.

whether the Friend virus is able to sustain, by itself, the committed precursor compartment without any stem cell contribution. Of course this speculation is no longer valid from the moment tumour cells appear since these cells have acquired permanent autonomous growth. The presence of stem cells is certainly not required for the multiplication of such cells, as experiments in our laboratory have shown (unpublished data).

In other leukaemias the role of the haemopoietic pluripotent stem cell remains completely unknown.

Fourth question: relation between tumour cell and transformed cell

Table 1 demonstrates that tumour cell populations and hyperbasophilic Friend cell populations consist of two different entities. But the relation between them is unknown. It is tempting to conclude that the tumour cell is an erythropoietin-responsive cell transformed by the virus into a different pathway. Another possibility would imply a second step of transformation which would occur in hyperbasophilic Friend cells. The tumour cell could also result from the transformation of a quite different haemopoietic cell, but the ability for in-vivo or in-vitro erythroid differentiation would suggest that the target cell for tumorigenicity belongs to the erythroid cell line (Figs. 1 and 2).

Although we are not able at the present time to say definitely that these tumour cells do appear later in the disease, we have never been able to detect them before the third week after infection (unpublished data). It is likely that the Friend tumour cells correspond in terms of neoplastic development to the numerous transplantable murine leukaemia cells of lymphoid, myeloid or plasma type. Whether, in these other murine leukaemias, a first step analogous to the hyperbasophilic Friend cell situation exists, remains to be determined. It seems particularly interesting to notice that in both situations (erythroblastosis and other leukaemias) tumour cell transformation requires a very much longer delay than the initial Friend-virus-induced transformation process.

CONCLUSION

The pathophysiological study of such erythroleukaemias allows some interesting and, possibly, fundamental conclusions.

The Friend and Rauscher viruses, which are known as non-transforming viruses *in vitro*, are able to transform target cells *in vivo* or *in vitro* in about 30 hours. The target cell for this particular oncogenic function is not the haemopoietic pluripotent stem cell but an erythroid-committed precursor closely related to or identical with the erythropoietin-responsive cell. The transformation is followed by a virus-induced 'malignant' erythropoiesis which develops without the physiological hormone for erythropoiesis, i.e. erythropoietin.

This morphological transformation appears to be more like a derepression of the cell population size control mechanisms than a true cancerous transformation.

The *tumour* cell transformation seems to occur much later and is characterised by acquisition of an infinite lifespan, an acquired ability to give rise to subcutaneous tumours or cell lines *in vitro* and by a restricted ability to differentiate spontaneously towards the erythroid cell line. During the leukaemic process the pluripotent haemopoietic stem cells appear to be infected and some of their properties modified. But they remain true pluripotent stem cells that preserve their main properties following infection.

These conclusions raise several important questions.

(1) Does the 'transformation' induced in fibroblasts by oncogenic viruses like Rous sarcoma virus correspond to the first or the second step of the neoplastic development found in Friend leukaemia?

(2) Is the Friend virus requirement for a specific differentiated cell to express its oncogenic potential a peculiarity of leukaemia viruses?

(3) Is this two-step (or more) process in leukaemia development particular to Friend virus or more general?

(4) How does a virus initiate the erythroid maturation process and therefore haemoglobin synthesis?

The author thanks Drs O. Gallien-Lartigue, M. Hirsch, and J. Wood for reading the manuscript; Dr F. Zajdela for valuable discussion and Mrs N. Barat, P. Bucau-Varlet, M. Charon, F. Moreau, O. Pierre and D. Reuter for their excellent and efficient assistance.

REFERENCES

Abelson, H. T. & Rabstein, L. S. (1969). A new tumor inducing variant of Moloney leukemia virus. *Proceedings of the American Association for Cancer Research*, **10**, 1 (Abstr.).

Abelson, H. T. & Rabstein, L. S. (1970). Lymphosarcoma: virus induced thymic-independent disease in mice. *Cancer Research*, **30**, 2213–19.

Adkins, A. T. & Fernbach, D. J. (1965). Thrombocytopenic purpura following rubella. *Journal of the American Medical Association*, **193**, 243–5.

Axelrad, A. A., Cinader, B., Koh, S. W. & Van der Gaag, H. C. (1976). Tumor colony formation by Friend virus-infected cells in immunosuppressed mice. *Cancer Research*, **36**, 28–32.

Axelrad, A. A. & Steeves, R. A. (1964). Assay for Friend leukemia virus: rapid quantitative method based on enumeration of macroscopic spleen foci in mice. *Virology*, **24**, 513–18.

Barski, G., Barbieri, D., Koo Youn, J. & Hue, G. (1973). Low-leukemogenic variants of Rauscher leukemia virus obtained in long-term cultures at supra-optimal temperature. *International Journal of Cancer*, **12**, 55–65.

Bassin, R. H., Simons, P. J., Chesterman, F. C. & Harvey, J. J. (1968). Murine sarcoma virus (Harvey) characteristics of focus formation in mouse embryo cell embryo cell cultures and virus production by hamster tumour cells. *International Journal of Cancer*, **3**, 265–72.

Bateman, J. C. (1950). Leukemoid reactions to transplanted mouse tumors. *Journal of the National Cancer Institute*, **11**, 671–82.

Becker, A. J., McCulloch, E. A., Siminovitch, L. & Till, J. E. (1965). The effect of differing demands for blood cell production in DNA synthesis by hemopoietic colony-forming cells of mice. *Blood*, **26**, 296–308.

Becker, A. J., McCulloch, E. A. & Till, J. E. (1963). Cytological demonstration of the clonal nature of spleen colonies derived from transplanted mouse marrow cells. *Nature, London*, **197**, 452–4.

Bennett, M., Steeves, R. A., Cudkowicz, G., Mirand, E. A. & Russell, L. B. (1968). Mutant *Sl* alleles of mice affect susceptibility to Friend spleen focus-forming virus. *Science*, **162**, 564–5.

Bentvelzen, P. (1972). Evidence against the neoplastic nature of virus induced erythroblastosis in mice. *European Journal of Cancer*, **8**, 584–5.

Bergsagel, D. E. & Valeriote, F. A. (1968). Growth characteristics of a mouse plasma cell tumor. *Cancer Research*, **28**, 2187–96.

Biggs, P. M., Milne, B. S., Graf, T. & Bauer, H. (1973). Oncogenicity of non-transforming mutants of avian sarcoma viruses. *Journal of general Virology*, **18**, 399–403.

Blackett, N. M. & Millard, R. E. (1973). Differential effect of Myleran on two normal haemopoietic progenitor cell populations. *Nature, London*, **244**, 300–1.

Blank, K. J., Steeves, R. A. & Lilly, F. (1976). The Fv-2^r resistance gene in mice: its effect on spleen colony formation by Friend virus transformed cells. *Journal of the National Cancer Institute*, **57**, 925–30.

Blankstein, L. A. & Levy, S. B. (1976). Changes in histone f2a2 associated with proliferation of Friend leukaemic cells. *Nature, London*, **260**, 638–40.

Boggs, D. R., Malloy, E., Boggs, S. S., Chevernick, P. A. & Lee, R. E. (1977). Kinetic studies of a tumor-induced leukemoid reaction in mice. *Journal of laboratory and clinical Medicine*, **89**, 80–92.

Boiron, M., Levy, J. P., Lasneret, J. & Oppenheim, S. (1965). Pathogenesis of Rauscher leukaemia. *Journal of the National Cancer Institute*, **35**, 865–84.

Bradley, T. R. & Metcalf, D. (1966). The growth of mouse bone marrow cells *in vitro. Australian Journal of experimental Biology and medical Science*, **44**, 287–300.

BRODSKY, I., DENNIS, L. H. & KAHN, S. B. (1966). Erythropoiesis in Friend leukemia: red blood cell survival and ferro-kinetics. *Cancer Research*, **26**, 1887–92.

BRODSKY, I. & DIMITROV, N. V. (1973). Platelet kinetics and adenosine triphosphatase activity in AKR mice. *Journal of the National Cancer Institute*, **50**, 997–1001.

BRODSKY, I., KAHN, S. B., ROSS, E. M., PETKOV, G. & BRAVERMAN, S. D. (1967). Prelymphoid leukemia phase of Rauscher virus infection. *Journal of the National Cancer Institute*, **38**, 779–87.

BRODSKY, I., ROSS, E. M., KAHN, S. B. & PETKOV, G. (1968). The effect of a leukemia virus on thrombopoiesis. *Cancer Research*, **28**, 2406–12.

BROMMER, E. J. P. (1972). The role of stem cell in Rauscher murine leukemia. Thesis, University of Rotterdam, Medical Faculty, The Netherlands.

BROMMER, E. J. P. & BENTVELZEN, P. (1970). Transplantation of haematopoietic tissue from virus induced murine leukaemias. In *XIIIth International Congress on Hematology (Munich), Abstract Volume*, p. 198. Munich: J. F. Lehmanns Verlag.

BROMMER, E. J. P. & BENTVELZEN, P. (1974). The haematopoietic stem cells in Rauscher virus-induced erythroblastosis of BALB/c mice. *European Journal of Cancer*, **10**, 827–33.

BROWN, W. M. & AXELRAD, A. A. (1976). Effect of Friend leukemia virus on megakaryocytes and platelets in mice. *International Journal of Cancer*, **18**, 764–73.

BROXMEYER, H. E., KOLTUN, L., LOBUE, J., FREDRICKSON, T. N. & GORDON, A. S. (1975). Granulopoiesis in 'preleukemic' mice with anemia induced by Rauscher leukemia virus, variant a. *Journal of the National Cancer Institute*, **55**, 1123–7.

BRUCE, W. R. & VAN DER GAAG, H. A. (1963). Quantitative assay for the number of murine lymphoma cells capable of proliferation *in vivo*. *Nature, London*, **199**, 79–80.

BUCHHAGEN, D., PINCUS, T., STUTMAN, O. & FLEISSNER, E. (1976). Leukemogenic activity of murine type C viruses after long term passage *in vitro*. *International Journal of Cancer*, **18**, 835–42.

BUFFETT, R. F. & FURTH, J. A. (1959). A transplantable reticulum-cell sarcoma variant of Friend's viral leukemia. *Cancer Research*, **19**, 1063–9.

CAMISCOLI, J. F., LOBUE, J., GORDON, A. S., ALEXANDER, P., SCHULTZ, E. F. & WEITZ-HAMBURGER, A. (1972). Absence of plasma erythropoietin in mice with anemia induced by Rauscher leukemia virus. *Cancer Research*, **32**, 2843–4.

CEGLOWSKI, W. S., CAMPBELL, B. P. & FRIEDMAN, H. (1975). Immunosuppression by leukemia viruses. XI. Effect of Friend leukemia virus on humoral immune competence of leukemia-resistant C57Bl/6 mice. *Journal of Immunology*, **114**, 231–6.

CEGLOWSKI, W. S. & FRIEDMAN, H. (1968). Immunosuppressive effects of Friend and Rauscher virus-induced erythroblastosis of BALB/c mice. *Journal of the National Cancer Institute*, **40**, 983–95.

CHAMORRO, A. (1962). Origine et transmission de la leucémie de Friend. *Bulletin du Cancer*, **49**, 399–415.

Chamorro, A. (1967). Separation par centrifugation différentielle de deux agents distincts de la leucémogénèse de la souris. *Comptes rendus de l'Académie des Sciences, Paris, Series D*, **265**, 649–51.

Chazan, R. & Haran-Ghera, N. (1976). The role of the thymus subpopulation in 'T' leukemia development. *Cellular Immunology*, **23**, 356–75.

Chesterman, F. C., Harvey, J. J., Dourmashkin, R. R. & Salaman, M. H. (1966). The pathology of tumors and other lesions induced in rodents by virus derived from a rat with Moloney leukemia. *Cancer Research*, **26**, 1759–68.

Chevalier, C. & Frindel, E. (1973). The evolution of number of bone marrow stem cells in mice with L1210 leukaemia. *Biomedicine*, **19**, 177–9.

Chevalier, C., Gaillard, N. & Frindel, E. (1974). The number of *in vitro* stem cells in AKR leukemic mice. *Blood*, **44**, 743–50.

Chirigos, M. A. & March, R. W. (1966). Reversal of antiviral activity (Friend virus) of drug and irradiation by syngeneic spleen cells. *Antimicrobial Agents and Chemotherapy*, **6**, 489–96.

Clarke, B. J., Axelrad, A. A. & Housman, D. (1976). Friend spleen focus-forming virus production *in vitro* by a nonerythroid cell line. *Journal of the National Cancer Institute*, **57**, 853–9.

Clarke, B. J., Axelrad, A. A., Shreeve, M. M. & McLeod, D. L. (1975). Erythroid colony induction without erythropoietin by Friend leukemia virus *in vitro*. *Proceedings of the National Academy of Sciences, USA*, **72**, 3556–60.

Clerici, E., Mocarelli, P., Villa, M. L. & Natale, N. (1971). Influence of Ehrlich ascites carcinoma on repopulating ability of mouse bone marrow cells. *Journal of the National Cancer Institute*, **47**, 555–60.

Cole, R. K. & Furth, J. (1941). Experimental studies on the genetics of spontaneous leukemia in mice. *Cancer Research*, **1**, 957–65.

Cox, K. O. & Keast, D. (1973). Rauscher virus infection, erythrocyte clearance studies and autoimmune phenomena. *Journal of the National Cancer Institute*, **50**, 941–6.

Croizat, H. & Frindel, E. (1972). Bone marrow stem cell response to tumour grafts. *Revue européenne d'Etudes cliniques et biologiques*, **17**, 200–4.

Dalton, A., Law, L. W., Moloney, J. B. & Manaker, R. A. (1961). An electron microscopic study of series of murine lymphoid neoplasms. *Journal of the National Cancer Institute*, **27**, 747–91.

Danon, D., Jerushalmy, Z. & Devries, A. (1959). Incorporation of influenza virus in human blood platelets *in vitro*. *Virology*, **9**, 719–22.

Dawson, P. J. & Fieldsteel, A. H. (1974). Reticulum cell sarcoma induced in mice by Rauscher virus. *Journal of the National Cancer Institute*, **52**, 1805–8.

Dawson, P. J., Fieldsteel, A. H. & Bostick, W. L. (1963). Pathologic studies of Friend virus leukemia and the development of a transplantable tumor in BALB/c mice. *Cancer Research*, **33**, 349–54.

Decleve, A., Travis, M., Weissmann, I. L., Lieberman, M. & Kaplan, H. S. (1975). Focal infection and transformation in situ of thymus cell subclasses by a thymotropic murine leukemia virus. *Cancer Research*, **35**, 3585–95.

De Harven, E. & Friend, C. (1958). Electron microscope study of a cell-free induced leukemia of the mouse: A preliminary report. *Journal of biophysical and biochemical Cytology*, **4**, 151–6.

De Harven, E. & Friend, C. (1966). Origin of the viremia in murine leukemia. *National Cancer Institute Monographs*, **22**, 79–101.

Delmonte, L. & Liebelt, R. A. (1965). Granulocytosis promoting extract of mouse tumor tissue: partial purification. *Science*, **148**, 521–3.

Delmonte, L., Liebelt, A. G. & Liebelt, R. A. (1966). Granulopoiesis and thrombopoiesis in mice bearing transplanted mammary cancer. *Cancer Research*, **26**, 149–59.

Dennis, L. H. & Brodsky, I. (1965). Thrombocytopenia induced by the Friend leukemia virus. *Journal of the National Cancer Institute*, **35**, 993–9.

Dicke, K. A., Platenburg, M. G. C. & Van Bekkum, D. W. (1971). Colony formation in agar: 'in vitro' assay for haemopoietic stem cells. *Cell and Tissue Kinetics*, **4**, 463–77.

Dunn, T. B., Moloney, J. B., Green, A. W. & Arnold, B. (1961). Pathogenesis of the virus induced leukemia in mice. *Journal of the National Cancer Institute*, **26**, 189–221.

Ebert, P. S., Maestri, N. E. & Chirigos, M. A. (1972). Erythropoietic responses of mice to infection with Rauscher leukemia virus. *Cancer Research*, **32**, 41–7.

Eckner, R. J. (1975). Continuous replication of Friend virus complex (spleen focus-forming virus-lymphatic leukemia-inducing virus) in mouse embryo fibroblasts. Retention of leukemogenecity and loss of immunosuppressive properties. *Journal of experimental Medicine*, **142**, 936–48.

Fedoroff, S. (1967). Proposed usage of animal tissue culture terms. *Journal of the National Cancer Institute*, **48**, 1531–3.

Feldman, D. G. & Gross, L. (1964). Electron microscopic study of the distribution of the mouse leukemia virus (Gross), and of tissue from mice with virus-induced leukemia. *Cancer Research*, **24**, 1760–83.

Feldman, D. G. & Gross, L. (1966). Electron microscopic study of the distribution of the mouse leukemia virus (Gross) in organs of mice and rats with virus-induced leukemia. *Cancer Research*, **26**, 412–26.

Feldman, D. G. & Gross, L. (1967). Electron microscopic study of the distribution of the mouse leukemia virus (Gross) in genital organs of virus-injected C3Hf mice and of AK mice. *Cancer Research*, **27**, 1513–27.

Fenner, F. (1976). Classification and nomenclature of viruses. *Intervirology*, **7**, 1–116.

Fey, F. (1969). Studies on the pathogenesis and mechanism of hematologic diversification by re-isolation of the myeloid leukemia virus (Graffi). *Acta haematologica*, **42**, 65–75.

Fieldsteel, A. H., Dawson, P. J. & Bostick, W. L. (1961). Quantitative aspects of Friend leukemia virus in various murine hosts. *Proceedings of the Society for experimental Biology and Medicine*, **108**, 825–9.

Fiore-Donati, L. & Chieco-Bianchi, L. (1964). Influence of host factors on development and type of leukemia induced in mice by Graffi virus. *Journal of the National Cancer Institute*, **32**, 1083–1107.

Forteza-Vila, J., Seidel, H. J. & Calvo, W. (1972). Virus Replikation in unreifen granulopoetischen Zellen bei Maüsen mit Rauscher-leukämie. *Acta haematologica*, **47**, 168–74.

Foulds, L. (1969). In *Neoplastic development*, vol. 1, pp. 41–75. New York & London: Academic Press.

FREDRICKSON, T., TAMBOURIN, P. E., WENDLING, F., JASMIN, C. & SMADJA, F. (1975). Target cell of the polycythemia-inducing Friend virus: studies with Myleran. *Journal of the National Cancer Institute*, **55**, 443–6.

FREI, J. V. (1976). Some mechanisms operative in carcinogenesis. A review. *Chemico-biological Interactions*, **13**, 1–25.

FRIEDMAN, H. (1974). Effects of the Friend leukemia virus on cells of the immune system. *Israel Journal of medical Sciences*, **10**, 1052–74.

FRIEND, C. (1956). The isolation of a virus causing a malignant disease of the hematopoietic system in adult Swiss mice. *Proceedings of the American Association for Cancer Research*, **2**, 106 (Abstr.).

FRIEND, C. (1957). Cell-free transmission in adult Swiss mice of a disease having the character of a leukemia. *Journal of experimental Medicine*, **105**, 307–18.

FRIEND, C. & HADDAD, J. R. (1959). Local tumor formation with transplants of spleen or liver of mice with a virus-induced leukemia. *Proceedings of the American Association for Cancer Research*, **3**, 21 (Abstr.).

FRIEND, C. & HADDAD, J. R. (1960). Tumor formation with transplants of spleen or liver from mice with virus-induced leukemia. *Journal of the National Cancer Institute*, **25**, 1279–89.

FRIEND, C., PATULEIA, M. C. & DE HARVEN, E. (1966). Erythrocytic maturation *in vitro* of murine (Friend) virus-induced leukemic cells. *National Cancer Institute Monograph*, **22**, 505–22.

FRIEND, C., SCHER, W., HOLLAND, J. G. & SATO, T. (1971). Hemoglobin synthesis in murine virus-induced leukemic cells *in vitro*. Stimulation of erythroid differentiation by dimethyl sulfoxide. *Proceedings of the National Academy of Sciences, USA*, **68**, 378–82.

FRINDEL, E. & CHEVALIER, C. (1975). Measurement of the number of bone marrow multipotential stem cells in AKR leukaemic mice. *Biomedicine*, **23**, 166–7.

FRINDEL, E. & TUBIANA, M. (1967). Durée du cycle cellulaire au cours de la croissance d'une ascite expérimentale de la souris C3H. *Comptes rendus de l'Académie des Sciences, Paris, Series D*, **265**, D, 829–31.

GALLIEN-LARTIGUE, O., TAMBOURIN, P., WENDLING, F. & ZAJDELA, F. (1969). Spontaneous hematopoietic recovery of Friend virus-infected mice after heavy X irradiation. *Journal of the National Cancer Institute*, **42**, 1061–8.

GILLEPSIE, A. V. & ROWSON, K. E. K. (1968). The influence of sex upon the development of Friend virus leukemia. *International Journal of Cancer*, **3**, 867–75.

GOLDE, D. W., FAILLE, A., SULLIVAN, A. & FRIEND, C. (1976). Granulocytic stem cells in Friend leukemia. *Cancer Research*, **36**, 115–19.

GOLDWASSER, E. (1976). Erythropoietin. *Blut*, **33**, 135–40.

GRAF, T. (1973). Two types of target cells for transformation with avian myelocytomatosis virus. *Virology*, **54**, 398–413.

GRAF, T., FINK, D., BEUG, H. & ROYER-POKORA, B. (1977). Oncornavirus induced sarcoma formation obscured by rapid development of lethal leukemia. *Cancer Research*, **37**, 59–63.

GRAF, T., ROYER-POKORA, B., SCHUBERT, G. E. & BEUG, H. (1976). Evidence for the multiple oncogenic potential of cloned leukemia virus: *in vitro* and *in vivo* studies with avian erythroblastosis virus. *Virology*, **71**, 423–33.

GRAFFI, A. & BIELKA, H. (1963). In *Problèmes de cancérologie expérimentale*, pp. 454–5. Paris: Gauthier-Villars.

GRAFFI, A., FEY, F. & SCHRAMM, T. (1966). Experiments on the hematologic diversification of viral mouse leukemias. *National Cancer Institute Monograph*, **22**, 21–31.

GREZES, F. & SALOMON, J. C. (1972). Myoepithelioma with a leukemoid reaction in BALB/c mice. *Journal of the National Cancer Institute*, **49**, 1527–38.

GROSS, L. (1951). 'Spontaneous' leukemia developing in C3H mice following inoculation, in infancy, with Ak-leukemic extracts, or Ak embryos. *Proceedings of the Society for experimental Biology and Medicine*, **76**, 27–32.

GROSS, L. (1959). Effect of thymectomy on development of leukemia in C3H mice inoculated with 'passage' virus. *Proceedings of the Society for experimental Biology and Medicine*, **100**, 325–8.

GROSS, L. (1963). Pathogenic potency and host range of the mouse leukemia virus. *Acta haematologica*, **29**, 1–15.

HACKER, M., ROBERTS, D. & JACKSON, C. (1977). Depression of the platelet count after inoculation of mice with L1210 or L5178Y cells. *British Journal of Haematology*, **35**, 465–71.

HAJDIK, I., NIEWISCH, H. & MATIOLI, G. (1971). Hematopoietic stem cells in mice with erythroblastosis. *Journal of cellular Physiology*, **77**, 393–400.

HANKINS, D. & KRANTZ, S. B. (1974). Rapid *in vivo* assay for Friend polycythemia virus. *Journal of the National Cancer Institute*, **52**, 1223–9.

HANKINS, W. D. & KRANTZ, S. B. (1975). *In vitro* expression of erythroid differentiation induced by Friend polycythaemia virus. *Nature, London*, **253**, 731–2.

HARRISON, P. R. (1976). Analysis of erythropoiesis at the molecular level. *Nature, London*, **262**, 353–6.

HÄYRY, P., RAGO, D. & DEFENDI, V. (1970). Inhibition of phytohemagglutinin- and alloantigen-induced lymphocyte stimulation by Rauscher leukemia virus. *Journal of the National Cancer Institute*, **44**, 1311–19.

HAYS, E. F., CRADDOCK, C. G., HASKETT, D. & NEWELL, M. (1976). *In vitro* colony-forming cells in the marrow of leukemic and preleukemic mice. *Blood*, **47**, 603–10.

HOROSZEWICZ, J. S., LEONG, S. S. & CARTER, W. A. (1975). Friend leukemia: rapid development of erythropoietin independent hematopoietic precursors. *Journal of the National Cancer Institute*, **54**, 265–7.

HUSSEINI, S., FRIED, W., KNOSPE, W. H. & TROBAUGH, F. E. Jr (1976). Dynamics of leukemic and normal stem cells in leukemic RFM mice. *Cancer Research*, **36**, 1784–9.

IKAWA, Y., FURUSAWA, M. & SUGANO, H. (1973). Erythrocyte membrane specific antigens in Friend virus-induced leukemia cells. In *Unifying concepts of leukemia, Bibliotheca Haematologica No. 39*, ed. R. M. Dutcher & L. Chieco-Bianchi, pp. 955–67. Basel: Karger.

IKAWA, Y. & SUGANO, H. (1967). Spleen focus in Friend's disease: an electron microscopic study. *Gann*, **58**, 155–60.

IKAWA, Y., SUGANO, H. & OOTA, K. (1967). Spleen focus in Friend's disease: histological study. *Gann*, **58**, 61–7.

ITURRIZA, R. G. & SEIDEL, H. J. (1974). Stem cell growth and production of colony-stimulating factor in Rauscher virus-infected CBA/J mice. *Journal of the National Cancer Institute*, **53**, 487–92.

Jacobson, L. O., Goldwasser, E., Plzak, C. F. & Fried, W. (1957). Studies on erythropoiesis. IV. Reticulocyte response of hypophysectomized and polycythemic rodents to erythropoietin. *Proceedings of the Society for experimental Biology and Medicine*, **94**, 243–9.

Jerushalmy, Z., Kaminski, E., Kohn, A. & De Vries, A. (1963). Interaction of Newcastle disease virus (NDV) with megakaryocytes in cell culture of guinea-pig bone marrow. *Proceedings of the Society for experimental Biology and Medicine*, **114**, 687–90.

Jerushalmy, Z., Kohn, A. & De Vries, A. (1961). Interaction of myxoviruses with human blood platelets *in vitro*. *Proceedings of the Society for experimental Biology and Medicine*, **106**, 426–66.

Karnjanaprakorn-Yorsook, C. & Thomson, S. (1973). Development and regression of murine leukemia induced by Friend virus complex. *Proceedings of the Society for experimental Biology and Medicine*, **143**, 300–3.

Karnjanaprakorn-Yorsook, C. & Thomson, S. (1974). Host responses in Friend virus-induced leukemia. *Journal of the National Cancer Institute*, **53**, 407–13.

Kasuga, T. & Oota, K. (1962). Pathological characteristics of Friend disease. *Gann*, **8**, 251–65.

Kirsten, W. H. & Mayer, L. A. (1967). Morphologic responses to a murine erythroblastosis virus. *Journal of the National Cancer Institute*, **39**, 311–35.

Kirsten, W. H., Mayer, L. A., Wollmann, R. L. & Pierce, M. I. (1967). Studies on a murine erythroblastosis virus. *Journal of the National Cancer Institute*, **38**, 117–39.

Knyszynski, A. & Danon, D. (1976). Membrane characteristics of old and Rauscher leukemia virus infected mouse red blood cells. *Experimental Cell Research*, **100**, 303–8.

Kumar, V., Bennett, M. & Eckner, R. J. (1974). Mechanisms of genetic resistance to Friend virus leukemia in mice. *Journal of experimental Medicine*, **139**, 1093–1109.

Lai, M., Duesberg, P., Horst, J. & Vogt, P. K. (1973). Avian tumor virus RNA. A comparison of three sarcoma viruses and their transformation defective derivates by oligonucleotide fingerprint and DNA–RNA hybridization. *Proceedings of the National Academy of Sciences, USA*, **70**, 2266–70.

Lajtha, L. G. (1975). Haematopoietic stem cells. *British Journal of Haematology*, **29**, 529–35.

Lajtha, L. G., Pozzi, L. V., Schofield, R. & Fox, M. (1969). Kinetic properties of haemopoietic stem cells. *Cell and Tissue Kinetics*, **2**, 39–49.

Lajtha, L. G., Tanaka, P. & Testa, N. E. G. (1972). Stem cell kinetics in a myeloid leukaemia in mice. *British Journal of Haematology*, **23**, 259 (Abstr.).

Lappat, E. J. & Cawein, M. (1964). A study of the leukemoid response to transplantable A-280 tumor in mice. *Cancer Research*, **24**, 302–11.

Levy, S. B., Rubenstein, C. B. & Friend, C. (1976*a*). The spleen in Friend leukemia. I. Prolonged survival of leukemic mice after auto-implantation of spleen tissue. *Journal of the National Cancer Institute*, **56**, 1183–8.

Levy, S. B., Rubenstein, C. B. & Tavassoli, M. (1976*b*). The spleen in Friend leukemia. II. Non-leukemic nature of spleen stroma. *Journal of the National Cancer Institute*, **56**, 1189–96.

LIAO, S. K. & AXELRAD, A. A. (1975). Erythropoietin-independent erythroid colony formation *in vitro* by hemopoietic cells of mice infected with Friend virus. *International Journal of Cancer*, **15**, 467–82.

LILLY, F. & PINCUS, T. (1973). Genetic control of murine viral leukemogenesis. *Advances in Cancer Research*, **17**, 231–77.

LUDWIG, F. G., BOSTICK, W. L. & EPLING, M. L. (1964). Quantitative analysis of Friend's disease in two inbred strains of mice with emphasis on bone marrow response. *Cancer Research*, **24**, 1308–17.

McCULLOCH, E. A. & TILL, J. E. (1960). The radiation sensitivity of normal mouse bone marrow cells determined by quantitative transplantation into irradiated mice. *Radiation Research*, **13**, 115–25.

McCULLOCH, E. A. & TILL, J. E. (1964). Proliferation of hemopoietic colony-forming cells transplanted into irradiated mice. *Radiation Research*, **22**, 383–97.

McGARRY, M. P. & MIRAND, E. A. (1973). Altered responsiveness to erythropoietin in mice following infection with polycythemia-inducing Friend virus. *Experimental Hematology*, **1**, 174–82.

McGARRY, M. P. & MIRAND, E. A. (1977). Presence of target cells for Friend virus (FVP) in the liver and spleen of early postnatal mice. *Proceedings of the Society for experimental Biology and Medicine*, **154**, 295–8.

McGARRY, M. P., STEEVES, R. A., ECKNER, R. J., MIRAND, E. A. & TRUDEL, P. J. (1974). Isolation of a myelogenous leukemia-inducing virus from mice infected with the Friend virus complex. *International Journal of Cancer*, **13**, 867–78.

MAJUMDAR, S. K. & BILENKER, J. D. (1975). Failure to detect chromosome damage *in vivo* in Friend virus-infected leukemic mice. *Journal of the National Cancer Institute*, **54**, 503–5.

MARKOE, A. M. & OKUNEWICK, J. P. (1973*a*). Radioprotective effect of Rauscher leukemia virus in the SJL/J mouse. II. Development of radioresistance and the distribution of mortality. *Radiation Research*, **53**, 115–23.

MARKOE, A. M. & OKUNEWICK, J. P. (1973*b*). Radioprotective effect of Rauscher leukemia virus in the SJL/J mouse. III. Relationship of hematologic factors and splenomegally to the development of radioresistance. *Radiation Research*, **53**, 428–34.

MARKOE, A. M., OKUNEWICK, J. P., ERHARD, P. & PHILLIPS, E. (1973). Increased splenic transplantable colony-forming unit related to increased radio-resistance after Rauscher leukemia virus infection of the SJL/J mouse. *Journal of the National Cancer Institute*, **50**, 449–55.

MARTIN, G. S. (1970). Rous sarcoma virus: a function required for the maintenance of the transformed state. *Nature, London*, **227**, 1021–3.

MARTIN, G. S. & DUESBERG, P. A. (1972). The a subunit in the RNA of transforming avian tumor viruses. I. Occurrence in different virus strains. II. Spontaneous loss resulting in nontransforming variants. *Virology*, **47**, 494–7.

MATIOLI, G. (1973). Friend leukemic mouse stem cell reversion to normal growth in irradiated hosts. *Journal of the reticuloendothelial Society*, **14**, 380–6.

MATIOLI, G. (1974). Metastasis and growth of Friend tumor cells in irradiated syngeneic hosts. *Journal of the reticuloendothelial Society*, **15**, 282–96.

MAUPIN, B. (1966). Platelets and viruses. *Hemostase*, **6**, 301–13.

METCALF, D., FURTH, J. & BUFFET, R. F. (1959). Pathogenesis of mouse leukemia caused by Friend virus. *Cancer Research*, **19**, 52–8.

MICKLEM, H. S. & LOUTIT, J. F. (1966). In *Tissue grafting and radiation*, pp. 171–3. New York & London: Academic Press.

MILAS, L. & BASIC, I. (1972). Stimulated granulocytopoiesis in mice bearing fibrosarcoma. *European Journal of Cancer*, **8**, 309–13.

MIRAND, E. A. (1967*a*). Erythropoietin-like effect of a polycythemia virus. *Proceedings of the Society for experimental Biology and Medicine*, **125**, 562–5.

MIRAND, E. A. (1967*b*). Virus-induced erythropoiesis in hypertransfused polycythemic mice. *Science*, **156**, 832–3.

MIRAND, E. A. (1976). Autonomous erythropoiesis induced by a virus. *Seminars in Hematology*, **13**, 49–56.

MIRAND, E. A., PRENTICE, T. C. & HOFFMANN, J. G. (1961). Effect of Friend virus in Swiss and DBA/1 mice on ^{59}Fe uptake. *Proceedings of the Society for experimental Biology and Medicine*, **106**, 423–6.

MIRAND, E. A., STEEVES, R. A., AVILA, L. & GRACE, J. T. (1968*a*). Spleen focus formation by polycythemic strains of Friend leukemia virus. *Proceedings of the Society for experimental Biology and Medicine*, **127**, 900–4.

MIRAND, E. A., STEEVES, R. A., LANGE, R. D. & GRACE, J. T. (1968*b*). Virus-induced polycythemia in mice: erythropoiesis without erythropoietin. *Proceedings of the Society for experimental Biology and Medicine*, **128**, 844–9.

MOHRI, N. (1962). Pathological study on the leukemoid reaction. *Acta pathological japonica*, **12**, 35–57.

MOLONEY, J. B. (1960). Biological studies on a lymphoid leukemia virus extracted from sarcoma 37. I. Origin and introductury investigations. *Journal of the National Cancer Institute*, **24**, 933–51.

NASRALLAH, A. G. & MCGARRY, M. P. (1976). *In vivo* distinction between a target cell for Friend virus (FVP) and murine hematopoietic stem cells. *Journal of the National Cancer Institute*, **57**, 443–5.

NIEBURG, H. E. & GOLDBERG, A. F. (1968). Changes in polymorphonuclear leucocytes as a manifestation of malignant neoplasia. *Cancer*, **22**, 35–42.

NOOTER, K. & BENTVELZEN, P. (1976), *In vitro* transformation of murine erythroid cells by Rauscher leukemia virus. *Cancer Letters*, **1**, 155–60.

NOOTER K. & GHIO, R. (1975). Hormone-independent *in vitro* erythroid colony formation by bone marrow cells from Rauscher virus-infected mice. *Journal of the National Cancer Institute*, **55**, 59–64.

NOOTER, K., VAN DEN BERG, K., DE VRIES, J. & BENTVELZEN, P. (1976). Transfection of mouse bone marrow cells with Rauscher virus proviral DNA. *European Journal of Cancer*, **12**, 633–7.

ODAKA, T. (1969). Effect of bone marrow graft on the susceptibility of mice to Friend leukemia virus. *Japan Journal of experimental Medicine*, **39**, 99–100.

OKUNEWICK, J. P. & ERHARD, P. (1973). Assessment of the radioprotective capacity of spleen cells transplanted from mice with Rauscher viral leukemia. *Radiation Research*, **53**, 241–7.

OKUNEWICK, J. P., FULTON, D. & MARKOE, A. M. (1972). Radioprotective effect of Rauscher leukemia virus in the SJL/J mouse. *Radiation Research*, **49**, 521–9.

OKUNEWICK, J. P., MARKOE, A. M., ERHARD, P. & PHILLIPS, E. L. (1973). Growth rate and ^{59}Fe uptake in the progeny of Rauscher leukemia virus-infected splenic colony forming units following transplantation. *Journal of laboratory and clinical Medicine*, **81**, 489–96.

OKUNEWICK, J. P. & PHILLIPS, E. L. (1973). Changes in marrow and spleen CFU compartments following leukemia virus injection: comparison of Friend and Rauscher virus. *Blood*, **42**, 885–92.

OKUNEWICK, J. P. & PHILLIPS, E. L. (1974). Colony-forming unit suicide in normal and Rauscher leukemic mice given tritiated thymidine *in vivo*. *Experimental Hematology*, **2**, 9–15.

OKUNEWICK, J. P., PHILLIPS, E. L. & BROZOVICH, B. J. (1974). Antiserum inactivation of hematopoietic colony forming stem cells from mice with Rauscher leukemia. *Experientia*, **30**, 1470–1.

OKUNEWICK, J. P., PHILLIPS, E. L. & BROZOVICH, B. J. (1976*a*). Effect of antiserum on transplantable hematopoietic colony-forming units during Rauscher leukemia development. *American Journal of Hematology*, **1**, 443–52.

OKUNEWICK, J. P., PHILLIPS, E. L. & BROZOVICH, B. J. (1976*b*). Effect of Rauscher leukemia development on DNA synthesis by hematopoietic CFU-s. *Experimental Hematology*, **4**, 143–50.

OKUNEWICK, J. P., PHILLIPS, E. L. & ERHARD, P. (1972). Increase in number of splenic transplantable colony-forming units in the SJL/J mice after infection with Rauscher leukemia virus. *Journal of the National Cancer Institute*, **49**, 1101–6.

OSKI, F. A. & NAIMAN, J. L. (1966). Effect of live measles vaccine on the platelet count. *New England Journal of Medicine*, **275**, 352–6.

OSTERTAG, W., MELDERIS, H., STEINHEIDER, G., KLUGE, N. & DUBE, S. (1972). Synthesis of mouse haemoglobin and globin mRNA in leukaemic cell cultures. *Nature New Biology, London*, **239**, 231–4.

PARR, I. B. & ROWSON, K. E. K. (1970). Erythrocyte osmotic fragility in Friend virus-infected mice. *European Journal of Cancer*, **6**, 411–15.

PENNELLI, N., CHIECO-BIANCHI, L., TRIDENTE, G. & FIORE-DONATI, L. (1968). Virus particles in bone marrow megakaryocytes and susceptibility of mice to virus-induced leukemia. *International Journal of Cancer*, **3**, 390–6.

PERIES, J., LEVY, J. P., BOIRON, M. & BERNARD, J. (1964). Multiplication of Rauscher virus in cultures of mouse kidney cells. *Nature, London*, **203**, 672–3.

PLAYFAIR, J. H. L. & COLE, L. J. (1965). Quantitative studies on colony-forming units in isogenic radiation chimeras. *Journal of cellular and comparative Physiology*, **65**, 7–18.

PLUZNIK, D. H. & SACHS, L. (1964). Quantitation of a murine leukemia virus with a spleen colony assay. *Journal of the National Cancer Institute*, **33**, 535–46.

PLUZNIK, D. H., SACHS, L. & RESNITZKY, P. (1966). The mechanism of leukemogenesis by the Rauscher leukemia virus. *National Cancer Institute Monographs*, **22**, 3–14.

RAUSCHER, F. J. (1962). A virus-induced disease of mice characterized by erythrocytopoiesis and lymphoid leukemia. *Journal of the National Cancer Institute*, **29**, 515–43.

REISSMANN, K. R. & ITO, K. (1966). Selective eradication of erythropoiesis by actinomycin D as the result of interference with hormonally controlled effector pathway of cell differentiation. *Blood*, **28**, 201–12.

REISSMANN, K. R. & SAMORAPOOMPCHIT, S. (1970). Effect of erythropoietin on proliferation of erythroid stem cell in the absence of transplantable colony forming units. *Blood*, **36**, 287–96.

REISSMANN, K. R. & UDUPA, K. B. (1972). Effect of erythropoietin on proliferation of erythropoietin responsive cells. *Cell and Tissue Kinetics*, **5**, 481–9.

RIFE L. L., SIZEMORE, D. B. & MATIOLI, G. (1975). Induction of differentiation instabilities in Friend erythroleukemia. *Journal of the reticuloendothelial Society*, **18**, 97–106.

RILEY, V. (1968). Lactate dehydrogenase in the normal and malignant state in mice and the influence of a benign enzyme-elevating virus. In *Methods in cancer research*, vol. IV, ed. H. Busch, pp. 493–618. New York & London: Academic Press.

ROSENBERG, N. & BALTIMORE, D. (1976). A quantitative assay for transformation of bone marrow cell by Abelson murine leukemia virus. *Journal of experimental Medicine*, **143**, 1453–63.

ROSENBERG, N., BALTIMORE, D. & SCHER, C. D. (1975). *In vitro* transformation of lymphoid cells by Abelson murine leukemia virus. *Proceedings of the National Academy of Sciences, USA*, **72**, 1932–6.

ROSSI, G. B., CUDKOWICZ, G. & FRIEND, C. (1970). Evidence for transformation of spleen cells one day after infection of mice with Friend leukemia virus. *Journal of experimental Medicine*, **131**, 765–81.

ROSSI, G., CUDKOWICZ, G. & FRIEND, C. (1973). Transformation of spleen cells three hours after infection *in vivo* with Friend leukemia virus. *Journal of the National Cancer Institute*, **50**, 249–54.

ROSSI, G. B., DE HARVEN, E., HADAD, I. R. & FRIEND, C. (1971). Studies on Friend virus-induced viremia in lethally irradiated mice with or without hematopoietic repopulation. *International Journal of Cancer*, **7**, 303–12.

ROSSI, G. B. & FRIEND, C. (1967). Erythrocytic maturation of (Friend) virus-induced leukemic cells in spleen clones. *Proceedings of the National Academy of Sciences, USA*, **58**, 1373–80.

ROWSON, K. E. K. & PARR, I. (1970). A new virus of minimal pathogenicity associated with Friend virus. I. Isolation by end point dilution. *International Journal of Cancer*, **5**, 96–102.

SANFORD, K. S. (1974). Biologic manifestations of oncogenesis *in vitro*: a critique. *Journal of the National Cancer Institute*, **53**, 1481–5.

SASSA, S., TAKAKU, F. & NAKAO, K. (1968). Regulation of erythropoiesis in the Friend leukemia mouse. *Blood*, **31**, 758–65.

SCHER, C. D., SCOLNICK, E. M. & SIEGLER, R. (1975). Induction of erythroid leukaemia by Harvey and Kirsten sarcoma viruses. *Nature, London*, **256**, 225–6.

SCHER, C. D. & SIEGLER, R. (1975). Direct transformation of 3T3 cells by Abelson murine leukaemia virus. *Nature, London*, **253**, 729–31.

SCHOOLEY, J. C., GARCIA, J. F., CANTOR, L. N. & HAVENS, V. W. (1968). A summary of some studies on erythropoiesis using anti-erythropoietin immune serum. *Annals of the New York Academy of Sciences*, **149**, 266–80.

SCHOOLMAN, H. M., SPURRIER, W., SCHWARTZ, S. O. & SZANTO, P. B. (1957).

Studies in leukemia. VII. The induction of leukemia virus in Swiss mice by means of cell-free filtrates of leukemic mouse brain. *Blood*, **12**, 694–700.

Scolnick, E. M., Howk, R. S., Anisowicz, A., Peebles, P. T., Scher, C. D. & Parks, W. P. (1975). Separation of sarcoma virus-specific and leukemia virus-specific genetic sequences of Moloney sarcoma virus. *Proceedings of the National Academy of Sciences, USA*, **72**, 4650–4.

Scolnick, E. M. & Parks, W. P. (1974). Harvey sarcoma virus: a second murine type C sarcoma virus with rat genetic information. *Journal of Virology*, **13**, 1211–19.

Scolnick, E. M., Rands, E., Williams, D. & Parks, W. P. (1973). Studies on the nucleic acid sequences of Kirsten sarcoma virus: a model for formation of a mammalian RNA-containing sarcoma virus. *Journal of Virology*, **12**, 458–63.

Seidel, H. J. (1972*a*). Die Blutzellbildung bei der Rauscher-Leukämie (Mäusestamm BALB/c) und ihre Beeinflussung durch Hypertransfusion. *Zell Krebsforschung*, **77**, 155–65.

Seidel, H. J. (1972*b*). Pancytopenia in CBA mice after Rauscher virus infection. *Journal of the National Cancer Institute*, **48**, 959–64.

Seidel, H. J. (1973*a*). Target cell characterization for Rauscher leukemia virus in vivo. In *Unifying concepts of leukemia, Bibliotheca Haematologica 39*, ed. R. M. Dutcher & L. Chieco-Bianchi, pp. 935–42. Basel: Karger.

Seidel, H. J. (1973*b*). Das verhalten hämopoetischer Stammzellen bei Mäusen mit Virus-leukämie. I. Milzkolonieversuche am Mäusestamm CBA nach Infektion mit dem Rauscher Virus. *Zell Krebsforschung*, **79**, 123–34.

Seidel, H. J. (1973*c*). Hemopoietic stem cells in mice with virus induced leukemia. II. Studies with C3H mice after Rauscher virus infection. *Zell Krebsforschung*, **80**, 229–37.

Seidel, H. J. (1973*d*). Stem cell in the peripheral blood of Rauscher leukemic mice. *Experientia*, **29**, 607 (Abstr.).

Sendo, F., Kodama, T. & Kobayashi, H. (1973). Leukemoid reaction in BALB/c mice bearing transplantable tumor. *Gann*, **64**, 297–300.

Shabad, L. M. (1973). Precancerous morphological lesions. *Journal of the National Cancer Institute*, **50**, 1421–8.

Siegel, B. V. & Morton, J. I. (1973). Immunologic stimuli in relation to leukemogenesis: the hematopoietic stem cell as target cell for Rauscher leukemia virus. In *Conference on virus tumorigenesis and immunogenesis*, ed. W. Ceglowski & H. Friedman, p. 271. New York & London: Academic Press.

Siegel, B. V. & Morton, J. I. (1976). Tumor virus effects on immunocyte precursor cells. Hemopoietic stem cell behavior and leukemogenic susceptibility. *Annals of the New York Academy of Sciences*, **276**, 442–54.

Siegler, R. S., Zajdel, S. & Lane, I. (1972). Pathogenesis of Abelson virus induced murine leukemia. *Journal of the National Cancer Institute*, **48**, 189–218.

Silvestre, D., Levy, J. P., Leclerc, J. C. & Boiron, M. (1966). Etude ultrastructurale du cycle du virus de Rauscher chez la souris. *Pathologie–Biologie*, **14**, 559–64.

Siminovitch, L., McCulloch, E. A. & Till, J. E. (1963). The distribution of colony forming cells among spleen colonies. *Journal of cellular and comparative Physiology*, **62**, 327–36.

Siminovitch, L., Till, J. E. & McCulloch, E. A. (1964). Decline in colony-forming ability of marrow cells subjected to serial transplantation into irradiated mice. *Journal of cellular and comparative Physiology*, **64**, 23–32.

Sinkovics, J. G., Bertin, B. A. & Howe, C. D. (1966). Occurrence of low leukemogenic but immunizing mouse leukemia virus in tissue culture. *National Cancer Institute Monographs*, **22**, 349–68.

Sklar, M. D., Shevach, E. M., Green, I. & Potter, M. (1975). Transplantation and preliminary characterization of lymphocyte surface markers of Abelson virus-induced lymphoma. *Nature, London*, **253**, 550–2.

Smadja-Joffe, F., Jasmin, C., Malaise, E. P. & Bournoutian, C. (1973). Study of the cellular proliferation kinetics of Friend leukemia. *International Journal of Cancer*, **11**, 300–13.

Smadja-Joffe, F., Klein, B., Kerdiles, C., Feinendegen, L. & Jasmin, C. (1976). Study of cell death in Friend leukaemia. *Cell and Tissue Kinetics*, **9**, 131–45.

Smith, L. H. & Vos, O. (1962). Sensitivity and protection of bone marrow cells X-irradiated *in vitro*. *International Journal of radiation Biology*, **5**, 461–70.

Soule, H. D. & Arnold, W. J. (1970). Murine myeloproliferative virus in cell culture. *Journal of the National Cancer Institute*, **45**, 253–62.

Steeves, R. A. (1975). Spleen focus-forming virus in Friend and Rauscher leukemia virus preparations. *Journal of the National Cancer Institute*, **54**, 289–97.

Steeves, R. A., Bennett, M., Mirand, E. A. & Cudkowicz, G. (1968). Genetic control by the W locus of susceptibility to (Friend) spleen focus-forming virus. *Nature, London*, **218**, 372–4.

Steeves, R. A., Mirand, E. A., Bulba, A. & Trudel, P. J. (1970). Spleen foci and polycythemia in C57Bl mice infected with host-adapted Friend leukemia virus. *International Journal of Cancer*, **5**, 346–56.

Steeves, R. A., Mirand, E. A. & Thomson, S. (1969). Properties of transformed hemopoietic cells in mice infected with the Friend virus complex. In *Comparative leukemia research, Bibliotheca Haematologica 36*, ed. R. M. Dutcher, pp. 624–33. Basel: Karger.

Stehelin, D., Guntaka, R. V., Varmus, H. E. & Bishop, S. M. (1976*a*). Purification of DNA complementary to nucleotide sequences required for neoplastic transformation of fibroblasts by avian sarcoma viruses. *Journal of molecular Biology*, **101**, 349–65.

Stehelin, D., Varmus, H. E., Bishop, S. M. & Vogt, P. K. (1976*b*). DNA related to the transforming gene(s) of avian sarcoma viruses is present in normal avian DNA. *Nature, London*, **260**, 170–3.

Stephenson, J., Axelrad, A., McLeod, D. & Schreeve, M. M. (1971). Induction of colonies of hemoglobin-synthesizing cells by erythropoietin *in vitro*. *Proceedings of the National Academy of Sciences, USA*, **68**, 1542–6.

Sundelin, P. & Adam, L. R. (1967). DNA, RNA and hemoglobin cytophotometry of maturing erythroid cells from liver nodules in Rauscher virus infected mice. *International Journal of Cancer*, **2**, 544–50.

Takada, A., Takada, Y. & Ambrus, J. L. (1971). Proliferation of spleen cells from mice infected with Friend virus in the spleens of unirradiated and irradiated mice. *Experientia*, **21**, 315–16.

TAMBOURIN, P. E. (1970). Antigènes viraux et cellulaires sur les membranes des cellules du système hematopoïétique dans la leucémie murine de Friend. Thesis, University of Paris.

TAMBOURIN, P. E., GALLIEN-LARTIGUE, O., WENDLING, F. & HUAULME, D. (1973). Erythrocyte production in mice infected by the polycythaemia-inducing Friend virus or by the anaemia-inducing Friend virus. *British Journal of Haematology*, **24**, 511–24.

TAMBOURIN, P. & WENDLING, F. (1971). Malignant transformation and erythroid differentiation by polycythaemia-inducing Friend virus. *Nature New Biology*, **234**, 230–3.

TAMBOURIN, P. E. & WENDLING, F. (1975). Target cell for oncogenic action of polycythaemia-inducing Friend virus. *Nature, London*, **256**, 320–2.

TAMBOURIN, P., WENDLING, F., BARAT, N. & ZAJDELA, F. (1969). Influence de différents facteurs d'homéostase érythropoïétiques sur l'évolution de la leucémie de Friend. *Nouvelle Revue française de'Hématologie*, **9**, 461–84.

TANAKA, T. & CRAIG, A. W. (1970). Cell free transmission of murine myeloid leukaemia. *European Journal of Cancer*, **6**, 329–33.

TANAKA, T. & LAJTHA, L. G. (1969). Characteristics of murine myeloid leukemia colonies in the spleen. *British Journal of Cancer*, **23**, 197–203.

TEMIN, H. M. (1974). On the origin of RNA tumor viruses. *Annual Review of Genetics*, **8**, 155–77.

THOMSON, S. (1969*a*). An assay method for target cell of Friend spleen focus-forming virus. *Proceedings of the American Association for Cancer Research*, **10**, 93 (Abstr.).

THOMSON, S. (1969*b*). A system for quantitative studies on interactions between Friend leukemia virus and hemopoietic cells. *Proceedings of the Society for experimental Biology and Medicine*, **130**, 227–31.

THOMSON, S. & AXELRAD, A. A. (1968). A quantitative spleen colony assay method for tumor cells induced by Friend leukemia virus infection in mice. *Cancer Research*, **28**, 2105–14.

THOMSON, S. & SHRIVANATAKUL, P. (1973). Factors influencing the assay for target cells of Friend spleen focus-forming virus. *Cancer Research*, **33**, 717–20.

THOMSON, S., SHRIVANATAKUL, P. & MIRAND, E. A. (1974). Studies on target cells of Friend spleen focus-forming virus in mice. *Proceedings of the Society for experimental Biology and medicine*, **145**, 1329–32.

TILL, J. E. (1966). Discussion. *Journal of cellular Physiology*, **67**, suppl. 1, 144 (Abstr.).

TILL, J. E. & MCCULLOCH, E. A. (1961). A direct measurement of the radiation sensitivity of normal mouse bone marrow cells. *Radiation Research*, **14**, 213–22.

TILL, J. E., MCCULLOCH, E. A. & SIMINOVITCH, L. (1964). A stochastic model of stem cell proliferation based on the growth of spleen colony-forming cells. *Proceedings of the National Academy of Sciences, USA*, **51**, 29–36.

TROXLER, D. H., PARKS, W. P., VASS, W. C. & SCOLNICK, E. M. (1977). Isolation of a fibroblast nonproducer cell line containing the Friend strain of the spleen focus-forming virus. *Virology*, **76**, 602–15.

UDUPA, K. B., OKAMURA, H. & REISSMANN, K. R. (1972). Granulopoiesis during Myleran-induced suppression of transplantable hematopoietic stem cells. *Blood*, **39**, 317–25.

Urso, P. & Congdon, C. C. (1957). The effect of the amount of isologous bone marrow injected on the recovery of haemopoietic organs, survival and body weight after lethal irradiation injury in mice. *Blood*, **12**, 251–60.

Van Beek, H. J., Van't Hull, E. & Van Griesven, L. J. L. D. (1976). Modification of hemopoietic stem cell of BALB/c mice by Rauscher leukemia virus. *Experimental Hematology*, **4**, 151–60.

Van Griensven, L. J. L. D., Van Beek, H. J. & Van't Hull, E. (1974). The nature of hemopoietic stem cells from Rauscher leukemia virus infected BALB/c mice. *Biomedicine*, **21**, 334–7.

Van Griensven, L. J. L. D., Van't Hull, E. & De Vries, M. J. (1973). The *in vitro* growth of bone marrow cell from Rauscher leukemia virus infected mice. *Biomedicine*, **19**, 138–41.

Van Griensven, L. J. L. D., Van't Hull, E., Van Beek, H. J., Buurman, W. A. & De Vries, M. J. (1975). The relation between the proliferative activity and the differentiation pattern of bone marrow cells from Rauscher leukemia virus infected BALB/c mice. *Biomedicine*, **22**, 393–8.

Vogel, H., Niewisch, H. & Matioli, G. (1968). The self renewal probability of hemopoietic stem cells. *Journal of cellular Physiology*, **72**, 221–8.

Vogt, P. K. (1971). Spontaneous segregation of nontransforming viruses from cloned sarcoma viruses. *Virology*, **46**, 939–46.

Weitz-Hamburger, A., Fredrickson, T. N., Lobue, J., Hardy, W. D., Camiscoli, J. F., Ferdinand, P., Gallicchio, V. & Gordon, A. S. (1975). Stimulation of erythropoietic differentiation in BALB/c mice infected with Rauscher leukemia virus. *Journal of the National Cancer Institute*, **55**, 1171–5.

Weitz-Hamburger, A., Fredrickson, T. N., Lobue, J., Hardy, W. D., Ferdinand, P. & Gordon, A. S. (1973). Inhibition of erythroleukemia in mice by induction of hemolytic anemia prior to infection with Rauscher leukemia virus. *Cancer Research*, **33**, 104–11.

Wendling, F., Jullien, P. & Tambourin, P. E. (1973). Hematopoietic CFU in mice infected by the polycythaemia-inducing Friend virus. II. Survival of lethally irradiated mice grafted with spleen or bone marrow cells from infected donors. *Radiation Research*, **55**, 177–88.

Wendling, F., Tambourin, P. E., Barat, N. & Zajdela, F. (1969). Différences de sensibilité vis à vis de l'actinomycine D du tissu pathologique induit par le virus de Friend et du tissu érythropoïétique normal. *Comptes rendus de l'Académie des Sciences, Paris, Series D*, **268**, 2222–5.

Wendling, F., Tambourin, P. E. & Bucau-Varlet, P. (1974). Capacité de prolifération des cellules souches (CFU-s) de la rate de souris leucémiques porteuses de la leucémie de Friend. *Comptes rendus de l'Académie des Sciences, Paris, Series D*, **278**, 2385–8.

Wendling, F., Tambourin, P. E., Gallien-Lartigue, O. & Charon, M. (1974). Comparative differentiation and numeration of CFU-s from mice infected either by the anemia or polycythemia-inducing strains of Friend viruses. *International Journal of Cancer*, **13**, 454–62.

Wendling, F., Tambourin, P. E. & Jullien, P. (1972). Hematopoietic CFU in mice infected by the polycythemia inducing Friend virus. I. Number of CFU and differentiation pattern in the spleen colonies. *International Journal of Cancer*, **9**, 554–66.

Wendling, F., Tambourin, P. E. & Jullien, P. (1973*a*). Hematopoietic colony-forming units in mice infected by the polycythaemia-inducing Friend virus. III. Pattern of blood recovery in irradiated mice grafted with normal or infected spleen cells. *Journal of the National Cancer Institute*, **51**, 179–89.

Wendling, F., Tambourin, P. & Jullien, P. (1973*b*). Hematopoietic CFU in mice infected by the polycythemia-inducing Friend virus. IV. Pattern of blood recovery in irradiated mice grafted with normal or infected bone marrow cells. *Biomedicine*, **18**, 521–9.

Wodinsky, I., Swiniarski, J. & Kensler, C. J. (1967). Spleen colony studies of leukemia L1210. I. Growth kinetics of lymphocytic L1210 cells *in vivo* as determined by spleen colony assay. *Cancer Chemotherapy Reports*, **51**, 415–21.

Wollmann, R. L., Pang, E. J., Evans, A. E. & Kirsten, W. H. (1970). Virus-induced hemolytic anemia in mice. *Cancer Research*, **30**, 1003–10.

Wright, B. S. & Lasfargues, J. C. (1965). Long term cultivation of the Rauscher murine leukemia virus in tissue culture. *Journal of the National Cancer Institute*, **35**, 319–27.

Wyke, J. A. (1975). Temperature sensitive mutants of avian sarcoma viruses. *Biochemica et biophysica Acta*, **417**, 91–121.

Yokoro, K. & Thorell, B. (1966). Cytology and pathogenesis of Rauscher virus disease in splenectomized mice. *Cancer Research*, **26**, 536–43.

Yoshikura, H., Hirokawa, Y., Ikawa, Y. & Sugano, H. (1969). Infections by non-leukemogenic Friend leukaemia virus obtained after prolonged cultivation *in vitro*. *International Journal of Cancer*, **4**, 636–40.

Yoshikura, H., Hirokawa, Y., Yamada, M. & Sugano, H. (1967). Production of Friend leukemia virus in a mouse lung cell line. *Japanese Journal of medical Science and Biology*, **20**, 225–36.

Zajdela, F. (1962). Contribution á l'étude de la cellule de Friend. *Bulletin du Cancer*, **49**, 351–73.

Zajdela, F., Tambourin, P. E., Wendling, F. & Pierre, O. (1968). Formation de particules virales sur la membrane d'érythrocytes de souris injectées avec le virus de Friend. *Comptes rendus de l'Académie des Sciences, Paris, Series D*, **267**, 2394–6.

Epithelial proliferative subpopulations

CHRISTOPHER S. POTTEN

Paterson Laboratories, Christie Hospital and
Holt Radium Institute, Withington, Manchester M20 9BX, UK

All adult mammalian tissues continually lose cells. This loss may be infrequent, as a result, for instance, of the occasional accident that may befall a cell, or it may be a more extensive programmed cell loss, e.g. through differentiation and maturation. In a stable steady state tissue this loss has to be compensated for by a precisely balanced cell replacement. This could be achieved by two types of replacement mechanism (Fig. 1): the straight replacement of lost cells by stem cell divisions (*a*) with or (*b*) without a series of amplifying divisions in the cells committed to differentiation. These *committed cells* with their limited life and possible transitory amplifying cell divisions resulting in a temporary cell replacement, represent a 'running out' population dependent on ultimate replacement by the stem cells. The *stem cells* are therefore the cells ultimately responsible for all cell replacement, including their own, within an adult tissue during the animal's lifetime (Lajtha, 1964, 1967; Barnes & Loutit, 1967; Fig. 1).

Differentiation is defined as the initiation of new, or a change in existing, biosynthetic processes, i.e. a change in the repression–activation pattern of the genome with a consequent rearrangement of the functional options open to that cell. It is evident from the appearance of new cellular products. The initial stages of differentiation would require exquisitely sensitive techniques to detect the novel constituents. *Maturation* is the quantitative change in the level, concentration or numbers of these novel cellular constituents (Lajtha & Schofield, 1974). Thus differentiation is a qualitative, while maturation is a quantitative, function.

It is clear that in the straight replacement scheme the stem cell cycle rate will be that determined for the proliferative population

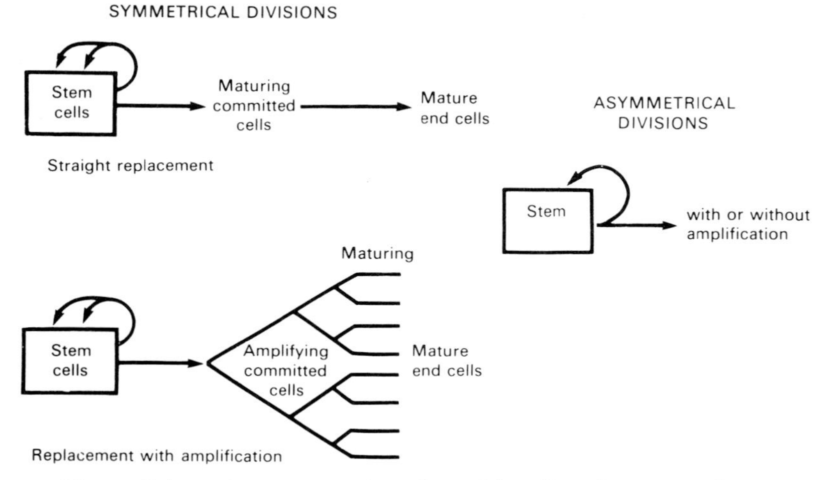

Fig. 1. Schematic representation of possible cell replacement schemes.

as a whole, while in the amplification scheme described the stem cells will divide less frequently. The amplifying committed cells may become more differentiated as they progress and form different sublines. The speed and number of the amplification divisions may be variable, depending on the demand for cells. The potential for division may be large, being rarely expressed under steady state conditions, but this potential may be released in severe cell-loss situations or if the cells are placed in suitable in-vitro conditions. The younger committed cells might be expected to have a greater division potential than the older cells. However, they will all eventually exhaust their potential and require replacement themselves. The ability of a cell to form many descendants (a clone or colony) is itself, therefore, not a good indicator of stem cell activity, but it represents the only quantitative assay available for epithelial tissues at present. Clearly the longer cell division is maintained within the clone (the bigger the clone) the younger the committed cells that are being assayed will be. The only ultimate test for stem cells, however, is whether the clone itself contains further clonogenic cells, i.e. whether it has produced further stem cells.

THE SYMMETRY OF STEM CELL DIVISIONS

Stem cell divisions could be either symmetrical (both daughters being stem cells) or asymmetrical (one stem and one committed cell). Symmetrical divisions will always produce two stem cells. The fate of these daughters may depend on some external factors or their position within the tissue. Symmetrical division must occur during regeneration and may occur in some normal tissues. However, it would be unlikely in the extreme situation where the stem cells' distribution is such that each 'tissue unit of proliferation' has only one stem cell. Under these circumstances an asymmetrical stem cell division is essential to ensure local stability. In epithelial tissues this may be the case. In skin the epidermal proliferative units (Potten, 1974) may have one or a very low number of stem cells and a similar situation may exist in the tongue papilla (Hume & Potten, 1976). The number of stem cells in small intestinal crypts is believed to be higher but preliminary experiments suggest that asymmetrical divisions may also occur here.

Asymmetrical division may have another significance in that a model has been put forward requiring such division in stem cells to preserve the integrity of the stem cell genome. This proposes the transfer of any damaged DNA to the committed cells, which are soon lost (Cairns, 1975). Since DNA replication is an error-prone process stem cells would benefit if they selectively segregated new and old DNA strands at division; the stem cells retaining the error-free parental template strands. It would, of course, be of even greater advantage if the cell could check the template strands as well and, by means of sister chromatid exchange, transfer any damaged portions to the newly synthesised strands which would then be discarded to the committed cells. Though the mechanics of the process were not discussed it would provide a means of ensuring unequal division. So far evidence for this hypothesis is lacking. However, preliminary experiments (C. S. Potten, W. J. Hume & P. Reid, unpublished data) in both intestinal crypts and tongue papillae provide data consistent with the idea that a few cells in the crypt, or tongue papilla, are selectively handling their DNA at mitosis.

I should like to consider how stem cells are distributed in epithelial tissue by looking at three regions of surface epithelia; intestinal mucosa, tongue epithelium and epidermis, considering them together as three examples of surface epithelia rather than as three separate tissues. The radiobiological data suggest that only a few of the proliferative cells in both skin and intestine can undergo extensive regenerative divisions, indicating that the amplifying committed cell scheme operates. In skin and tongue the presence of few stem cells per unit of proliferation suggests an asymmetric division process. Taken as a whole, the data for all three epithelial regions, together with their three-dimensional topography and the cell flow patterns that are a consequence of that topography, suggest that these regions are best explained by the asymmetric amplifying scheme.

SMALL INTESTINE

Cell proliferation in the small intestine is restricted to a centrally located band of cells (about 10 cell layers deep) within small 'balloon shaped' structures, the crypts (Fig. 2). Within this band the cells divide rapidly (every 10–14 hours) and the product of this division activity matures and moves out of the crypt, onto the villus and is eventually lost (2–3 days later) from the villus tip. Each villus is served by several crypts and each crypt may provide cells for more than one villus. If one considers the three-dimensional aspects of the crypt itself it is clear that, unless there is extensive, constant up-and-down cell position changing in the crypt (for which there is no evidence) then the cells at the top of the proliferative band must be pushed out by the cell division adjacent and below. This 'pushing up' process must occur within all the cell layers. Thus many of the cells within this rapidly dividing region must be lost and therefore represent a 'running out' population requiring constant replacement from the cells ultimately responsible for the entire crypt proliferation. These must be positioned near the crypt base. However, small intestinal crypts, like those from some other regions, contain a mature end cell population also at the crypt base, the Paneth cells. Their presence indicates that some cell replacement is needed both down and up the crypt.

There are three other observations on the crypt base cells consistent with the hypothesis that these are the crypt stem cells. These I should like to summarise. They will be dealt with in greater detail in the next paper (Wright, this volume). First, cell kinetic studies indicate the presence of a more slowly dividing (cycle times of 20 hours) cell population at the boundary between the Paneth cells and

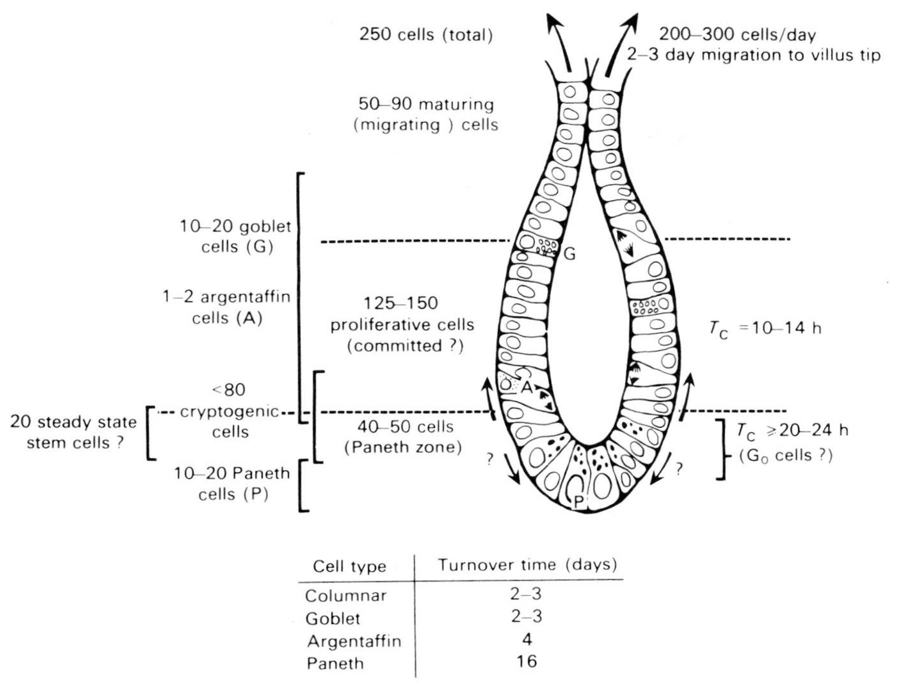

Cell type	Turnover time (days)
Columnar	2–3
Goblet	2–3
Argentaffin	4
Paneth	16

Fig. 2. Schematic representation of small intestinal crypt cell populations, cell topography and cell kinetics. (Data drawn from Potten & Hendry, 1975; Potten, 1975*a*, 1976*a*; Cheng & Leblond, 1974; Leblond & Cheng, 1976.)

the proliferating cells (Cairnie, Lamerton & Steel, 1965; Potten, Kovacs & Hamilton, 1974; Al-Dewachi, Wright, Appleton & Watson, 1975). The number of these slowly cycling cells is difficult to determine but might be about 20 per crypt (Potten *et al.*, 1974). This number is consistent with the existence of a single ring of cells above the Paneth cells. These crypt base presumptive stem cells also exhibit different thymidine pool characteristics since they label more heavily after a single exposure to tritiated thymidine ([^{3}H]TdR) than the other crypt cells (C. S. Potten, S. E. Al-Barwari & J. Searle, unpublished data; Potten, Al-Barwari, Hume & Searle, 1977). They also exhibit particularly marked circadian variations in proliferation (Potten *et al.*, 1977).

Secondly, some of the proliferative cells at the crypt base, unlike the cells higher up the proliferative band, possess an extreme sensitivity to radiation damage as evidenced by apoptosis (Kerr, Wyllie & Currie, 1972). This crypt base cell sensitivity is evident after either [^{3}H]TdR incorporation (Cheng & Leblond, 1974; Leblond & Cheng, 1976; C. S. Potten, S. E. Al-Barwari & J. Searle, unpublished data) or external X- or γ-irradiation (Fig. 3) (C. S. Potten, S. E. Al-Barwari & J. Searle, unpublished data). These radiation-killed cells appear first in the crypt base, where they are phagocytosed by the neighbouring crypt base cells, and appear later at higher cell positions. By following the fate of [^{3}H]TdR-induced labelled phagosomes as they migrate Cheng & Leblond (1974) have come to the conclusion that they are ingested by stem cells which are present in small numbers at the crypt base and are responsible for replacement of all four cell types in the crypt: columnar, goblet, argentaffin and Paneth. In contrast, hydroxyurea (HU) kills more cells more efficiently in the upper regions of the crypt where many are in S phase (C. S. Potten, S. E. Al-Barwari & J. Searle, unpublished data). After this type of crypt cell destruction by HU in the rat the crypt base cells respond rapidly by shortening their cell cycle and this region appears to be responsible for crypt repopulation (Al-Dewachi, Wright, Appleton & Watson, 1977; Wright, this volume).

Thirdly, the number of cells per crypt capable of regenerating the crypt within 3–4 days after severe radiation depletion can be estimated by using split-doses of radiation (Hendry & Potten, 1974; Potten & Hendry, 1975). The errors on these estimates are fairly large and several assumptions have to be made but the results of several experiments consistently provide values of about 80 cells per crypt (Hendry & Potten, 1974; Potten & Hendry, 1975; C. S. Potten, unpublished data). However, higher estimates have recently been obtained using a different indirect method (Masuda, Withers, Mason & Chen, 1977). The reason for the difference, which is unlikely to be significant considering the errors inherent in both techniques, is not clear at present. The split-dose assay provides values that are higher by a factor of 4 than the estimates of about 20 stem cells obtained by histologic and labelling methods, but lower by a factor of 2 than the total number of proliferative cells. This suggests some early committed cells are capable of some regeneration. This is also supported by the fact that treatment with S-phase specific agents before irradiation selectively kills cells in S, most of which are in the mid- and upper crypt regions, and this results in a greater crypt destruction than after radiation alone: i.e. some of the S phase cells killed by HU were capable of some regeneration (Boarder & Blackett, 1976).

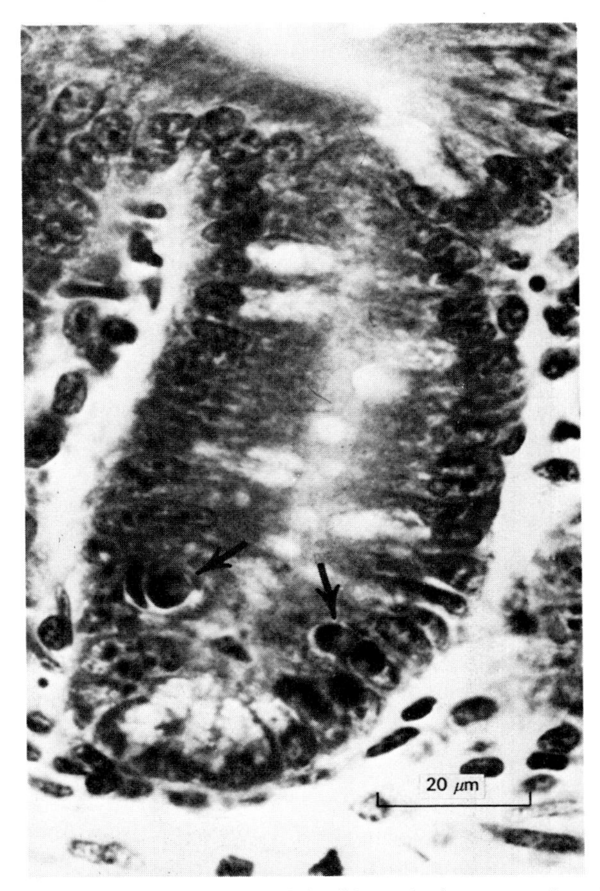

Fig. 3. Haematoxylin and eosin stained small intestinal crypt section, 6 hours after 25 rads of X-rays (300 kVp at 30 rad/min). Several apoptotic cells can be seen particularly in the lower third of the crypt (arrows).

TONGUE EPITHELIUM

The dorsal surface of the tongue is covered with numerous small projections, the filiform papillae. The structure and mode of cell replacement within these papillae can be compared with the small intestine (Fig. 4). Proliferation is restricted to a single layer of cells with some regions (the deeper lying regions) showing rapid proliferation and others (the more distal regions) showing none. Recent work (Hume & Potten, 1976) shows that the papilla can be regarded as having two major parts, the anterior and posterior aspects. Each of these consists of a highly ordered column of mature end cells which represent the historical record of several days' cell

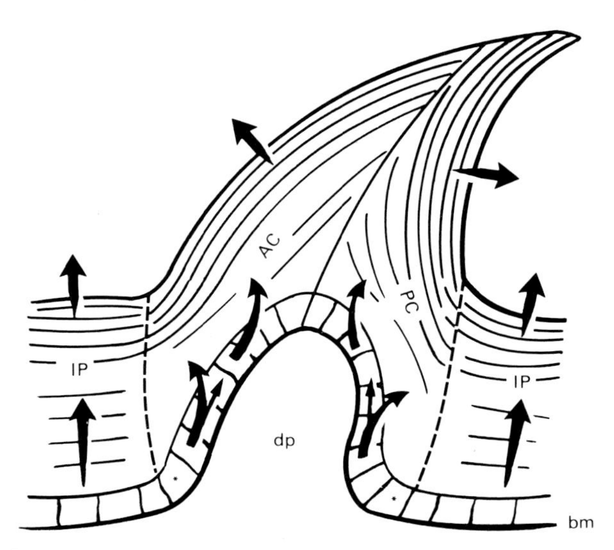

Fig. 4. Organisation and suggested cell flow pattern for the anterior filiform papilla. Large arrows represent desquamation from the four modified columns of differentiated cells. Cell production is restricted to the cells on the basement membrane (bm), which is folded up beneath the papilla to form a dermal papilla (dp). The cell flow patterns are indicated by the small arrows. The presumed locations of the papilla column stem cells are indicated by the asterisks. The papilla is composed of four columns, an anterior (AC) and posterior (PC) column with two collar columns at right-angles. There is an interpapillary (IP) region between adjacent papilla. (From Hume & Potten, 1976.)

production by the area of the basal layer beneath each column. The actual situation is complicated by the extremely complex three-dimensional shape of the papilla and the presence of other less prominent lateral columns, as well as by an interpapillary region. The boundaries of the two major columns can be traced to the basal layer and the cell proliferation studied in relation to these boundaries. What features does this system have in common with the intestine? First, there are (*a*) regions where the basal cells apparently never divide (top of the dermal papilla) and (*b*) regions of high basal cell division activity (base of the dermal papilla). The major difference is that the movement and maturation of committed cells in the tongue papilla occur away from the basement membrane (unlike intestine) as well as along it distally (as in intestine). The complex topography and proliferation patterns suggest a cell flow pattern from the base of the dermal papilla up and into the columns. As in the intestine the cells ultimately responsible for this flow must lie at the base of it. The only other information available is that, as in the intestine, the cells at the presumptive stem cell position show a very marked circadian rhythm with probable cell cycle times of about 24 hours (W. J. Hume, unpublished data).

EPIDERMIS

The fact that epidermis from many regions in many species including man is arranged in a highly organised fashion is well documented (MacKenzie, 1969; Christophers, 1970, 1971; Menton & Eisen, 1971; Potten, 1974; Allen & Potten, 1974). Some regions lack this organisation (Allen & Potten, 1976) probably because of some functional necessity associated with the site rather than the rate of cell production within the region. In well organised regions such as rodent ear or dorsal epidermis the six to 20 mature end cells (corneocytes) are thin cells with a large roughly hexagonal surface area arranged into precise columns with a minimal area of overlap and a regular alternation at the column boundaries. These columns can be seen in suitable sectioned material (Fig. 5) or in the other axis from the surface in suitable thin epidermal sheets in the light

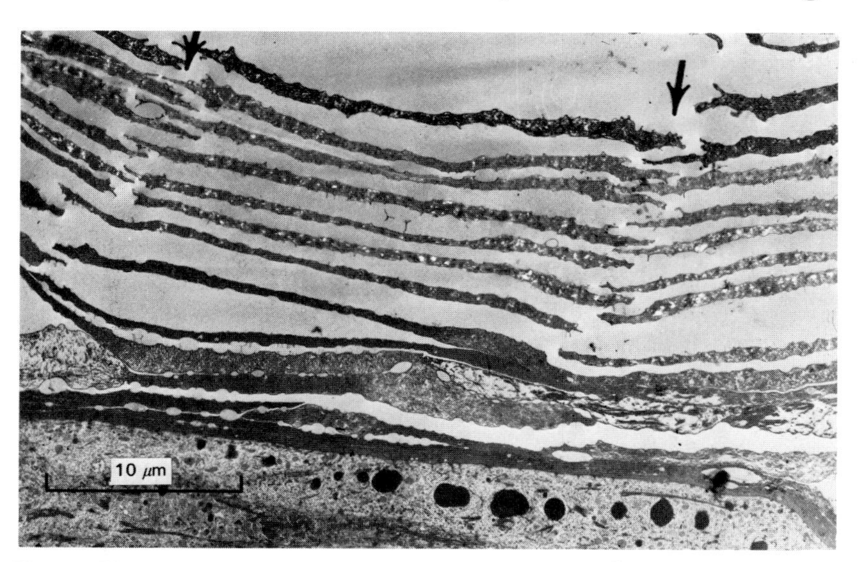

Fig. 5. Electron micrograph of the cornified cell layers of mouse ear epidermis showing precise cornified cell edge alignment (arrows) with minimal but regular alternating overlap. Ear placed in short-term culture conditions to expand the cornified region.

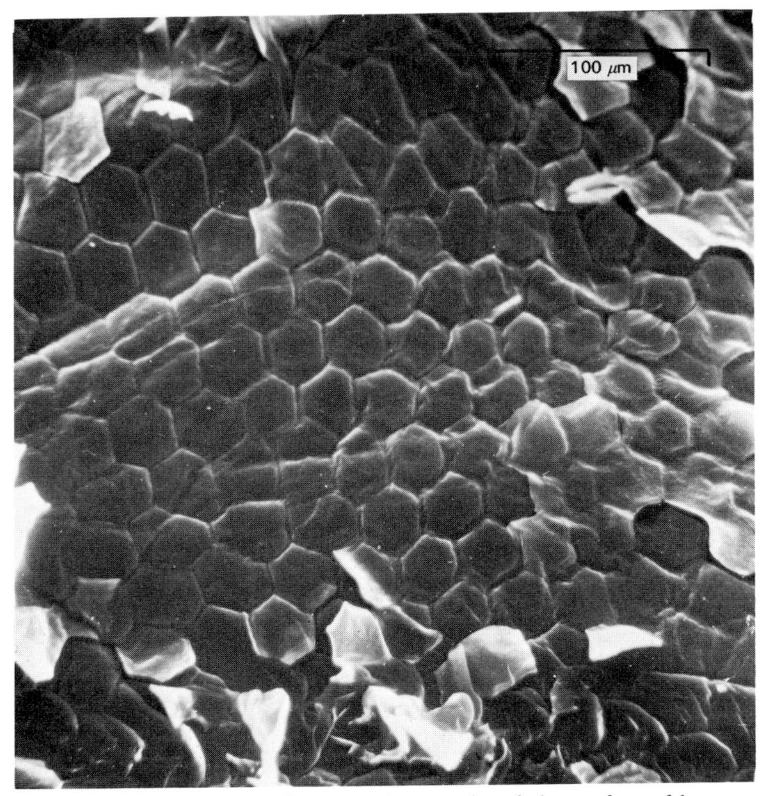

Fig. 6. Low-power scanning electron micrographs of the surface of hamster ear showing regular hexagonal arrangement of cornified cells with occasional individual cell detachment.

microscope or directly in the scanning electron microscope (Figs. 6 and 7). In transmission electron microscopy the column edges can be traced to the basal layer which is a single layer of cells arranged in a flat sheet. Cell migration and maturation in this case are predominantly away from the basal layer. No cell division occurs in a suprabasal position. Cells migrate suprabasally, flatten and increase their surface area ten-fold while aligning themselves perfectly into the columns. It is clear that as a general rule these columns must be maintained by the group of 10 basal cells that lies beneath the columns, from which derives the concept of discrete Epidermal Proliferative Units or EPUs (Potten, 1974; Fig. 8). Each EPU in mouse dorsum consists of six to 10 mature end cells (corneocytes), three maturing end cells (spinous and granular cells)

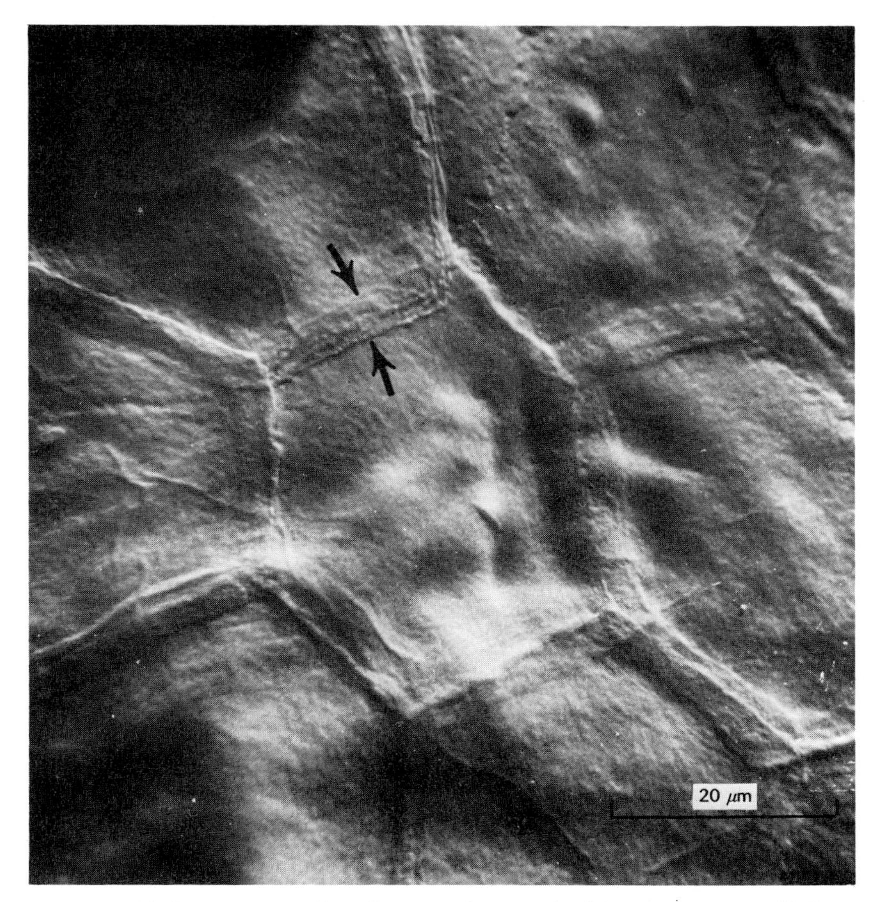

Fig. 7. High-power scanning electron micrograph of hamster ear cornified cells showing hexagonal shape and extent of overlap (arrows).

and 10 basal cells. Within the basal component there are probably two or three peripherally positioned early maturing end cells (awaiting the signal for migration) with one centrally positioned Langerhans cell and approximately one amelanotic melanocyte for every three EPUs. There are six to eight proliferative basal cells producing about one new cell between them each day (Potten, 1974; Allen & Potten, 1974; Potten, 1975*b*, 1976*b*). The 10 basal cells tend to be positioned according to a fairly regular pattern, with six to seven cells having a peripheral position relative to the column boundaries and three to four cells centrally positioned with a smaller internuclear distance than the peripheral cells. Cell migration into

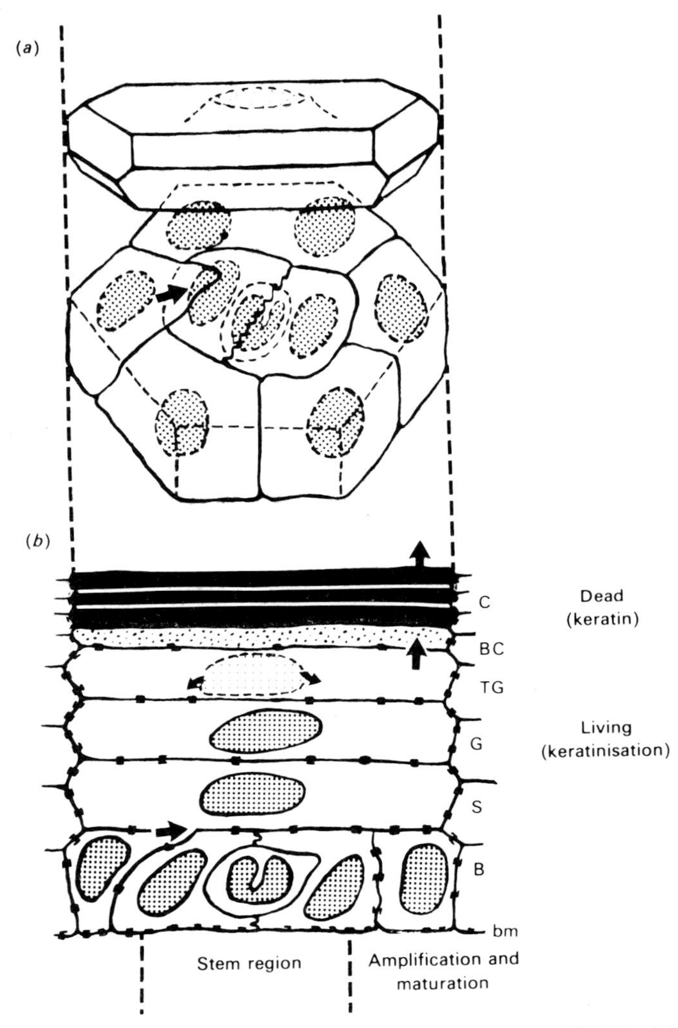

Fig. 8. Schematic representation of the Epidermal Proliferative Unit (EPU) with its basal cells (B) (nine in this case), one of which is beginning to migrate. There are three maturing cell layers, spinous (S), granular (G), and top granular (TG). In the last the nucleus is degraded releasing material that can be utilised by the basal layer. The upper layers are dead plates of keratin, the cornified cells (C), the lowest of which is the basal cornified (BC). bm, basal membrane. (*a*) is the surface view of the basal and spinous cells, (*b*) the section view. (From Potten, 1974; Allen & Potten, 1974; Allen & Potten, 1977.)

the epidermal columns occurs from a peripheral position which suggests that the peripheral region is more mature than the central region. Repeated labelling studies have shown that the peripheral cells become labelled more rapidly than the central ones. After stimulation by wounding, the central region responds dramatically and rapidly (Potten, 1974). The central region also appears to exhibit a well defined circadian rhythm (Potten *et al.*, 1977). When the columnar organisation (EPUs) and the basal cell topography is considered it seems likely that each EPU is discrete with its own stem cell complement and that this is probably centrally positioned. Once the EPU and its columns are laid down it is difficult to see how the number of EPUs can change (i.e. how EPUs could be lost or, more particularly, inserted) without first destroying the organisation through a transient hyperplasia with its apparently random organisation. This implies that the EPU must be laid down with a division potential to last the life of the animal, provided it is neither wounded nor needs to increase its skin surface area.

In epidermis the most striking evidence for different proliferative subpopulations comes from radiobiological experiments. An analysis of the extrapolated origin of all radiation dose–response curves suggests that only a fraction of the basal cells are capable of extensive epithelial regeneration (Potten, 1977; C. S. Potten, S. E. Al-Barwari & J. Searle, unpublished data), while an analysis of those data where the actual origin can be estimated suggests that only 2–7 % of the basal cells are capable of regeneration (clonogenic) (Potten & Hendry, 1973).

Further evidence can be found if one considers the radiation production of histologically dead or dying cells (apoptotic or pycnotic cells). Over a fairly wide dose-range where more than 99 % of the epidermal cells are known to be reproductively sterilised there is no clear dose-dependence in the yield of apoptotic cells. The apoptotic yield rarely rises above a few per cent (Potten, 1977; C. S. Potten, S. E. Al-Barwari & J. Searle, unpublished data). The data are consistent with the hypothesis that the epidermis contains a small subpopulation of sensitive regenerative cells most of which are killed by doses in the range studied and appear as apoptoses. In the intestine there also appears to be a highly radiosensitive subpopulation which in this case was positioned in the presumptive stem cell region.

In both intestine and epidermis the radiobiological estimates of regenerative cells may overestimate stem cells because some early committed cells may be capable of many divisions and the formation of a temporary clone. The extent of this contamination is unknown but this may partially explain the different results obtained with skin experiments, the widely differing clone growth rates reported (Al-Barwari & Potten, 1976) and the discrepancies as the result of using different techniques in the crypt 'stem' cell estimates.

In all three regions there are cell kinetic indications of cellular subpopulations; regions of maturing cells and regions with differing cell kinetics. In intestine and skin there are also strong indications of radiobiologically distinct subpopulations. In epidermis and tongue there is considerable tissue organisation with mature cell columns. A comparison of the cycle kinetics and probable numbers of stem and committed cells for skin and intestine is given in Fig. 9.

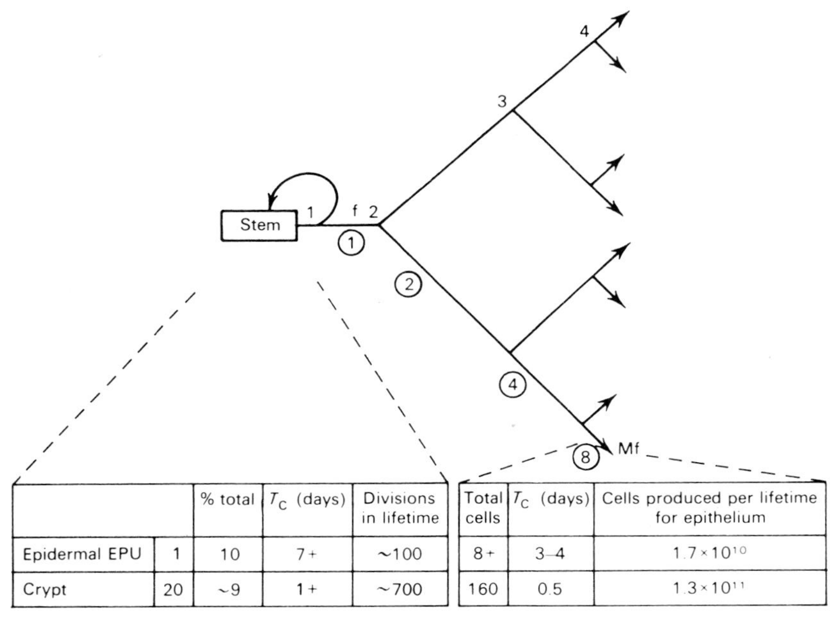

		% total	T_C (days)	Divisions in lifetime
Epidermal EPU	1	10	7+	~100
Crypt	20	~9	1+	~700

Total cells	T_C (days)	Cells produced per lifetime for epithelium
8+	3–4	1.7×10^{10}
160	0.5	1.3×10^{11}

Fig. 9. Schematic representation of the probable amplification divisions that follow an epithelial stem cell division. The cell divisions are numbered 1 to 4 (the most likely number of amplification divisions is 3 to 4) and the numbers of cells produced (for each stem cell division) are shown encircled. Some parameters relevant to two epithelia are shown. M, number of committed cells produced for each stem cell division; f, flux out of stem cell compartment. (Data from Potten, 1974, 1975*a*, *b*.)

STEM CELL POSITIONS

In tongue and intestine the presumptive stem cell regions are in the deepest (most proximal) part of the epithelia. In this position the cells may be protected from the turbulence of extensive cell movement, from some of the factors that control differentiation and maturation, and also from the potentially harmful environment. The stem cells may also be somewhat protected by their slow cycle rate (less time in the error-susceptible stages of DNA synthesis and mitosis). They may also benefit from their periods of apparent inactivity by using this time for reorganising themselves, recognising their status (position), or repairing deficiencies. In the haemopoietic system the stem cells normally cycle very slowly and may in fact need time between divisions to restitute their full stem cell potential (L. G. Lajtha & C. W. Gilbert, unpublished data). The epidermis differs from the intestine and tongue in that it is a flat sheet of cells with its presumptive stem cells regularly spaced within the sheet. However, it is possible that even in epidermis the true stem cells may be placed in a protected position tucked deep in the hair follicles or rete-pegs, thus bringing the epidermis even more in line with the tongue and intestinal arrangements. In this case the EPUs must be laid down with an early committed cell with sufficient division potential to last the lifetime of the animal provided no extensive regeneration is required. This could explain why (1) a recent radiobiological approach (Al-Barwari & Potten, 1976) provides a dose–response curve which extrapolates to an origin more consistent with the number of follicles than the number of EPUs; (2) during regeneration the columnar organisation is broken down through a transient hyperplasia which enables new early committed cells to be laid down in EPUs and why in severely damaged regions the regenerated epithelium has very deep rete-peg or follicle-like structures (Potten, 1977); (3) many early foci of regeneration are seen to be associated with follicles (Al-Barwari & Potten, 1976), an observation reported for several wounded states; and (4) those regions lacking follicles or other associated appendages tend to have an undulating or folded basement membrane.

In conclusion there are indications that epithelial tissues operate via an asymmetrical amplifying scheme. There are many similarities between the three epithelial regions discussed here. The precise location and number of stem cells are uncertain in some regions and

Table 1. *Properties of epithelial stem cells*

Specific location within the tissue
No clear histological features
Slowly cycling (G_0) (few in number)
Characteristic thymidine pools
Strong circadian rhythm
Characteristic sensitive radiation response (apoptotic)
Asymmetric division (selective segregation)?

complicated by the fact that the early committed cells may possess some degree of clonogenicity. The numbers of stem cells might in fact be very low, particularly if selective segregation is regarded as a characteristic feature of stem cells. Some of the properties that may be attributable to epithelial stem cells are shown in Table 1.

This work was supported by grants from the Medical Research Council and the Cancer Research Campaign. I am grateful to Professor Lajtha for his helpful comments, to S. E. Al-Barwari and W. J. Hume for the use of their unpublished data and to Dr T. Allen for his help with the electron microscopy.

REFERENCES

Al-Barwari, S. E. & Potten, C. S. (1976). Regeneration and dose–response characteristics of irradiated mouse dorsal epidermal cells. *International Journal of Radiation Biology*, **30**, 201–16.

Al-Dewachi, H. S., Wright, N. A., Appleton, D. R. & Watson, A. J. (1975). Cell population kinetics in the mouse jejunal crypt. *Virchows Archiv*, **18B**, 225–43.

Al-Dewachi, H. S., Wright, N. A., Appleton, D. R. & Watson, A. J. (1977). The effect of a single injection of hydroxyurea on cell population kinetics in the rat jejunum. *Cell and Tissue Kinetics*, **10**, 203–14.

Allen, T. D. & Potten, C. S. (1974). Fine structural identification and organisation of the epidermal proliferative unit. *Journal of Cell Science*, **15**, 291–319.

Allen, T. D. & Potten, C. S. (1976). Ultrastructural site variations in mouse epidermal organisation. *Journal of Cell Science*, **21**, 341–59.

Allen, T. D. & Potten, C. S. (1977). Significance of cell shape in tissue architecture. *Nature, London*, **264**, 545–7.

Barnes, W. H. & Loutit, J. F. (1967). Haemopoietic stem cells in the peripheral blood. *Lancet*, **ii**, 1138–41.

Boarder, T. A. & Blackett, N. M. (1976). The proliferative status of intestinal epithelial clonogenic cells: sensitivity to S phase specific cytotoxic agents. *Cell and Tissue Kinetics*, **9**, 589–96.

CAIRNIE, A. B., LAMERTON, L. F. & STEEL, G. G. (1965). Cell proliferation studies in the intestinal epithelium of the rat. I. Determination of the kinetic parameters. *Experimental Cell Research*, **39**, 528–38.

CAIRNS, J. (1975). Mutation selection and the natural history of cancer. *Nature, London*, **255**, 197–200.

CHENG, H. & LEBLOND, C. P. (1974). Origin, differentiation and renewal of the four main epithelial cell types in the mouse small intestine. V. Unitarian theory of the origin of the four epithelial cell types. *American Journal of Anatomy*, **141**, 537–62.

CHRISTOPHERS, E. (1970). Eine neue Methode zur Darstellung des Stratum Corneum. *Archiv für klinische und experimentelle Dermatologie*, **237**, 717–22.

CHRISTOPHERS, E. (1971). Cellular architecture of the stratum corneum. *Journal of investigative Dermatology*, **57**, 241–6.

HENDRY, J. H. & POTTEN, C. S. (1974). Cryptogenic cells and proliferative cells in intestinal epithelium. *International Journal of Radiation Biology*, **25**, 583–8.

HUME, W. J. & POTTEN, C. S. (1976). The ordered columnar structure of mouse filiform papillae. *Journal of Cell Science*, **22**, 149–60.

KERR, J. F. R., WYLLIE, A. H. & CURRIE, A. R. (1972). Apoptosis: a basic biological phenomenon with wide-ranging implications in tissue kinetics. *British Journal of Cancer*, **26**, 239–57.

LAJTHA, L. G. (1964). Recent studies in erythroid differentiation and proliferation. *Medicine (Baltimore)*, **43**, 625–33.

LAJTHA, L. G. (1967). Stem cells and their properties. In *Proceedings of VII Canadian Cancer Research Conference*, pp. 31–9. Oxford: Pergamon Press.

LAJTHA, L. G. & SCHOFIELD, R. (1974). On the problem of differentiation in haemopoiesis. *Differentiation*, **2**, 313–20.

LEBLOND, C. P. & CHENG, H. (1976). Identification of stem cells in the small intestine of the mouse. In *Stem cells of renewing cell populations*, ed. A. B. Cairnie, P. K. Lala & D. G. Osmond, pp. 7–31. New York & London: Academic Press.

MACKENZIE, I. C. (1969). Ordered structure of the stratum corneum of mammalian skin. *Nature, London*, **222**, 881–2.

MASUDA, K., WITHERS, H. R., MASON, K. A. & CHEN, K. Y. (1977). Single dose–response curves of murine gastrointestinal crypt stem cells. *Radiation Research*, **69**, 65–75.

MENTON, D. N. & EISEN, A. Z. (1971). Structure and organisation of mammalian stratum corneum. *Journal of Ultrastructure Research*, **35**, 247–64.

POTTEN, C. S. (1974). The epidermal proliferative unit: the possible role of the central basal cell. *Cell and Tissue Kinetics*, **7**, 77–88.

POTTEN, C. S. (1975*a*). Kinetics and possible regulation of crypt cell populations under normal and stress conditions. *Bulletin du Cancer*, **62**, 419–30.

POTTEN, C. S. (1975*b*). Epidermal cell production rates. *Journal of investigative Dermatology*, **65**, 488–500.

POTTEN, C. S. (1976*a*). Small intestinal crypt stem cells. In *Stem cells of renewing cell populations*, ed. A. B. Cairnie, P. K. Lala & D. G. Osmond, pp. 79–84. New York & London: Academic Press.

POTTEN, C. S. (1976*b*). Identification of clonogenic cells in the epidermis and the structural arrangement of the epidermal proliferative unit (EPU). In *Stem cells of renewing cell populations*, ed. A. B. Cairnie, P. K. Lala & D. G. Osmond, pp. 91–102. New York & London: Academic Press.

Potten, C. S. (1977). The cellular and tissue response of skin to single doses of ionising radiation. *Current Topics in Radiation Research*, in press.

Potten, C. S., Al-Barwari, S. E., Hume, W. J. & Searle, J. (1977). Circadian rhythms of presumptive stem cells in three different epithelia of the mouse. *Cell and Tissue Kinetics*, **10**, 557–68.

Potten, C. S. & Hendry, J. H. (1973). Clonogenic cells and stem cells in epidermis. *International Journal of Radiation Biology*, **24**, 537–40.

Potten, C. S. & Hendry, J. H. (1975). Differential regeneration of intestinal proliferative cells and cryptogenic cells after irradiation. *International Journal of Radiation Biology*, **27**, 413–24.

Potten, C. S., Kovacs, L. & Hamilton, E. (1974). Continuous labelling studies on mouse skin and intestine. *Cell and Tissue Kinetics*, **7**, 271–83.

The cell population kinetics of repopulating cells in the intestine

NICHOLAS A. WRIGHT
Dept. of Pathology, University of Oxford, Harkness Laboratories, Radcliffe Infirmary, Oxford OX2 6HE, UK

It will be noted that the term 'stem cell' is carefully avoided in the title to this article; this is because, in the small bowel at least, there has been some confusing use of the term. For example, Cheng & Leblond (1974*b*; Leblond & Cheng, 1976) use the term exclusively to mean those cells which have the ability to undergo divisions throughout the life of the organism, and give rise to progeny which ultimately undergo differentiation and to others which are similar to themselves. Other workers have reserved the term for those cells which are capable of repopulating the crypt after death of proliferative cells, produced, for example, by irradiation or chemotherapeutic insult. These latter cells are often referred to as 'clonogenic' cells. The point of contention is, do those cells which feed the proliferative and differentiating compartments in physiological circumstances also exclusively constitute the clonogenic cell pool? Do the proliferating or differentiating progeny of Cheng & Leblond's stem cells also have clonogenic properties?

Consequently one must distinguish between 'functional' stem cells and 'potential' stem cells; functional stem cells are responsible for population of the normal crypt, while potential stem cells are capable of repopulating a crypt in abnormal circumstances, i.e. they are clonogenic. Boarder & Blackett (1976) have suggested that these may be the same population, different populations, or overlapping populations. It is the purpose of this article to distinguish between these possibilities and, in doing so, to assess the proliferative status of clonogenic crypt cells.

FUNCTIONAL STEM CELLS

The crypts of the small and large bowel are closed systems (Cairnie, 1976). Most cell production occurs in the lower two-thirds of the crypt, in the proliferative compartment (Fig. 1), and differentiating cells migrate through the maturation compartment onto the surface, probably as a result of population pressure induced by mitotic activity, and are lost into the bowel lumen. Hence those cells which act as functional stem cells should be housed in the crypt base, where they are in a position to feed the proliferative compartment and to maintain their own number without being swept away in the escalator.

Cheng & Leblond (1974*a*; Leblond & Cheng, 1976) have studied

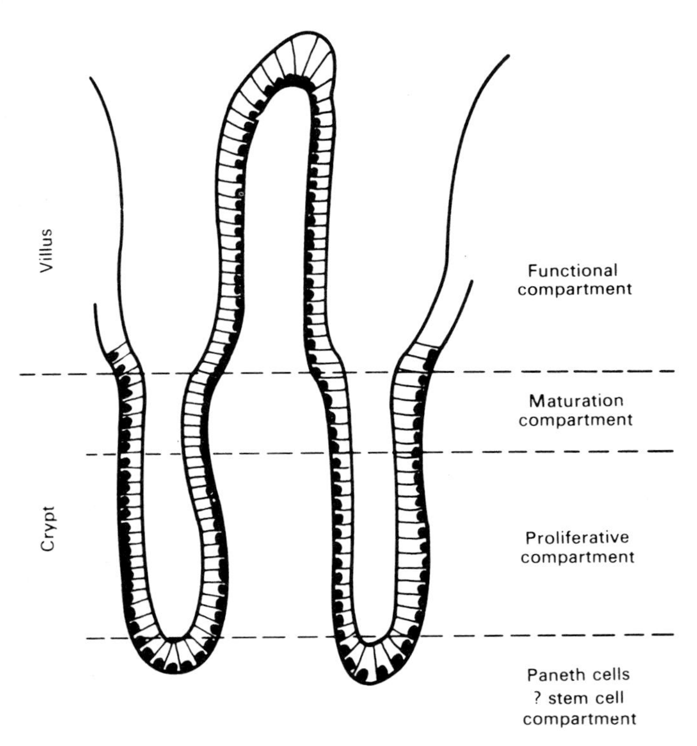

Fig. 1. The small bowel crypt divided into the several kinetic compartments. Functional stem cells are found at the crypt base, in the Paneth cell zone; cells then pass into the proliferative compartment, and losing the capacity to divide, enter the maturation compartment. After traversing the villi, they are lost into the bowel lumen.

the ultrastructure of crypt base columnar cells in the small bowel of the mouse and noted that they share certain morphological characteristics with the stem cells of other renewal systems; these are small size, diffuse nuclear chromatin structure, and few cytoplasmic organelles but many free ribosomes. These crypt base cells were able to elaborate mucous, enteroendocrine and Paneth cell granules. Cheng & Leblond (1974*b*) also reported that a dose of 2 μCi/g body weight of [^{3}H]thymidine killed some crypt base columnar cells; adjacent crypt base columnar cells took up the nuclear debris to produce a labelled phagosome. Using this as a marker, phagosomes were subsequently seen in upper crypt and villus columnar cells, and also in mucous cells, and in enteroendocrine and Paneth cells. They concluded that basal crypt columnar cells were functional stem cells for all small bowel cells. This latter evidence is, of course, rather dependent upon basal columnar cells being the only phagocytic cells in the crypt, and in the rat, after cell death induced by hydroxyurea, mid-crypt columnar cells appear to be phagocytic (N. A. Wright, unpublished observations). However, there is certainly a strong case to be made out for crypt base columnar cells being the functional stem cells for the small bowel crypt.

Analogies with another renewal system, the bone marrow, has led to the suggestion that functional crypt stem cells are few in number; Potten (Kovacs & Potten, 1973; Potten, 1976), using crypt squash techniques, reports that mouse small bowel crypts contain some 20–30 undifferentiated cells of non-Paneth nature, out of a total crypt population of some 250 cells. Other investigators (see for example Withers, 1976), concerned to satisfy the Hayflick hypothesis, and further extending the bone marrow analogy, have suggested that functional crypt stem cells are slowly cycling. Since the response of functional stem cells to cytotoxic chemotherapy depends largely on their proliferative rate, it is perhaps worth considering the evidence for the concept of longer cycling crypt base columnar cells.

Cairnie, Lamerton & Steel (1965) measured the cell cycle time of cells at different levels in the rat jejunal crypt, and showed that cells in basal cell positions had extended cell cycle times compared with those in the above proliferative compartment. These results were confirmed by Al-Dewachi, Wright, Appleton & Watson (1974), and are summarised in Table 1. It is possible to obtain an estimate of proliferative rates at each cell position using the metaphase arrest

Table 1. *Variation in kinetic parameters in rodent intestine*

	Fraction labelled mitosis (FLM) data (mean cell cycle time ±S.E. in hours)								
Cell positions	1–4	5–8	9–12	13–16	17–20	20+	Whole crypt		
Rat jejunum	15.5±0.3	12.3±0.2	11.2±0.2	10.8±0.2	11.0±0.2	10.7±0.1	11.32±0.15		
Rat colon	59.5±1.4	46.7±3.2	40.8±1.2	41.8±1.8	50.9±1.8	—	53.5±2.1		
	Metaphase arrest data (apparent cell cycle time in hours)								
Cell position	1	2	3	4	5	6	7	8	Whole crypt
Rat jejunum[a]	30	22	15	10	10	10	11	10	14
Mouse jejunum[a]	—	—	31	21	16	17	18	15	19
Cell position	1–4	5–8	9–12	13–16	17–20	21–24	Whole crypt		
Rat colon	210	65	50	44	37	34	58		

[a] With Paneth cells removed from the population.

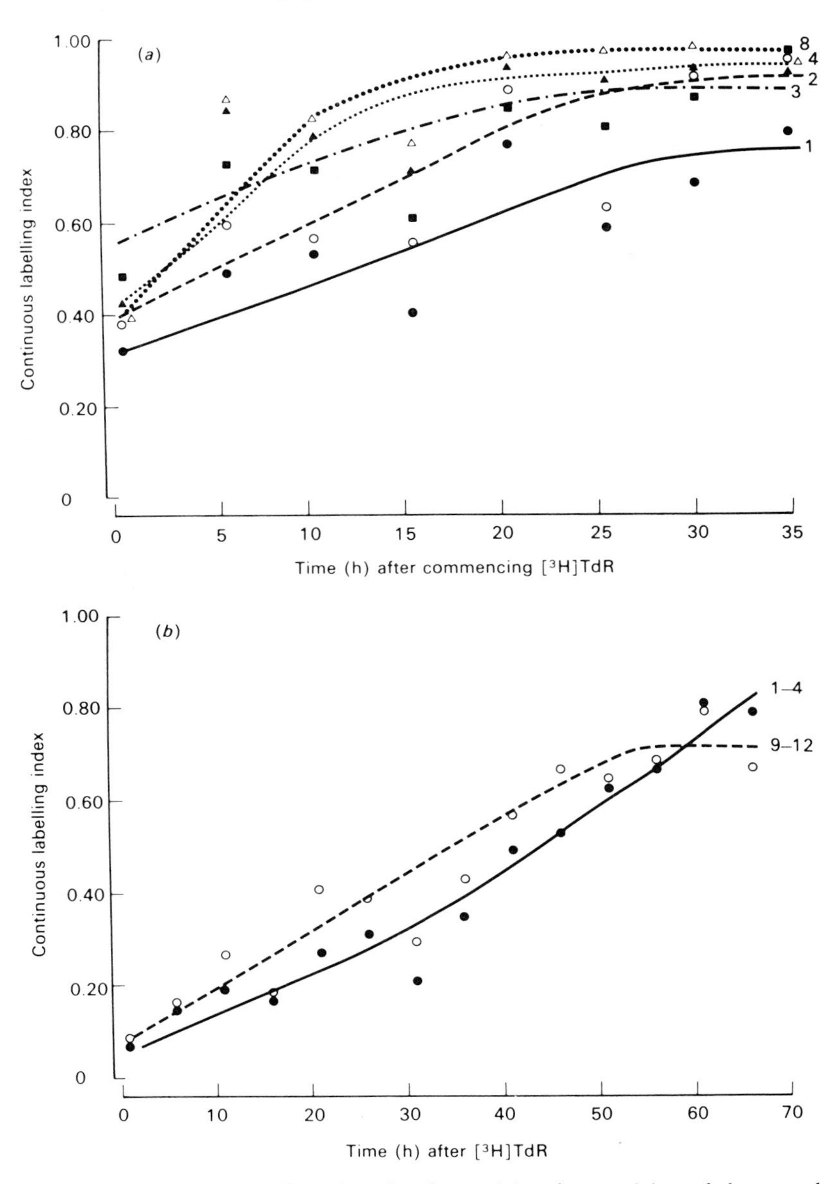

Fig. 2. Continuous labelling data for the rat jejunal crypt (*a*), and the rat colonic crypt (*b*). Animals were injected with 0.5 μCi [^{3}H]TdR per g body weight every 5 hours to label all cells entering S phase. Lines are hand-optimised, and the time taken to achieve a plateau is related to $t_{g1}+t_{g2}+t_m$. Numbers at ends of lines indicate cell positions. Note in (*a*) the decreasing $t_{g1}+t_{g2}+t_m$ period with cell positions up the crypt. In (*b*) cell positions 9–12 achieve a plateau before 1–4.

technique with vincristine, which measures the birth rate, and allows the calculation of an apparent cell cycle time; assuming a growth fraction of unity, this then equals the cell cycle time (Wright, Morley & Appleton, 1972). Apparent cell cycle times for the basal crypt columnar cells in rat and mouse (Al-Dewachi, Wright, Appleton & Watson, 1975) jejunum are also shown in Table 1, together with values for the rat colon (J. P. Sunter, N. A. Wright & D. R. Appleton, unpublished data), compared with higher cell positions. Even when Paneth cells are excluded from the calculations, it is evident that cell production rates are lower in basal cell positions in both jejunum and colon. Taken together with the fraction labelled mitoses (FLM) measurements, which are independent of growth fraction considerations, these data support the concept of slower cycling basal cells, especially as the FLM technique will give preference to the faster cycling cells in cell position 4.

Continuous labelling data also indicate that basal cell positions show longer cell cycle times compared with proliferative cells in higher cell positions (Fig. 2). These results show that basal crypt cells in both small bowel and colon have relatively extended $t_{g1}+t_{g2}+t_{m}$; similar results have been reported by Potten, Kovacs & Hamilton (1974) and Cairnie (1976).

We may conclude that basal crypt cells in both the large and small intestine form a slowly cycling subpopulation compared with their progeny, the cells in the proliferating compartment.

POTENTIAL STEM CELLS

We have seen above that some 20–30 basal crypt cells are held to be functional stem cells in the mouse small bowel crypt; to answer our question about functional and potential stem cells being the same population or different populations, we must obviously have a reliable method for the determination of the number of clonogenic cells per crypt. It is probably true to say such a method does not presently exist.

Withers & Elkind (1969, 1970) developed the macrocolony and microcolony assay technique for measuring the survival of potential stem cells in small and large bowel crypts over a range of radiation doses. The survival curves obtained from these assay methods usually have a large shoulder, partly because of the many potential stem cells (multiplicity) present in each crypt. The way in which investigators have measured this number varies. Withers & Elkind (1970) simply assume a value which is approximately equal to the total number of proliferating cells, about 130 clonogenic cells per crypt. Hagemann, Sigdestad & Lesher (1971) also assumed that proliferative cells are potential stem cells and, using Poisson statistics, were able to explain their crypt survival data. Hagemann *et al.* employed a method based on the identification of labelled microcolonies as dpm/mg intestine after injection of [^{3}H]thymidine.

Hendry & Potten (Hendry & Potten, 1974; Hendry, 1975; Potten & Hendry, 1975) have used a different approach; employing split doses of irradiation they measured the recovery factor for crypt potential stem cells, and making a number of assumptions, they originally calculated a potential stem cell number of 44 per crypt; this was revised to 86 ± 31 potential stem cells per crypt (on the basis of a small change in the D_0 value), with a range of 20–120. It must be conceded that these very wide confidence limits are an obvious drawback to the method. It can be seen that the maximum estimate of 120 cells per crypt is close to the 130–160 potential stem cells per crypt assumed by Withers & Elkind (1970) and by Hagemann *et al.* (1971). On the other hand, if the lower part of the range is assumed to obtain, then the value corresponds with the estimate of the slowly cycling undifferentiated cells found in the crypt base. Potten & Hendry (1975) have in fact advanced the hypothesis that the clonogenic cells in the mouse small bowel crypt are indeed that very same slowly cycling subpopulation of basal crypt columnar cells; that is, that functional stem cells and potential stem cells are the same subpopulation. However, if we take the mean value of 80 cells reported by Potten & Hendry, it is evident that if a maximum of 30 cells are slowly cycling, then the remainder, some 50 cells, are rapidly proliferating and most probably found in the proliferative compartment.

It is thus apparent that methods which attempt to measure the number of clonogenic cells give no easy answer to the problem of identification of the potential crypt stem cells; assumptions about potential stem cell number can at best provide only circumstantial evidence concerning their identity.

THE PROLIFERATIVE STATUS OF POTENTIAL STEM CELLS

Some recent publications have cast some light on the proliferative state of crypt clonogenic cells, and have indicated that a considerable number of them may be rapidly proliferating. Thus Boarder & Blackett (1976) treated mice with high doses of [^{3}H]thymidine (1 mCi per mouse), and also with cytosine arabinoside, both of which kill proliferating cells in DNA synthesis. Soon afterwards clonogenic cell survival was assessed over a range of doses of irradiation, using the microcolony technique. There was no change in the D_0 value between the treated animals and irradiated controls, but 60 ± 3.5 % of clonogenic cells were killed by the thymidine suicide technique, and 85 ± 5 % were killed by the cytosine arabinoside. These results show that potential stem cells, as measured by the microcolony technique, are susceptible to agents which act on cells in the S phase. Assuming that the treatment with [^{3}H]thymidine and cytosine arabinoside does not alter the slope of the survival curve, and that the extrapolation number is unchanged, it can be concluded that a considerable portion of potential stem cells is rapidly proliferating; furthermore the slowly cycling cells cannot make up a large proportion of the potential stem cells.

Of course these results cannot be interpreted as meaning that all rapidly proliferating cells are clonogenic, and nor do they give any insight concerning the clonogenic powers of the slowly cycling population. But they do lead us to question the role of slowly cycling basal crypt cells in crypt repopulation after death of proliferative cells. Theoretically, basal cells, because of their kinetic parameters, would be expected to survive preferentially cycle-specific damage. Consequently, an examination of their role in physiological responses, and also their own response to death of proliferative cells, becomes important.

KINETIC BEHAVIOUR OF BASAL CRYPT COLUMNAR CELLS IN PHYSIOLOGICAL RESPONSES

Several investigators have demonstrated diurnal variation of proliferative indices in small intestinal mucosa; in the mouse peaks of labelling and mitotic indices occur at 03.00 hours (Sigdestad & Lesher, 1970; Hendry, 1975). Hendry (1975) has reported that a fall in microcolony survival occurred at 03.00 hours. Hendry & Potten (1975) noted that basal crypt cells in the Paneth cell zone showed an increase in labelling index at that time, and the increase in labelling index noted at 03.00 hours was thought to be due to a greater proportion of slowly cycling cells entering DNA synthesis. These cells were held to be more radiosensitive, thus explaining the decrease in clonogenic cell survival, in accord with the general hypothesis that slowly cycling basal cells constitute the clonogenic crypt cells. Similar diurnal variation in microcolony number has been found by Boarder & Blackett (1976); both groups reported a decrease in the D_0 value at 03.00 hours.

Diurnal variation of proliferative indices has also been reported in the rat small intestine (Al-Dewachi, Wright, Appleton & Watson, 1976). In the rat jejunum, however, the peaks of labelling and mitotic indices coincide at 15.00 hours, while the nadirs also coincide during the dark schedule (Fig. 3); the curves are not significantly out of phase. Fig. 3(*a*) shows diurnal changes in labelling and mitotic indices for the whole crypt. Separate readings were also obtained for each cell position in the crypt. When the basal cells, which are known to be longer cycling, are grouped together (i.e. cell positions 1–4, see Table 1), no real evidence of non-random patterns emerged (Fig. 3*b*). However, in those cell positions immediately above the slowly cycling cells, positions 5–8, there was evidence of a non-random pattern which resembled the pattern shown for the whole crypt (Fig. 3*c*). The comparative results of an FLM curve started at 17.00 hours, and timed so that the second peak of the FLM curve coincided with the nadir of labelling and mitotic indices, are shown in Fig. 4. It can be seen that there is an apparent small decrease in the duration of the cell cycle time in most cell position groups, but not selectively in the lower cell positions.

It is interesting that the timing of diurnal variation in proliferative indices in the rat is almost diametrically opposite to that found in

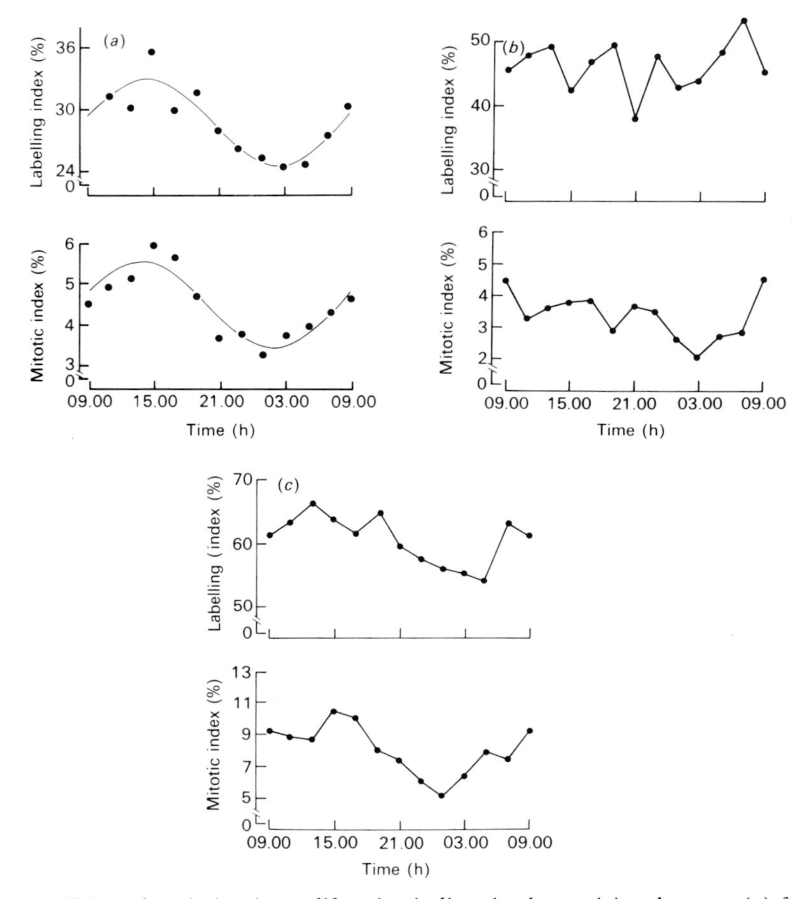

Fig. 3. Diurnal variation in proliferative indices in the rat jejunal crypt: (*a*) for the whole crypt; (*b*) for cell positions 1–4; and (*c*) for cell positions 5–8.

the mouse; similar timings were also found in the rat small bowel by Scheving, Burns & Pauly (1972), and such coincidental peaks in labelling and mitotic indices can only be explained by diurnal variation in transit times through both G_1 and G_2. An obviously important experiment would be to see if the diurnal changes in proliferative indices in the rat are also associated with diurnal changes in microcolony number which correspond with those seen in the mouse.

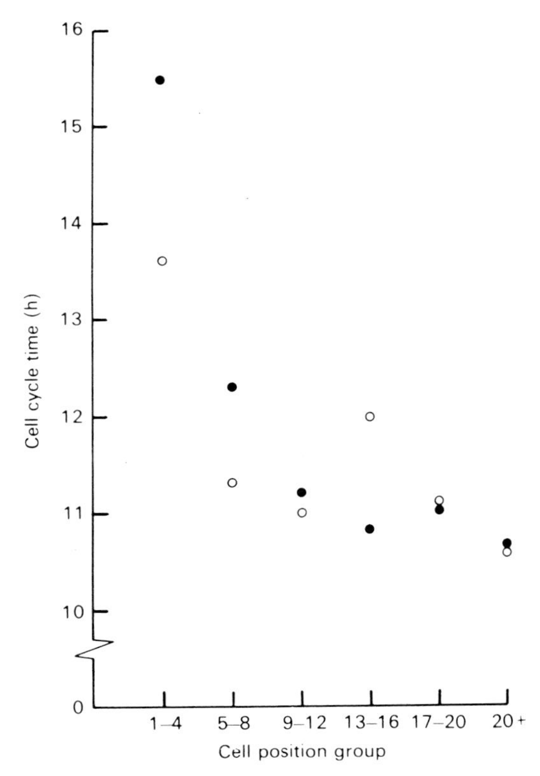

Fig. 4. The cell cycle time in the rat jejunal crypts. Solid circles, with the FLM started at 09.00 hours; open circles, with the FLM started at 17.00 hours, and timed so that the second peak coincides with the nadir of mitotic indices (see Fig. 3*a*).

THE RESPONSE OF THE BASAL CRYPT COLUMNAR CELLS TO PROLIFERATIVE CELL DEATH

As noted above, the subpopulation of basal crypt cells is possessed of comparatively long cell cycle times, is less likely to be in a sensitive phase when exposed to irradiation or chemotherapy, and is thus available for crypt repopulation. The response of these cells to proliferative cell death was explored by treating rats with 1840 mg/kg body weight of hydroxyurea, an agent which kills cells in DNA synthesis (Al-Dewachi, Wright, Appleton & Watson, 1977). Manifest damage to crypt cells was first seen at 2 hours after

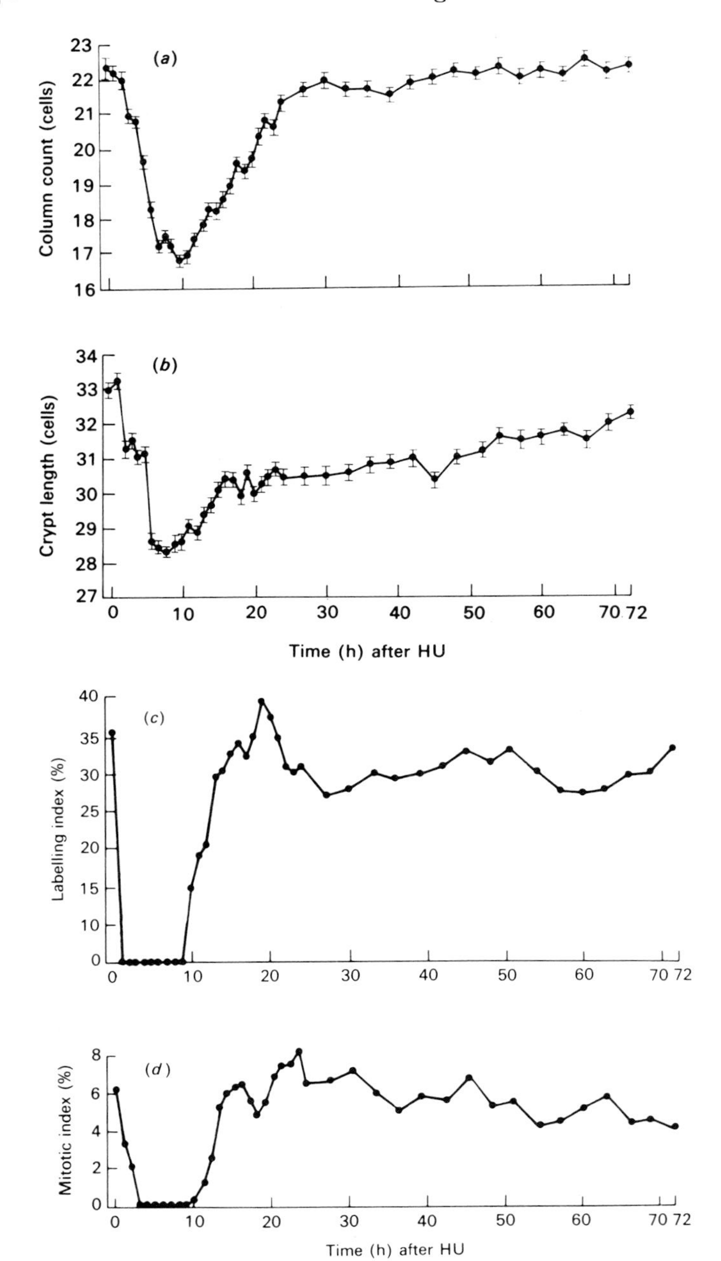

Fig. 5. Changes in (*a*) crypt column length, (*b*) column count, (*c*) flash labelling index and (*d*) mitotic index in the rat jejunal crypt after a single injection of hydroxyurea (HU) in a dosage of 1840 mg/kg body weight.

injection, and reached maximum severity at 7 hours. Changes appeared to be confined to the proliferative zone, and cells occupying the upper third of the crypt, and those in the bottom few cell positions, did not appear to be affected. Fig. 5 shows that hydroxyurea causes an immediate inhibition of DNA synthesis; a reduction in crypt length also occurred at 10 hours, reflecting the cell death induced by hydroxyurea. The labelling index had returned to control levels by 20 hours.

Changes in the distribution of labelled cells in the crypt with time after injection of hydroxyurea are shown in Fig. 6. From 10 to 12 hours low labelling indices are found at each cell position. At 13 hours there was a large increase in labelling indices in lower cell positions, and at 14, 15 and 16 hours the increased labelling indices in lower cell positions are maintained, while the labelling index in cell positions 9 and above increased to equal that of lower positions; at 20 hours after hydroxyurea a smooth curve is again realised.

That these changes in labelling index in basal cell positions are due to increased cell production rates was shown by measuring the rate of entry into mitosis at each cell position with vincristine; Fig. 7 shows that, at 15 hours after hydroxyurea, increased birth rates are found in lower cell positions, particularly positions 3 and 4. Furthermore, an FLM experiment started at 15 hours after hydroxyurea showed a considerable reduction in the cell cycle time for cell positions 1–4 (Fig. 8).

We conclude that the initial response to proliferative cell death is an increase in cell production rate in lower cell positions, produced by a decrease in the cell cycle time of the slowly cycling basal crypt cells. The increase in cell production rate augments the efflux into the proliferative compartment, replacing the dead cells. The repopulating cells also have a shorter cell cycle time (see Fig. 8).

It is of considerable interest to compare these results with those of Potten & Hendry (1975), who studied the differential response of clonogenic and proliferative compartment cells in mouse jejunal crypts after a dose of 900 rads of γ-irradiation. Immediately after irradiation the clonogenic cell number is reduced, but the number of clonogenic cells then *increases*, while proliferative cellularity is *decreasing*. This of course supports the concept that clonogenic and proliferative compartment cells are different populations. However, Potten & Hendry could not demonstrate differential increases in cell production rate (as indicated by labelling index measurements) in

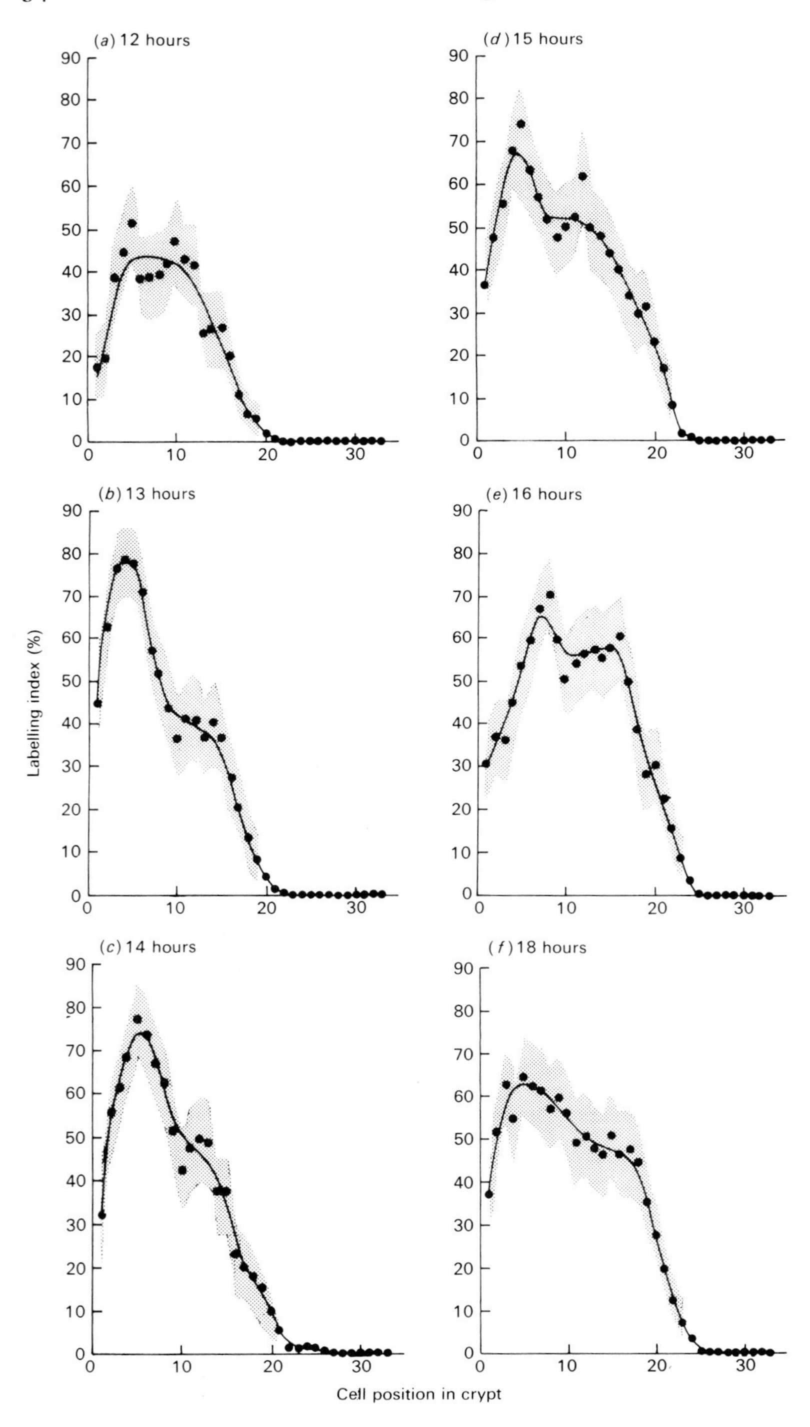
(a) 12 hours
(b) 13 hours
(c) 14 hours
(d) 15 hours
(e) 16 hours
(f) 18 hours
Labelling index (%)
Cell position in crypt
0
10
20
30
40
50
60
70
80
90

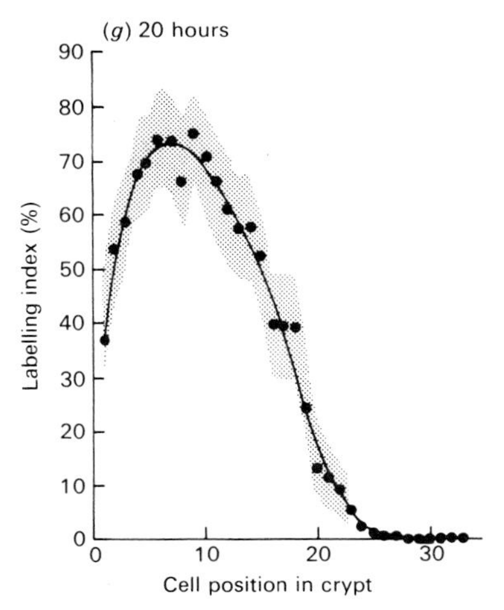

Fig. 6. A selection of labelling index distribution curves for the rat jejunal crypt at stated intervals after injection of hydroxyurea (1840 mg/kg body weight). Note the early and differential rise in labelling index in basal cell positions.

basal crypt cells (see Fig. 9) as was shown above for the rat treated with hydroxyurea.

These results may be compared with studies on the response of the mouse small bowel mucosa to a single injection of cytosine arabinoside (ara-C) (H. S. Al-Dewachi, N. A. Wright, D. R. Appleton & A. J. Watson, unpublished data). An injection of 10.8 mg per animal produces sufficient cell death to decrease the length of the crypt (Fig. 10), and inhibits DNA synthesis for 6 hours before the labelling index recovers and overshoots control levels. Fig. 11 shows a selection of labelling index distribution curves with time after ara-C injection. In this instance there is no evidence of a selective increase in cell production rate in basal cell positions; this dose of ara-C is very similar to that which Boarder & Blackett (1976) found would decrease clonogenic cell number by 82%. The mean cell cycle time for the rapidly proliferating cells was reduced from 12.4 hours to 10.1 hours.

Thus we have two instances where repopulating mechanisms appear to differ; in the rat with hydroxyurea, repopulation of the crypt originates from an increased cell production rate in slowly cycling basal crypt cells. In the mouse, with ara-C in the dose used, repopulation may be a function of the rapidly proliferating cells, which increased their cell production rate by a decrease in cell cycle

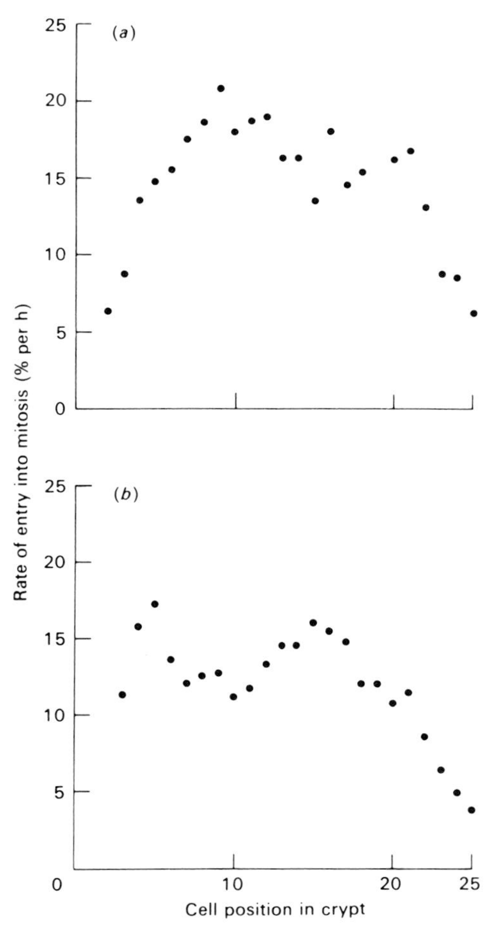

Fig. 7. The rate of entry into mitosis measured at each cell position in the rat jejunal crypt at (*a*) 15 hours and (*b*) 20 hours after a single injection of hydroxyurea (1840 mg/kg body weight). Vincristine was injected at 15 hours and 20 hours and the birth rate was calculated from readings made over 2.5 hours after injection of vincristine. Birth rates in basal cell positions are increased at 15 hours after hydroxyurea, in accord with the labelling index data.

time. Certainly no differential increase in the cell production rate of basal crypt cells occurred.

In the rat the mechanism of increased cell production by basal cells, and in the mouse by proliferative compartment cells, is likely to be mediated by decreases in cell cycle time. In the normal animal there is no real evidence of resting G_0 cells (Wright, Al-Dewachi,

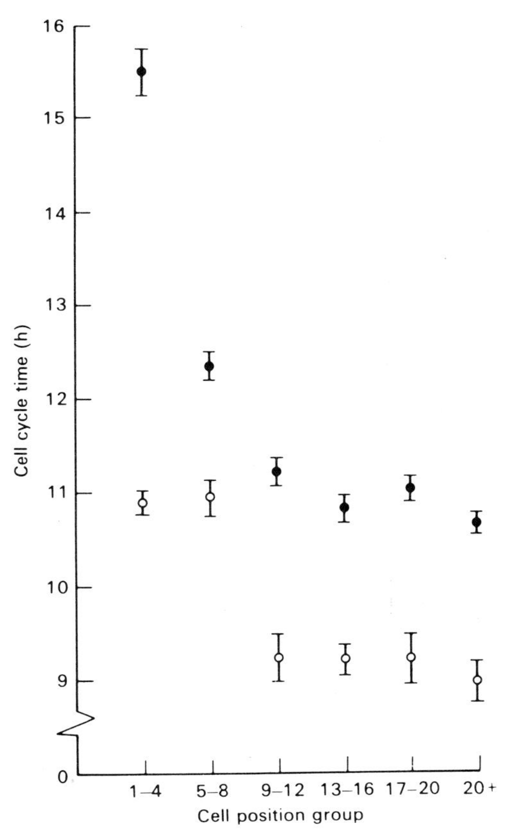

Fig. 8. The changes in the duration of the cell cycle time ±S.E. at 15 hours after a single injection of hydroxyurea, as measured by the FLM technique. Note the prominent decrease in cell cycle time in cell positions 1–4 in the treated group of animals (open circles), compared with control animals (solid circles).

Appleton & Watson, 1975), although in abnormal conditions, for example in starvation, cells in the proliferative compartment appear to enter a resting phase, from which they are ejected upon refeeding (Al-Dewachi, Wright, Appleton & Watson, 1974; H. S. Al-Dewachi, N. A. Wright, D. R. Appleton & A. J. Watson, unpublished data).

These observations lead to the concept that *the repopulating mechanism may depend on the severity of damage, i.e. the demands made upon the renewal system.* Although as yet we have no evidence as to which treatment produced the larger cell kill, a possible interpretation is that the damage produced by ara-C can be made good by the repopulating ability of the rapidly proliferating cells, whereas

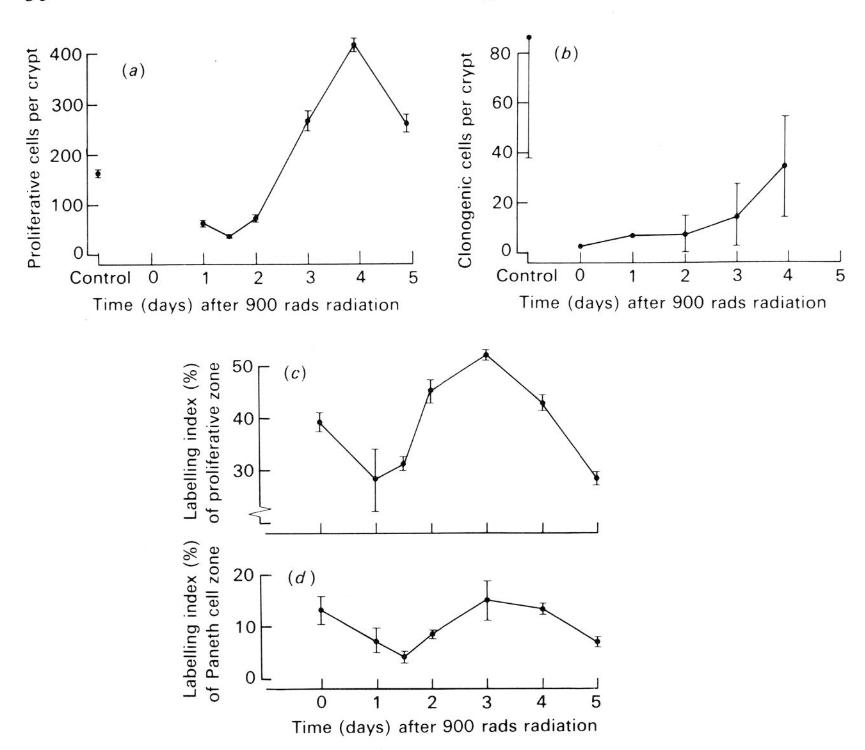

Fig. 9. Changes in (*a*) cellularity of the proliferative zone, (*b*) number of clonogenic cells, (*c*) labelling index of the proliferative zone, and (*d*) labelling index of the Paneth cell (functional stem cell) zone, per mouse jejunal crypt after a dose of 900 rads of γ-radiation. (Redrawn from Potten & Hendry, 1975.)

with hydroxyurea in the rat, death of rapidly proliferating repopulating cells may be so severe that the crypt repopulation has perforce to take place from the surviving basal crypt cells. This of course is sheer speculation, and more studies relating the degree of cell kill to repopulating mechanisms are needed. However, these results indicate that crypt repopulation may be a property of both rapidly and slowly cycling crypt cells.

Do slowly cycling crypt cells occur in man?

There have been no differential studies published on proliferative rates in human intestinal crypts, but Fig. 12 shows the rate of entry into mitosis (as measured with vincristine) plotted against cell position in the human jejunal crypt. It can be seen that cell production rates in basal cell positions are low compared with cells in

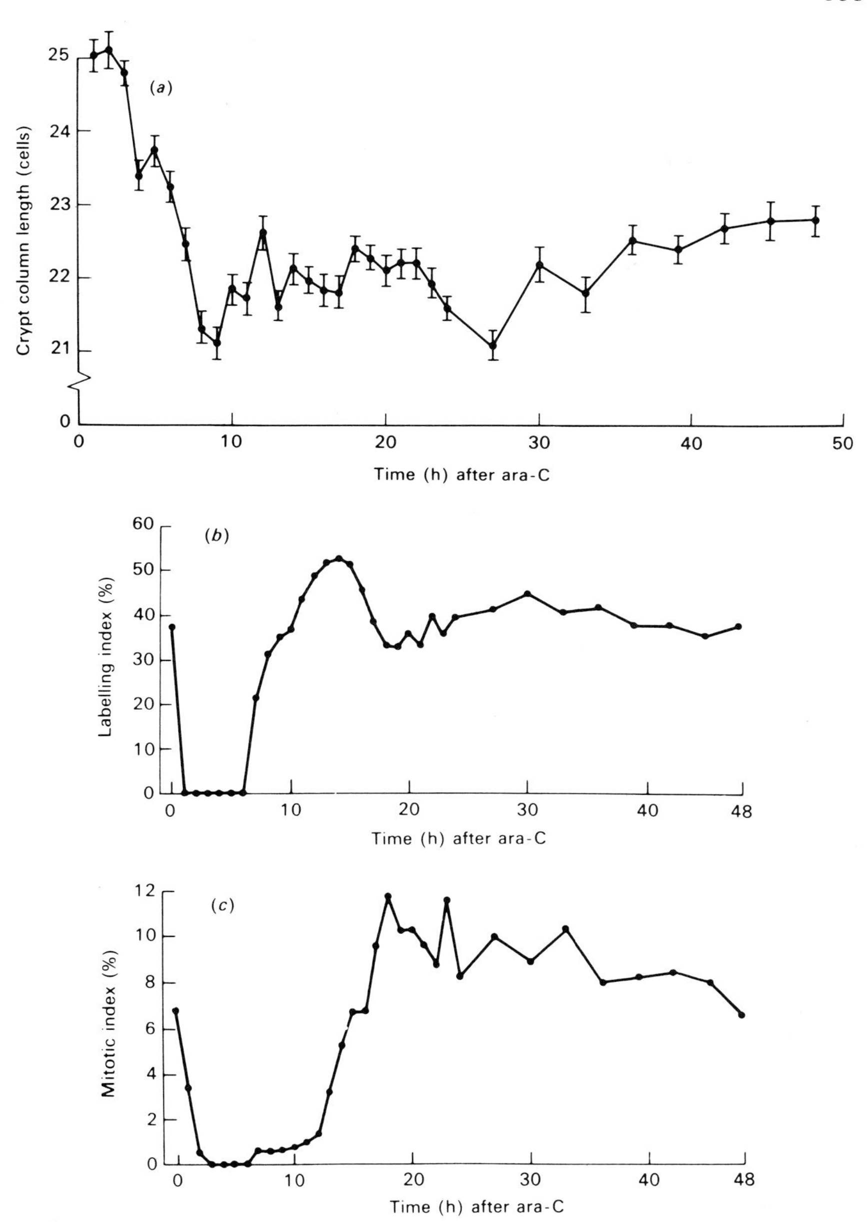

Fig. 10. Changes (*a*) in crypt column length, (*b*) flash labelling index, and (*c*) mitotic index, in the mouse jejunal crypt after a single injection of 10.8 mg per mouse of cytosine arabinoside (ara-C).

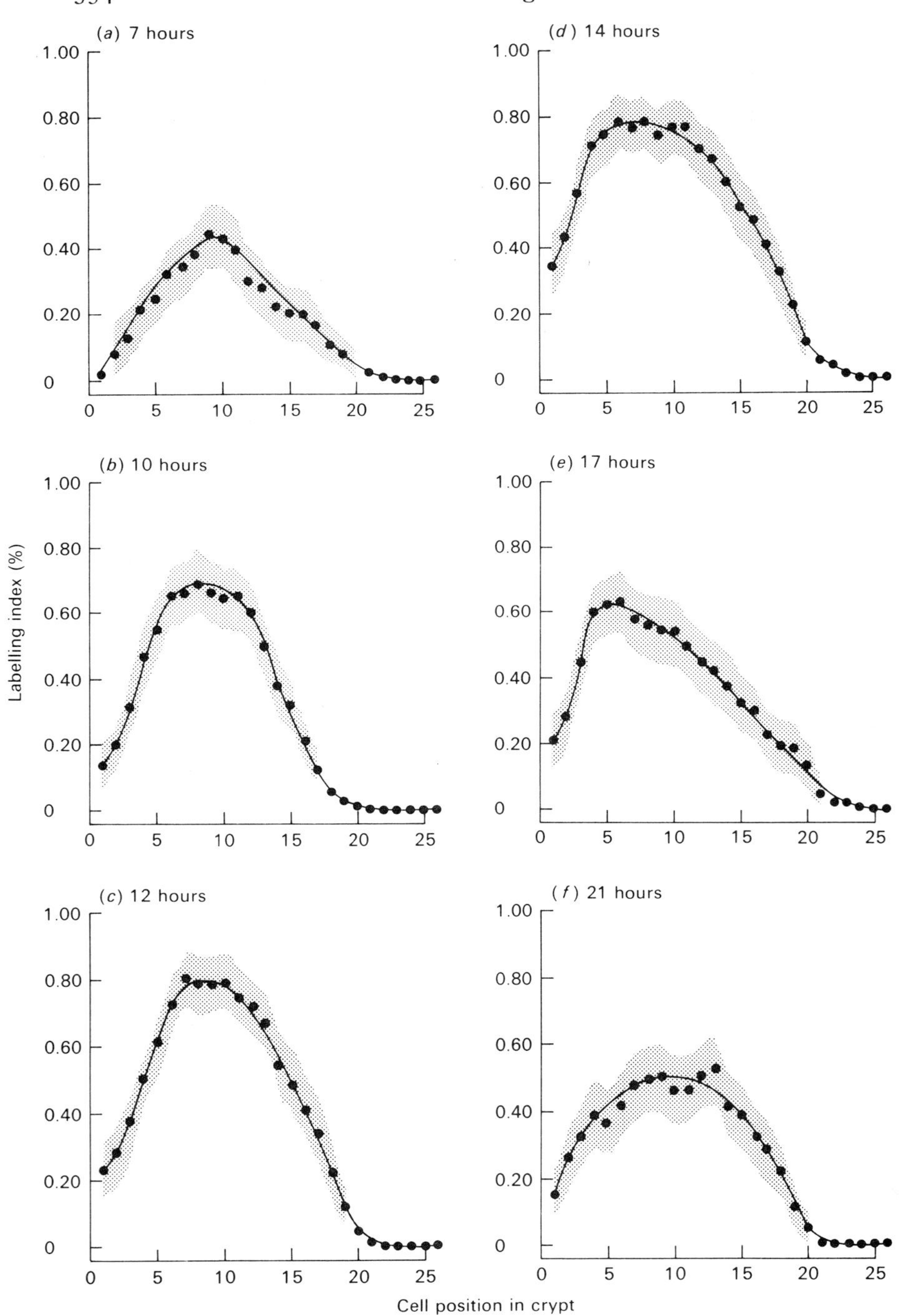

(a) 7 hours
(b) 10 hours
(c) 12 hours
(d) 14 hours
(e) 17 hours
(f) 21 hours
Labelling index (%)
Cell position in crypt
1.00
0.80
0.60
0.40
0.20
0
0
5
10
15
20
25

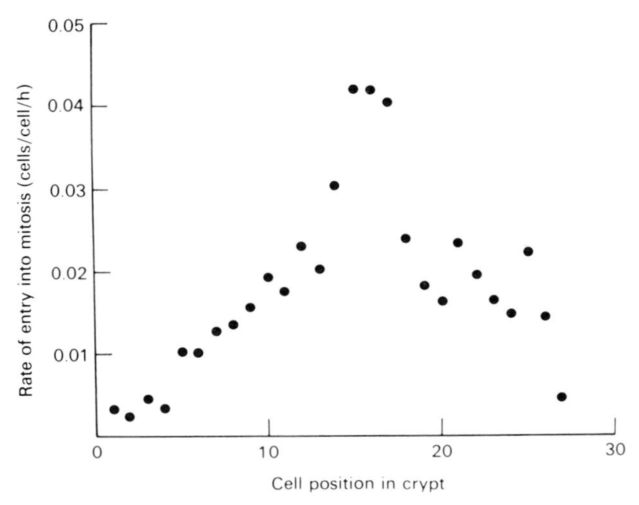

Fig. 12. Rates of entry into mitosis plotted as a function of cell position in the human jejunal crypt. The results are derived from metaphase accumulation lines produced for each cell position, and collected over 2.5 hours after injection of vincristine sulphate. The results are shown for an individual showing normal stereomicroscopical and histological appearances. (For details of patient and techniques see Wright, Watson, Morley, Appleton, Marks & Douglas, 1973.)

the proliferative compartment. These findings support the concept of slowly cycling basal crypt cells in human small bowel.

Possible role of slowly cycling basal crypt cells in carcinogenesis

One puzzling and as yet unanswered question is the problem of carcinogenesis in closed renewal systems such as the large bowel crypts. Rapidly proliferating cells have comparatively short transit times from birth in the proliferative compartment; consequently any carcinogenic effect has a strictly limited period of action. However, slowly cycling functional stem cells in the crypt base are self-renewing, and hence are available for continued action by carcinogenic agents. Consequently, carcinogenesis in the large bowel may be a primary process affecting basal functional stem cells. Circumstantial evidence for this concept comes from Tutton & Barkla

---

Fig. 11. A selection of labelling index distribution curves in the mouse jejunal crypt at stated intervals after a single injection of cytosine arabinoside (10.8 mg per mouse). In this instance there is no differential increase in labelling indices in basal crypt cells.

(1976) who demonstrated that cell production rates in dimethylhydrazine-induced adenocarcinomas of the rat colon are similar to those of basal crypt cells. However, any such hypothesis must explain the rarity of spontaneous carcinomas of the small bowel in man, even though slowly cycling basal crypt cells are present. Of interest in this respect is the fact that intestinal metaplasia of the small bowel type is a very common antecedent lesion of the intestinal type of gastric carcinoma.

CONCLUSION

In the intestine it is important to distinguish between functional stem cells which give rise to crypt cells in normal circumstances, and potential stem cells, which repopulate crypts after the death of proliferative cells. Current radiobiological studies offer conflicting evidence as to whether these constitute the same, different, or overlapping populations. Kinetic studies of crypt repopulation after death of proliferative cells induced by cytotoxic agents acting on S phase cells suggest that both slowly cycling basal crypt cells and rapidly proliferating cells have clonogenic properties, and it is possible that crypt repopulating mechanisms vary according to the degree of cell death.

This work was supported by the Cancer Research Campaign. The studies reported here were carried out in association with Dr H. S. Al-Dewachi, Dr D. R. Appleton and Dr A. J. Watson. I am grateful to Mrs Anne Reeve for typing the manuscript.

REFERENCES

Al-Dewachi, H. S., Wright, N. A., Appleton, D. R. & Watson, A. J. (1974). The cell cycle time in the jejunal crypt of the rat. *Cell and Tissue Kinetics*, **7**, 599–606.

Al-Dewachi, H. S., Wright, N. A., Appleton, D. R. & Watson, A. J. (1975). Cell population kinetics in the mouse jejunal crypt. *Virchows Archiv*, **18B**, 225–42.

Al-Dewachi, H. S., Wright, N. A., Appleton, D. R. & Watson, A. J. (1976). Studies on the mechanism of diurnal variation in the rat jejunal crypt. *Cell and Tissue Kinetics*, **9**, 459–67.

Al-Dewachi, H. S., Wright, N. A., Appleton, D. R. & Watson, A. J. (1977). The effect of a single injection of hydroxyurea on cell population kinetics in the rat jejunum. *Cell and Tissue Kinetics*, **10**, 203–13.

Boarder, T. A. & Blackett, N. M. (1976). The proliferative status of intestinal stem cells: sensitivity to S phase specific cytotoxic agents. *Cell and Tissue Kinetics*, **9**, 589–96.

CAIRNIE, A. B. (1976). Homeostasis in the small intestine. In *Stem cells of renewing cell populations*, ed. A. B. Cairnie, P. K. Lala & D. G. Osmond, pp. 67–77. New York & London: Academic Press.

CAIRNIE, A. B., LAMERTON, L. F. & STEEL, G. G. (1965). Cell proliferation studies in the intestinal epithelium of the rat. I. Determination of kinetic parameters. *Experimental Cell Research*, **39**, 528–38.

CHENG, H. & LEBLOND, C. P. (1974*a*). Origin, differentiation and renewal of the four main epithelial cell types in the mouse small intestine. I. Columnar cells. *American Journal of Anatomy*, **141**, 461–80.

CHENG, H. & LEBLOND, C. P. (1974*b*). Origin, differentiation and renewal of the four main epithelial cell types in the mouse small intestine. V. Unitarian theory of the origin of the four epithelial cell types. *American Journal of Anatomy*, **141**, 537–62.

HAGEMANN, R. F., SIGDESTAD, C. P. & LESHER, S. (1971). Intestinal crypt survival and total and per crypt levels of proliferative cellularity following irradiation. Single X-Ray exposures. *Radiation Research*, **46**, 533–46.

HENDRY, J. H. (1975). Diurnal variations in radiosensitivity of mouse intestine. *British Journal of Radiology*, **48**, 312–14.

HENDRY, J. H. & POTTEN, C. S. (1974). Cryptogenic and proliferative cells in intestinal epithelium. *International Journal of Radiation Biology*, **25**, 583–88.

HENDRY, J. H. & POTTEN, C. S. (1975). In *Paterson Laboratories Annual Report*. Christie Hospital and Holt Radium Institute, Manchester.

KOVACS, L. & POTTEN, C. S. (1973). An estimation of proliferative population size in stomach, jejunum and colon of DBA-2 mice. *Cell and Tissue Kinetics*, **6**, 125–33.

LEBLOND, C. P. & CHENG, H. (1976). Identification of stem cells in the small intestine of the mouse. In *Stem cells of renewing cell populations*, ed. A. B. Cairnie, P. K. Lala & D. G. Osmond, pp. 7–31. New York & London: Academic Press.

POTTEN, C. S. (1976). Small intestinal crypt stem cells. In *Stem cells of renewing cell populations*, ed. A. B. Cairnie, P. K. Lala & D. G. Osmond, pp. 79–84. New York & London: Academic Press.

POTTEN, C. S. & HENDRY, J. H. (1975). Differential regeneration of intestinal proliferative cells and cryptogenic cells after irradiation. *International Journal of Radiation Biology*, **27**, 413–24.

POTTEN, C. S., KOVACS, L. & HAMILTON, E. (1974). Continuous labelling studies on mouse skin and intestine. *Cell and Tissue Kinetics*, **7**, 271–83.

SCHEVING, L. E., BURNS, R. E. & PAULY, J. E. (1972). Circadian rhythms in mitotic activity and ^{3}H-thymidine uptake in the duodenum; effect of isoproterenol on the mitotic rhythm. *American Journal of Anatomy*, **135**, 311–18.

SIGDESTAD, C. P. & LESHER, S. (1970). Further studies on the circadian rhythm in the proliferative activity of mouse intestinal epithelium. *Experientia*, **26**, 1321–2.

TUTTON, P. J. M. & BARKLA, D. H. (1976). Cell proliferation in the descending colon of dimethylhydrazine treated rats and in dimethylhydrazine induced adenocarcinoma. *Virchows Archiv*, **21B**, 147–60.

WITHERS, H. R. (1976). Colony forming units in the intestine. In *Stem cells of renewing populations*, ed. A. B. Cairnie & P. K. Lala, pp. 33–40. New York & London: Academic Press.

WITHERS, H. R. & ELKIND, M. M. (1969). Radiosensitivity and fractionation response of crypt cells of mouse jejunum. *Radiation Research*, **38**, 598–613.

WITHERS, H. R. & ELKIND, M. M. (1970). Microcolony survival assay for cells of mouse intestinal mucosa exposed to radiation. *International Journal of Radiation Biology*, **17**, 261–7.

WRIGHT, N. A., AL-DEWACHI, H. S., APPLETON, D. R. & WATSON, A. J. (1975). Cell population kinetics in the rat jejunal crypt. *Cell and Tissue Kinetics*, **8**, 361–8.

WRIGHT, N. A., MORLEY, A. R. & APPLETON, D. R. (1972). Variation in the duration of mitosis in the crypts of Lieberkuhn of the rat; a cytokinetic study using vincristine. *Cell and Tissue Kinetics*, **5**, 351–62.

WRIGHT, N. A., WATSON, A. J., MORLEY, A. R., APPLETON, D. R., MARKS, J. & DOUGLAS, A. (1973). The cell cycle time in the flat (avillous) mucosa of the human small intestine. *Gut*, **14**, 603–6.

INDEX

Page numbers in italic type indicate reference to a table or figure.